ALLAN'S CIRCUITS PROBLEMS

Allan D. Kraus

Allan D. Kraus Associates, Beachwood, Ohio

New York   Oxford
OXFORD UNIVERSITY PRESS
2001

Oxford University Press

Oxford   New York
Athens   Auckland   Bangkok   Bogotá   Buenos Aires   Calcutta
Cape Town   Chennai   Dar es Salaam   Delhi   Florence   Hong Kong   Istanbul
Karachi   Kuala Lumpur   Madrid   Melbourne   Mexico City   Mumbai
Nairobi   Paris   Sao Paulo   Shanghai   Singapore   Taipei   Tokyo   Toronto   Warsaw

and associated companies in
Berlin   Ibadan

Published by Oxford University Press, Inc.
198 Madison Avenue, New York, New York 10016
http://www.oup-usa.org

ISBN: 0-19-514248-9

Printing (last digit): 9 8 7 6 5 4 3 2 1

Printed in the United States of America
on acid-free paper

CONTENTS

PREFACE

PURPOSE AND TOPICAL CONTENT

The problems contained in this manual are intended to be comprehensive and have been designed to allow the reader to obtain a rather strenuous workout in their solution. Several of the problems are given more than once so that the reader may observe alternate methods of solution.

The topical content is as follows:

1   INTRODUCTION AND BASIC CONCEPTS
2   KIRCHHOFF'S CURRENT AND VOLTAGE LAWS
3   NODAL AND LOOP ANALYSES
4   THE OPERATIONAL AMPLIFIER
5   SUPERPOSITION AND SOURCE TRANSFORMATIONS
6   THEVENIN, NORTON AND MAXIMUM POWER
       TRANSFER THEOREMS
7   INDUCTORS, CAPACITORS AND DUALITY
8   FIRST ORDER RL AND RC CIRCUITS
9   SECOND ORDER RL AND RC CIRCUITS
10 SINUSOIDAL STEADY STATE ANALYSIS BY
       PHASOR METHODS
11 SINUSOIDAL STEADY STATE POWER CALCULATIONS
12 THREE PHASE POWER SYSTEMS

NOMENCLATURE

Components are given by capital letters. Hence, R, L, and C designate resistance, inductance and capacitance in ohms (Ω), henries (H) and farads (F) respectively.

Instantaneous values are given by lower case roman letters. Hence, i, v, q, p and w indicate instantaneous current in amperes (A), volts, (V), coulombs (C), watts (W) and joules, (J).

Constant values (direct-current or DC) values are indicated by capital roman letters. Thus, I, V, Q, P and W indicate current in amperes (A), volts, (V), coulombs (C), watts (W) and joules, (J). Observe that an integral over a time period is represented by a capital letter such as the charge accumulated in coulombs or the energy dissipated or stored in Joules between two points in time.

Phasors in alternating current analysis are designated by a bold face letter and apply only to the voltage, current and charge phasors, $\mathbf{V}, \mathbf{I}$ and $\mathbf{Q}$. Notice that these phasors contain real and imaginary parts and when written in polar form, the magnitude of the phasor is designated by the capital roman letter, V, I or Q. Notice that the magnitudes, V, I or Q do *not* bear vertical rules which are reserved for determinants.

The impedance in Ω and the admittance in $\mho$ may also be written in polar form but because impedances and admittances are not phasors, their magnitudes are designated as $|Z|$ and $|Y|$ respectively. Observe that because the admittance is in mhos (not siemens), to avoid confusion, the instantaneous power, $\mathbf{S} = P + jQ$ in VA, in itself a complex number, is written with a magnitude, $|S|$ where P is the power in watts and Q is the reactive power in VARS.

Other nomenclature is apparent and is defined where employed.

A NOTE ON SOLVING PROBLEMS

1. Read the problem carefully and begin by drawing a neatly labeled circuit diagram. Be sure to leave enough space for the sought after quantities and in the case of alternating current analysis, the circuit diagram may well be for the circuit in the so-called "phasor domain."

2. Think about the problem and decide upon the solution strategy. There is more than one way to do a problem and there may be one best solution technique. For example, if two quantities are required to solve a problem employing superposition and each quantity must be determined from a mesh or node analysis, think about employing the mesh or nodal analysis directly to find the solution.

3. Be careful about what is required by the statement of the problem. If, in the problem in (2) above *requires* a solution by superposition, you can use mesh or node analysis to establish each component of the solution, but you *must* show that you are doing the problem by superposition.

4. Be careful when you are using complex numbers in the conversions between rectangular and polar form. This is a dog that frequently bites because, with just a little experience, there is a tendency to let the mind race and frequently this leads to a severe disappointment.

5. When you are finished solving the problem, ask youself, first, if you have done all that is required by the problem statement and, second, does you result make sense. It always helps to think about the "real world."

6. List your solution clearly with units. And remember, that a person grading your work tends to give more credit to a neat, well organized, effort as opposed to one that is sloppy and haphazard and that you are likely to lose ten to twenty percent of the credit for not including the units.

Allan D. Kraus
Beachwood, Ohio
January, 2001

PROBLEM STATEMENTS

CHAPTER ONE
INTRODUCTION AND
BASIC CONCEPTS

CHARGE AND CURRENT

1.1: How many electrons pass a given point in a conductor in 20 s if the conductor is carrying 10 A?

1.2: If 240 C pass through a wire in 12 s what is the current?

1.3: What current passes through a conductor if 3.2×10^{21} electrons flow through the conductor in 16 s?

1.4: If the current in a conductor varies in accordance with the relationship

$$i = 4t + 24(1 - e^{-4t})\, \text{A}$$

how much charge flows from time, $t = 0$ to time, $t = 5\,s$?

ENERGY, POWER, VOLTAGE, CURRENT AND CHARGE

1.5: How much work is done in moving 10 nC of charge a distance of 68 cm in the direction of a uniform electric field of having a field strength of $E = 80\,\text{kV/m}$?

1.6: A charge of 0.5 C is brought from infinity to a point. Assume that infinity is at 0 V and determine the voltage at the terminal point if 14.5 J is required to move the charge.

1.7: If the potential difference between two points is 125 V, how much work is required to move a 3.2 C charge?

1.8: How many coulombs can be moved from point-A to point-B if $\Delta V_{AB} = 440\,\text{V}$ and a maximum of 842 J can be expended?

1.9: If 1 horsepower (hp) is equal to 0.746 kW, how much energy does a 20 hp motor deliver in 20 min?

1.10: If a 150 W incandescent bulb operates at 120 V, how many coulombs and electrons flow through the bulb in 1 h?

1.11: If a light bulb takes 1.2 A at 120 V and operates for 8 h/day, what is the cost of its operation for 30 days if power costs \$0.21/kwh?

1.12: If 1 calorie (1 cal) is equal to 4.184 J and it takes 1000 cal to raise 1 kg of water 1°C. how much current is carried by a 120 V heater if it is used to heat 4.82 kg of water from 25°C to 45°C in 4 min?

3

$$P_3 = 1875\,\text{W}$$
$$P_4 = 1406.25\,\text{W}$$

and

$$P_{12} = 468.755\,\text{W}$$

Determine V_{30}, V_3, I_4 and I_{12}.

1.32: In a certain network containing four resistors, the resistors dissipate powers (with the subscript corresponding to the value of the resistor) of

$$P_5 = 45\,\text{W}$$
$$P_1 = 9\,\text{W}$$
$$P_3 = 12\,\text{W}$$

and

$$P_6 = 6\,\text{W}$$

Determine V_5, V_1, I_3 and I_6.

CONTROLLED SOURCES

1.33: In the circuit in Fig 1.3, if the resistor dissipates $2304\,\text{W}$, determine the value of β,

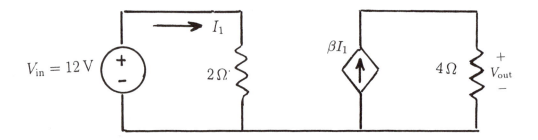

Figure 1.3.

1.34: In the circuit of Problem 1.33, determine the voltage gain, $|V_{\text{out}}/V_{\text{in}}|$ and the power gain, $|P_{\text{out}}/P_{\text{in}}|$

1.35: If $\mu = 10$ in Fig 1.4, determine the value of R_1 that is required to deliver a power of $25,600\,\text{W}$ to the $16\,\Omega$ resistor.

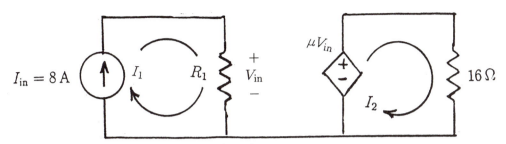

Figure 1.4.

1.36: Determine the current gain $|I_2/I_1|$ in the circuit of Problem 1.35.

1.37: In the circuit of Fig 1.5, $\beta = 8$ and $r_m = 8\,\Omega$. Determine the voltage gain, $|V_{\text{out}}/V_{\text{in}}|$ and the power gain, $|P_{\text{out}}/P_{\text{in}}|$.

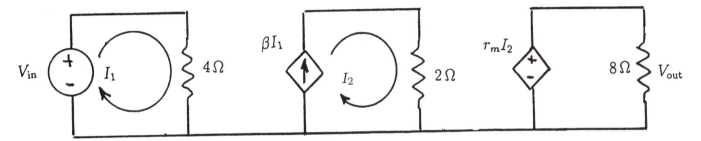

Figure 1.5.

1.38: In the circuit of Fig 1.6, $\mu = 8$ and $g_m = 0.5\,\mho$. Determine the current gain, $|I_3/I_{\text{in}}|$.

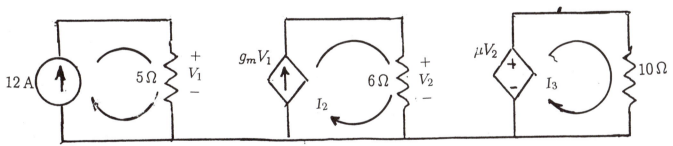

Figure 1.6.

10

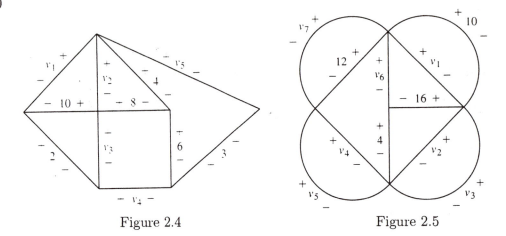

Figure 2.4   Figure 2.5

2.6: Determine v_1 through v_6 in the network of Fig 2.6.

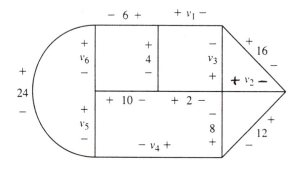

Figure 2.6

EQUIVALENT RESISTANCE PROBLEMS

2.7: Two resistors, R_1 and R_2 are connected in parallel. What is the value of the equivalent resistance if

(a) $R_1 = 20\,\Omega$ and $R_2 = 20\,\Omega$
(b) $R_1 = 20\,\Omega$ and $R_2 = 80\,\Omega$
(c) $R_1 = 20\,\Omega$ and $R_2 = \infty\,\Omega$
(d) $R_1 = 20\,\Omega$ and $R_2 = 0\,\Omega$
(e) $R_1 = 20\,\Omega$ and $R_2 = 50\,\Omega$

2.8: One of the resistances in the network shown in Fig 2.7 is blurred because of coffee spillage. If the equivalent resistance of the network is $R_{eq} = 20\,\Omega$, what is the value of the blurred resistor?

2.9: Determine the equivalent resistance looking into terminals *a-b* of the network shown in Fig 2.8.

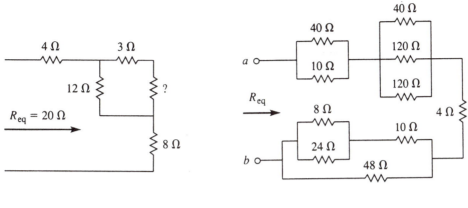

Figure 2.7   Figure 2.8

2.10: Determine the equivalent resistance looking into terminals *a-b* of the network shown in Fig 2.9.

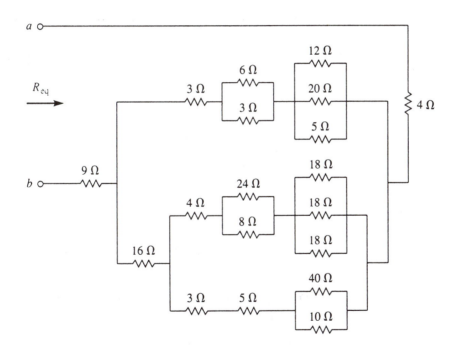

Figure 2.9

14

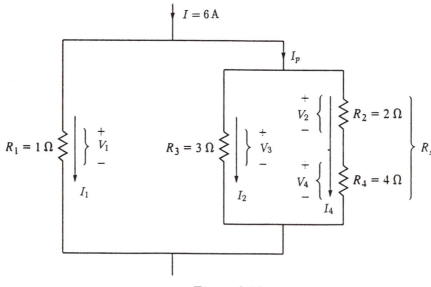

Figure 2.14

2.23: Find the current through and the voltage across each resistor in Fig 2.15.

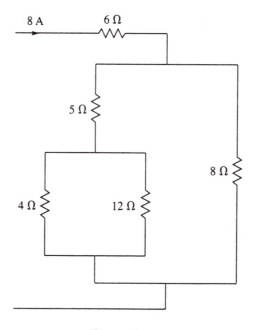

Figure 2.15

2.24: In the network of Fig 2.16, the resistance designators correspond to the resistance values. Determine the currents, I_1 and I_2 and the voltage across R_6 without using current and voltage division.

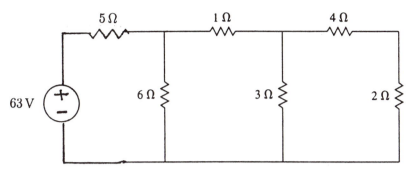

Figure 2.16

2.25: In the network of Problem 2.24, determine the currents, I_1 and I_2 and the voltage across R_6 by employing current and voltage division in place of KVL and KCL.

2.26: Find I_9, I_4 and V_7 in the network shown in Fig 2.17 without using voltage division.

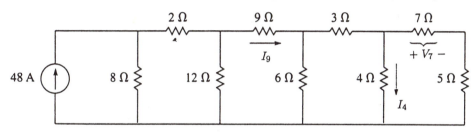

Figure 2.17

2.27: Find I_9, I_4 and V_7 in the network shown in Fig 2.17 without using current division.

2.28: Find I_3, I_4, V_{12} and V_{14} in the network shown in Fig 2.18.

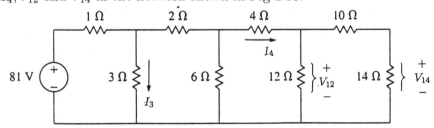

Figure 2.18

CHAPTER THREE
NODAL AND LOOP ANALYSES

SINGLE NODE PROBLEMS

3.1: Use nodal analysis to determine the voltage at point-1 (Node-1) in the network of Fig 3.1.

3.2: Use nodal analysis to determine the voltage at point-1 (Node-1) in the network of Fig 3.2.

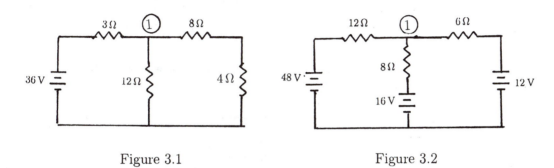

Figure 3.1   Figure 3.2

3.3: Use nodal analysis to determine the voltage at point-1 (Node-1) in the network of Fig 3.3.

3.4: Using nodal analysis, determine I_2 in the network of Fig 3.4.

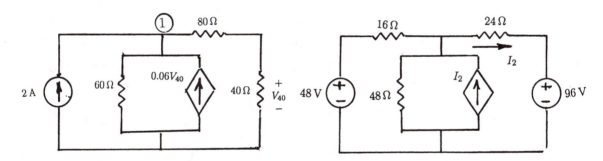

Figure 3.3   Figure 3.4

3.5: Use nodal analysis to determine the current flowing downward in the $8\,\Omega$ resistor in the network of Fig 3.5.

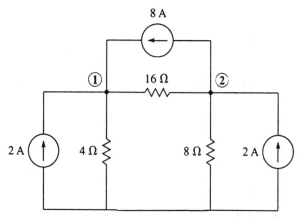

Figure 3.5

3.6: Determine the node voltages, V_1 and V_2, in the network shown in Fig 3.6.

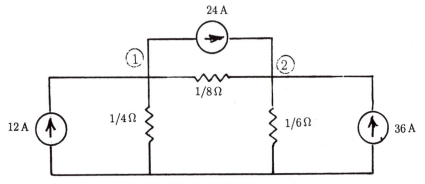

Figure 3.6

3.7: Use nodal analysis to determine the node voltages, V_1 and V_2 and then find the current through the $4\,\Omega$ resistor in the network of Fig 3.7.

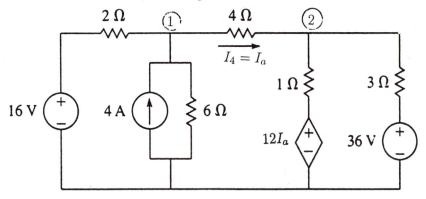

Figure 3.7

3.8: Write the node equations for the network shown in Fig 3.8 but do not attempt to solve them.

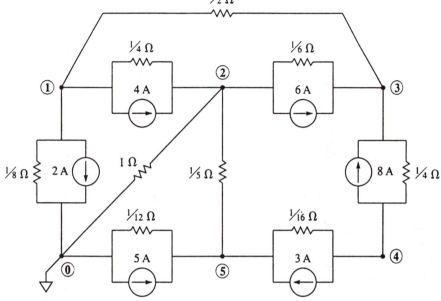

Figure 3.8

3.9: Determine the value of I_s that will produce a voltage of 36 V at node-2 in the network of Fig 3.9.

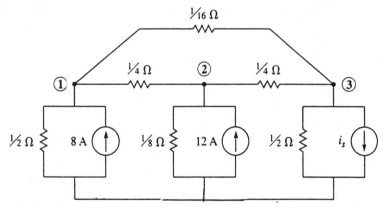

Figure 3.9

3.10: In the network of Fig 3.10, determine the current I_{40} using nodal analysis.

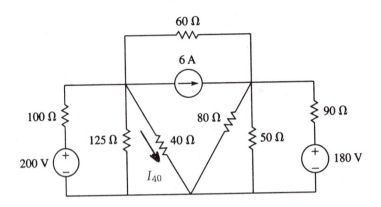

Figure 3.10

3.11: Determine the node voltages, V_1 and V_2 in the network shown in Fig 3.11.

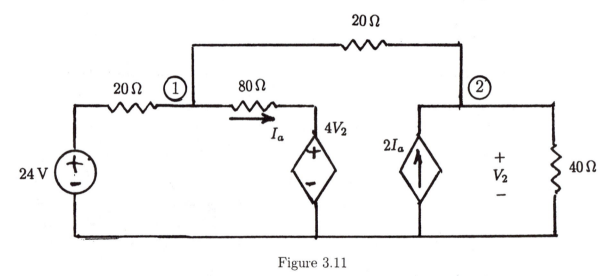

Figure 3.11

3.12: Use nodal analysis to determine the current in the $1/8\,\Omega$ resistor in the network of Fig 3.12.

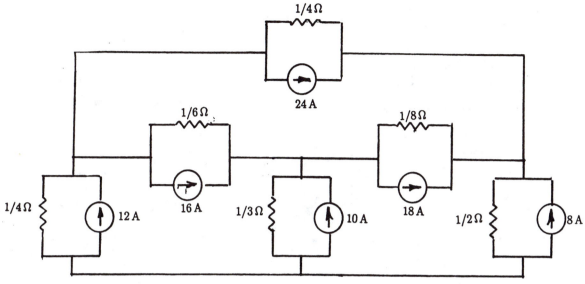

Figure 3.12

3.13: Determine the node voltages, V_1, V_2 and V_3 in the network of Fig 3.13.

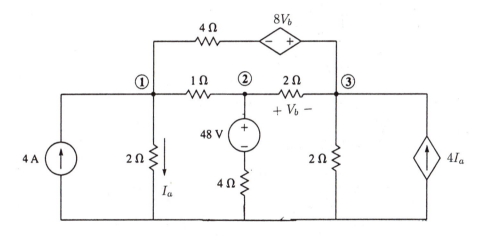

Figure 3.13

SUPERNODE PROBLEMS

3.14: Determine the node voltages, V_1, V_2 and V_3 in the network of Fig 3.14.

3.15: Determine the node voltages, V_1, V_2, V_3 and V_4 in the network of Fig 3.15.

24

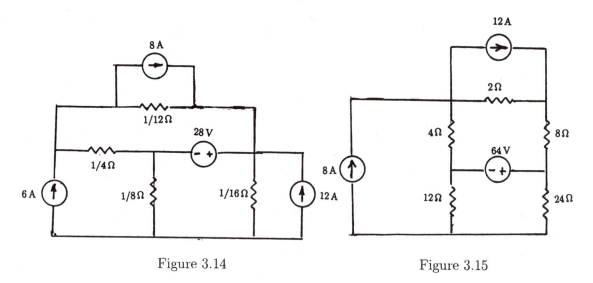

Figure 3.14   Figure 3.15

3.16: Determine the node voltages, V_1, V_2 and V_3 in the network of Fig 3.16.

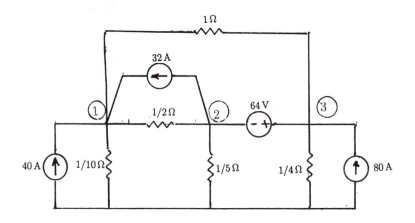

Figure 3.16

3.17: Determine the node voltages, V_1, V_2 and V_3 in the network of Fig 3.17.

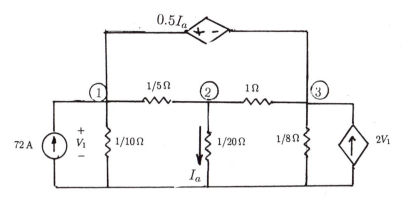

Figure 3.17

A LINK TO CHAPTER TWO

3.18: Figure 2.14 shows four resistors with subscripts corresponding to their resistance values. The right hand leg containing the $2\,\Omega$ and $4\,\Omega$ resistors is designated as R_s (s for series combination) and the current entering the parallel combination of the $3\,\Omega, 2\,\Omega$ and $4\,\Omega$ resistors is designated as I_p. If the current entering the network is $I = 6\,\text{A}$, determine the current through and the voltage across each resistor (This is a repetition of Problem 2.22).

3.19: In the network of Fig 2.16, the resistance designators correspond to the resistance values. Use nodal analysis to determine the currents, I_1 and I_2 and the voltage across R_6 (This is a repetition of Problem 2.24).

3.20: Use nodal analysis to determine I_3, I_4, V_{12} and V_{14} in the network shown in Fig 2.18 (This is a repetition of Problem 2.28).

3.21: Use nodal analysis to determine I_5 and V_{20}. Then determine the power dissipated in the $24\,\Omega$ resistor in Fig 2.22 (This is a repetition of Problem 2.32).

SINGLE LOOP ANALYSIS

3.22: Determine the current, I, in the network shown in Fig 3.18.

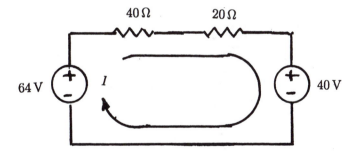

Figure 3.18

3.23: Determine the current, I, in the network shown in Fig 3.19.

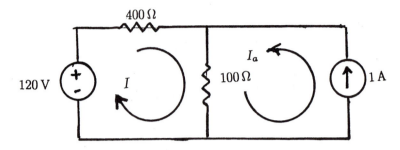

Figure 3.19

3.24: Determine the current, I, in the network shown in Fig 3.20.

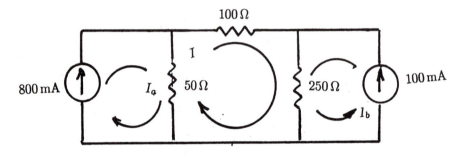

Figure 3.20

3.25: Determine the current, I, and the equivalent resistance seen by the 102 V source in Fig 3.21.

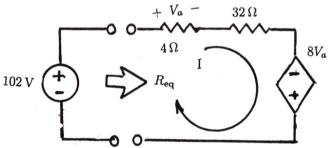

Figure 3.21

3.26: Use loop analysis to determine the current flowing in the $8\,\Omega$ resistor in Fig 3.5 (This is a repeat of Problem 3.5).

3.27: Determine the current, I, in the network shown in Fig 3.22.

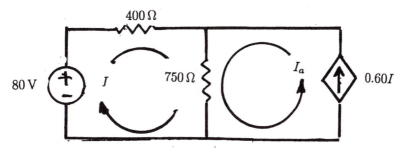

Figure 3.22

MULTIPLE LOOP ANALYSIS

3.28: Use loop analysis to determine the current flowing in the $4\,\Omega$ resistor in Fig 3.7 (This is a repeat of Problem 3.7).

3.29: Write the mesh equations for the network shown in Fig 3.23 but do not attempt to solve them.

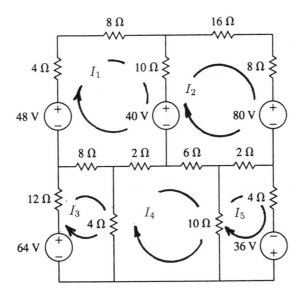

Figure 3.23

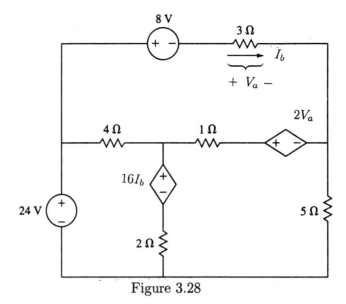

Figure 3.28

3.35: Problem 2.22 described a problem in which the network contained four resistors driven by a current of 6 A. The problem is repeated here with excitation by a current source of 6 A as shown in Fig 3.29. Use mesh analysis to determine the currents through and the voltage across each resistor. (This is a repeat of Problem 2.22 and Problem 3.18).

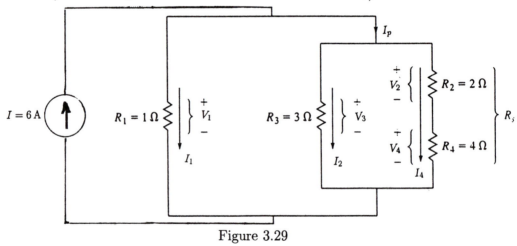

Figure 3.29

3.36: Use mesh analysis to determine the currents, I_1 and I_2 and the voltage across R_6 in the network of Fig 2.16 (This is a repeat of Problems 2.24 and 3.19).

3.37: Use mesh analysis to find I_3, I_4, V_{12} and V_{14} in the network shown in Fig 2.18 (This is a repeat of Problems 2.28 and 3.20).

3.38: Use mesh analysis to determine I_5, V_{20} and the power dissipated in the 24 Ω resistor in Fig 2.22 (This is a repeat of Problems 2.32 and 3.21).

CHAPTER FOUR

THE OPERATIONAL AMPLIFIER

ANALYSIS USING THE IDEAL OP-AMP MODEL

4.1: Determine the voltage gain and the input resistance for each of the ideal operational amplifier circuits of Fig 4.1.

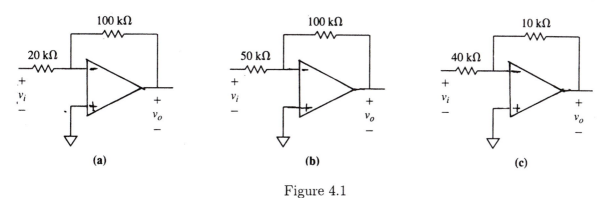

(a) (b) (c)

Figure 4.1

4.2: Determine the voltage gain and the input resistance for each of the ideal operational amplifier circuits of Fig 4.2.

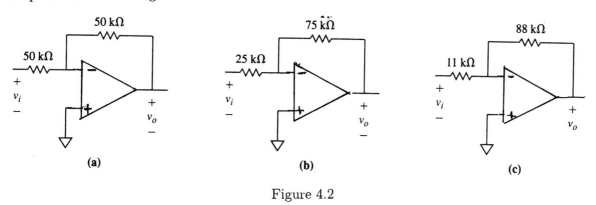

(a) (b) (c)

Figure 4.2

4.3: Determine the voltage gain of the ideal operational amplifier circuit of Fig 4.3.

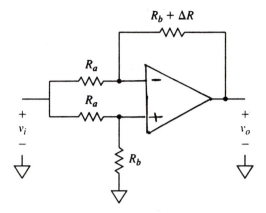

Figure 4.8

4.9: Figure 4.9 shows how an ideal operational amplifier can be put together by using resistors that have relatively small resistance values. If $R_1 = 2000$ Om, determine the value of a single feedback resistor to produce a gain of -1200 and then, with $R_1 = 2000\,\Omega$ and $R_b = 50\,\Omega$, determine the value of R_a to provide a gain of -1200.

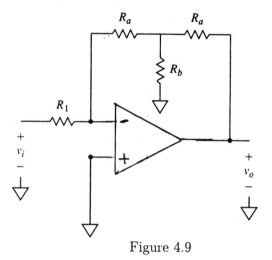

Figure 4.9

4.10: Figure 4.10 illustrates the use of an ideal operational amplifier as a *negative impedance converter*. Determine the input resistance.

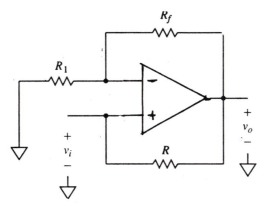

Figure 4.10

4.11: An ideal operational amplifier in the inverting configuraton is to have a gain of -125 and an input resistance as high as possible. If no resistance in the op-amp circuit is to have a value higher that $5\,\text{M}\Omega$, how is this design achieved by using just two resistors.

4.12: In the cascade of ideal operational amplifiers shown in Fig 4.11, if $v_i = 2\,\text{V}$ and $v_o = 30\,\text{V}$, determine the value of R.

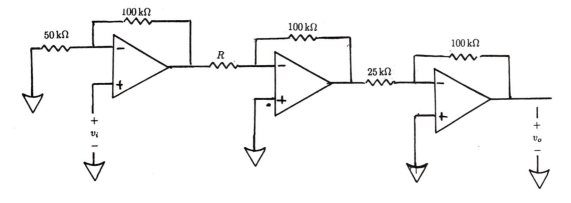

Figure 4.11

GENERAL SUMMING AMPLIFIER PROBLEMS (THE IDEAL OP-AMP MODEL)

4.13: Determine v_o in the ideal operational amplifier circuit shown in Fig 4.12.

NON IDEAL OP-AMPS: FINITE OPEN LOOP GAIN

4.19: If the operational amplifier in Fig 4.18 has a finite gain of $A = 500$, determine the output voltage, v_o, if $v_i = 4\,\text{V}$.

4.20: In problem 4.19, to determine the value of the gain, A to make $v_o = -18.5\,\text{V}$.

4.21: Determine the value of R_f required in Fig 4.19 to make the outout voltage, $v_o = 16\,\text{V}$ when $R_1 = 50\,\text{k}\Omega$, $V_i = -4\,\text{V}$ and the amplifier has an actual gain of $A = 400$.

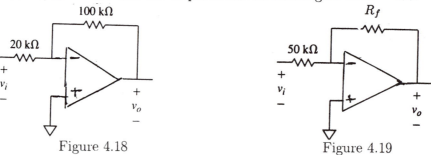

Figure 4.18   Figure 4.19

4.22: If the operational amplifier in Fig 4.20 has an actual gain of $A = 200$, determine the value of R_1 required to make $v_o = 18\,\text{V}$ with $v_i = -2\,\text{V}$.

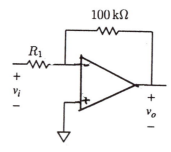

Figure 4.20

SATURATION EFFECTS

4.23: The input to an ideal operational amplifier in the inverting configuration is shown in Fig 4.21, If $R_f = 200\,\text{k}\Omega$, $R_1 = 50\,\text{k}\Omega$ and $V_{\text{sat}} = 12\,\text{V}$, determine and plot the output characteristic.

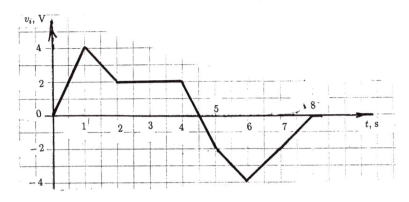

Figure 4.21

4.24: The operational amplifier of Problem 4.23 is subjected to a sinusoidal input of

$$v_i(t) = 3\sqrt{2}\sin 100t \text{ V}$$

Determine the time intervals for operation in both the positive and negative saturation regions.

4.25: Determine the extremes of operation of an operational amplifier with a saturation voltage of 8 V and a gain of 400. Then determine the output voltage when $v_d = v_+ - v_- = -6\,\text{mV}$.

4.26: An ideal operational amplifier has a saturation voltage of $V_{\text{sat}} = 8\,\text{V}$ and operates in the inverting configuration with feedback and input resistances of $R_f = 100\,\text{k}\Omega$ and $R_1 = 25\,\text{k}\Omega$, Determine v_o and $v_d = v_+ - v_-$ if (a) $v_i = 0.625\,\text{V}$, (b) $v_i = -3\,\text{V}$ and (c) $v_i = +4\,\text{V}$

SELECTION OF OPERATIONAL AMPLIFIER COMPONENTS

4.27: Design an ideal operational amplifier circuit to have the general input-output relationship

$$v_o = -(8v_{a1} + 4v_{a2} + 2v_{a3}) + (3v_{b1} + 6v_{b2} + 9v_{a3})$$

4.28: Use the ideal operational amplifier circuit developed in Problem 4.27 to determine the output voltage if all input voltages are taken at a nominal 1 V.

4.29: Design an ideal operational amplifier circuit to have the general input-output relationship

$$v_o = -(4v_{a1} + 2v_{a2} + v_{a3} + 8v_{a4}) + (5v_{b1} + 2v_{b2})$$

4.30: Use the ideal operational amplifier circuit developed in Problem 4.27 to determine the output voltage if all input voltages are taken at a nominal 1 V.

42

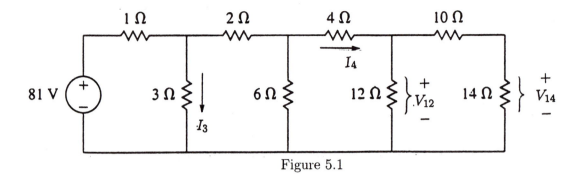

Figure 5.1

5.7: In the network of Fig 5.2, the resistance designators correspond to the resistance values. By assuming that $I_2 = 1\,A$, determine I_6 and V_3. (This is a repeat of Problem 2.24).

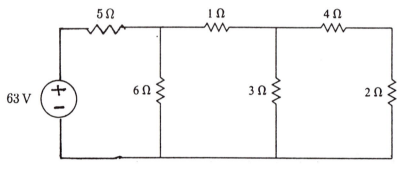

Figure 5.2

5.8: In the network of Fig 5.3, the resistance designators correspond to the resistance values. By assuming that $V_1 = 1\,V$, determine I_4, I_6 and V_2.

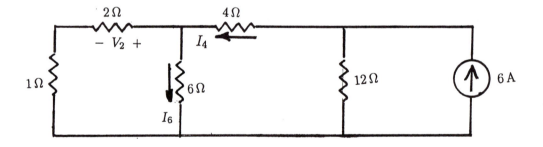

Figure 5.3

5.9: In the network of Fig 5.4, the resistance designators correspond to the resistance values. By assuming that $I_{10} = 1\,\text{A}$, determine I_2, I_4 and V_8.

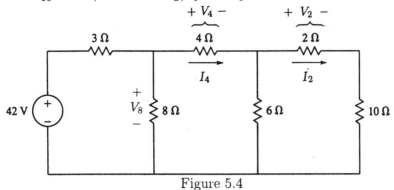

Figure 5.4

5.10: In the network of Fig 5.5, the resistance designators correspond to the resistance values. By assuming that $I_3 = 1\,\text{A}$, determine I_9, I_{40} and V_7. (This is a repeat of Problem 2.30).

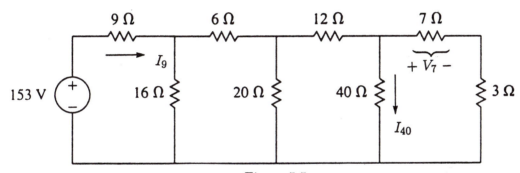

Figure 5.5

SOURCE TRANSFORMATION PROBLEMS

5.11: Transform the ideal voltage sources shown in Fig 5.6 to ideal current sources.

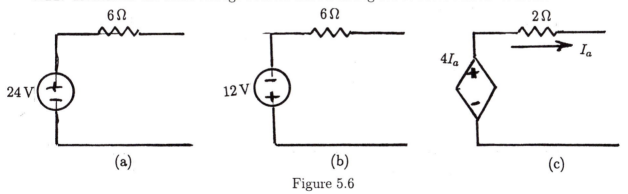

(a)   (b)   (c)

Figure 5.6

46

SUPERPOSITION PROBLEMS

5.16: Use superposition to determine the value of the current, I, in the network of Fig 5.11.

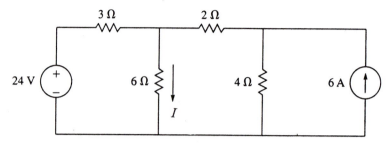

Figure 5.11

5.17: Use superposition to determine the value of the current, I, in the network of Fig 5.12.

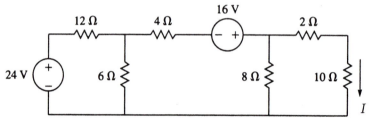

Figure 5.12

5.18: Use superposition to determine the value of the current, I, in the network of Fig 5.13.

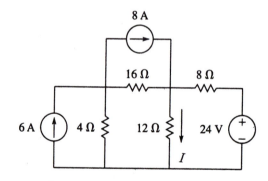

Figure 5.13

5.19: Use superposition to determine the value of the current, I, in the network of Fig 5.14.

5.20: Use superposition to determine the value of the voltage, V, in the network of Fig 5.15.

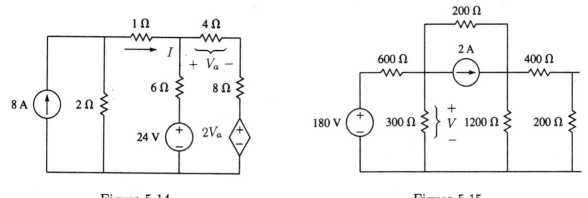

Figure 5.14   Figure 5.15

5.21: Use superposition to determine the value of the current, I, in the network of Fig 5.16.

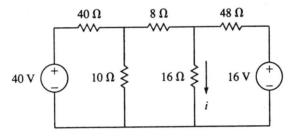

Figure 5.16

5.22: Use superposition to determine the value of the voltage, V, in the network of Fig 5.17.

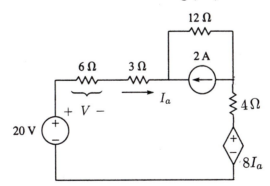

Figure 5.17

5.30: In the arrangement of Fig 5.23, $v_{i1} = 4\,$V, $v_{i2} = 8\,$V, $R_2 = 80\,$kΩ and $R_4 = 40\,$kΩ. Find the values of R_1 and R_3 needed to make $v_o = 4\,$V suject to the constraint that $R_1 = 1.25R_3$.

CHAPTER SIX

THEVENIN, NORTON AND
MAXIMUM POWER TRANSFER THEOREMS

THEVENIN AND NORTON THEOREMS FOR
NETWORKS WITHOUT CONTROLLED SOURCES

6.1: Use Thevenin's theorem to find the resistance that must be connected across terminals
a-b in Fig 6.1 in order for the resistor current to be 3 A.

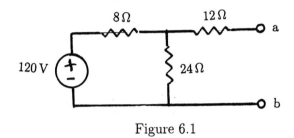

Figure 6.1

6.2: Determine the Thevenin equivalent network between terminals *a-b* in Fig 6.2.

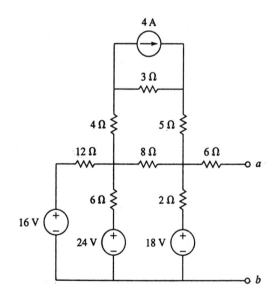

Figure 6.2

6.3: Use Thevenin's theorem to find the current, I, in the network of Fig 6.3.

6.4: Use Thevenin's theorem to find the current, I, flowing through the $600\,\Omega$ resistor in the network of Fig 6.4.

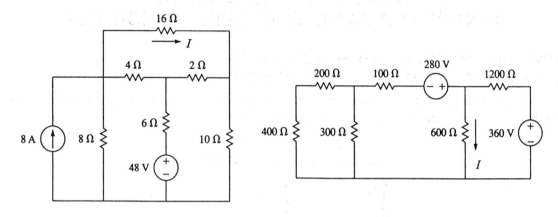

Figure 6.3   Figure 6.4

6.5: Determine the Norton equivalent circuit for terminals a-b in the network of Fig 6.5.

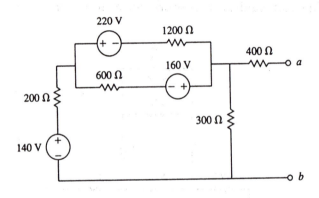

Figure 6.5

6.6: Use Thevenin's theorem to find the voltage across the $320\,\Omega$ resistor in the network of Fig 6.6.

6.7: Use the Thevenin's theorem or the Norton's theorem to determine the value of R that will allow a current of $1\,A$ to flow through the $2\,\Omega$ resistor in Fig 6.7.

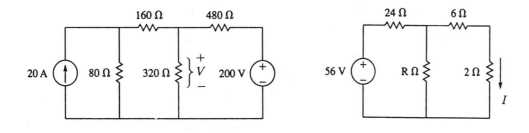

Figure 6.6

Figure 6.7

6.8: Use Norton's theorem to determine the current through the 10 Ω resistor in the network of Fig 6.8.

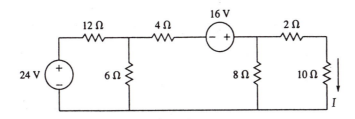

Figure 6.8

THEVENIN AND NORTON THEOREMS FOR NETWORKS WITH CONTROLLED SOURCES

6.9: Use Thevenin's theorem to determine the power dissipated by the 12 Ω resistor in Fig 6.9.

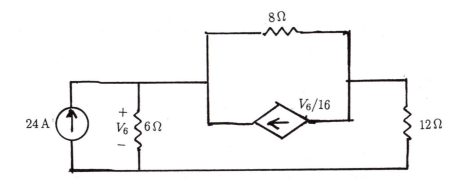

Figure 6.9

54

6.10: Determine the Norton equivalent for terminals *a-b* in the network of Fig 6.10.

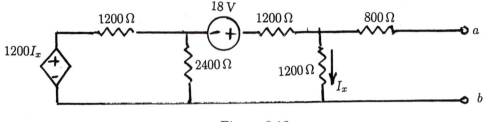

Figure 6.10

6.11: Use Thevenin's theorem to determine the current flowing through the 1.6 Ω resistor in Fig 6.11.

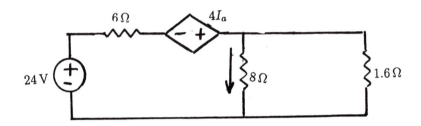

Figure 6.11

6.12: Use Norton's theorem to determine the current flowing through the right-hand 4 Ω resistor in Fig 6.12.

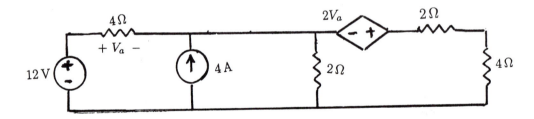

Figure 6.12

6.13: Use Thevenin's theorem to determine the current flowing through the 16 Ω resistor in Fig 6.13.

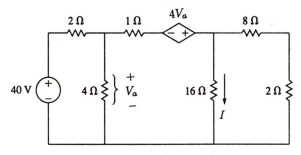

Figure 6.13

6.14: Use Thevenin's theorem to determine the power dissipated by a $300\,\Omega$ resistor connected across terminals a-b in Fig 6.14.

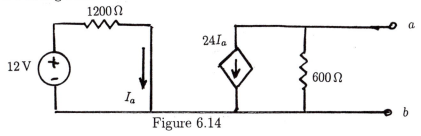

Figure 6.14

6.15: Use Thevenin's theorem to determine the current through the $6\,\Omega$ resistor in the network of Fig 6.15.

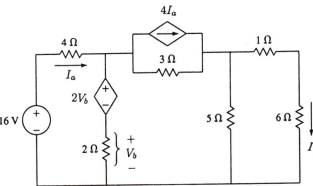

Figure 6.15

6.16: Determine the current through the $3\,\Omega$ resistor in Fig 6.16 by using superposition to find the Thevenin equivalent circuit.

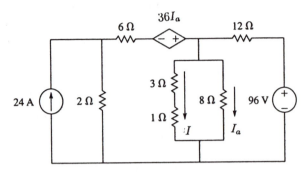

Figure 6.16

MAXIMUM POWER TRANSFER

6.17: In the network of Fig 6.17, determine the value of the load to be placed across terminals *a-b* in order for the load to draw maximum power. Then, determine the value of this power.

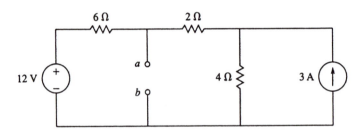

Figure 6.17

6.18: In the network of Fig 6.18, determine the value of the load to be placed across terminals *a-b* in order for the load to draw maximum power. Then, determine the value of this power.

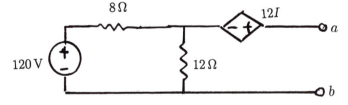

Figure 6.18

6.19: In the network of Fig 6.19, determine the value of the load to be placed across terminals a-b in order for the load to draw maximum power. Then, determine the value of this power.

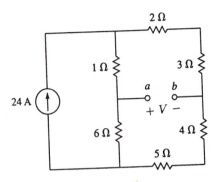

Figure 6.19

6.20: In the network of Fig 6.20, determine the value of the load to be placed across terminals a-b in order for the load to draw maximum power. Then, determine the value of this power.

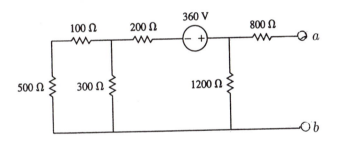

Figure 6.20

6.21: In the network of Fig 6.21, determine the value of the load to be placed across terminals a-b in order for the load to draw maximum power. Then, determine the value of this power.

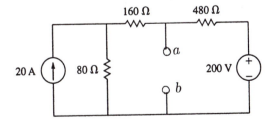

Figure 6.21

6.22: In the network of Fig 6.22, determine the value of the load to be placed across terminals *a-b* in order for the load to draw maximum power. Then, determine the value of this power.

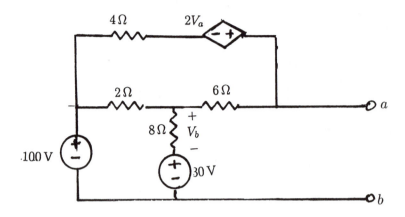

Figure 6.22

CHAPTER SEVEN
INDUCTORS, CAPACITORS AND DUALITY

INDUCTORS

7.1: Determine the voltage induced in a 250 mH inductor when the current changes at the rate of 50A/s.

7.2: Two inductors of 300 mH and 600 mH are connected in parallel. Determine the equivalent inductance.

7.3: If the current through an 80 mH inductor is

$$i = 20\sqrt{2}\sin 400t\,\text{A}$$

determine the power drawn by the inductor.

7.4: The current through a 400 mH inductor is given by

$$i = 4e^{-t} - 4e^{-2t}\,\text{A}$$

Determine
(a) the energy stored in the inductor at $t = 0\,s$,
(b) the energy stored in the inductor between $t = 0\,s$ and $t = 0.5\,s$,
(c) the voltage at $t = 0.5\,s$ and
(d) the instantaneous power at $t = 0.5\,s$

7.5: A parallel combination of two inductors ($L_1 = 1.2\,\text{H}$ and $L_2 = 0.60\,\text{H}$) is placed in series with two more inductors ($L_3 = 1.0\,\text{H}$ and $L_2 = 0.80\,\text{H}$). Determine the equivalent inductance.

7.6: The simple network shown in Fig 7.1 is connected to a current source at $t = 0$. The current provided by the source is $i = 4t\,\text{A}$.

At $t = 5\,s$, find
(a) the instantaneous power drawn by the resistor,
(b) the instantaneous power drawn by the inductor,
(c) the instantaneous power delivered by the source,

and between $0 \leq t \leq 4\,s$,

(d) the energy dissipated by the resistor
(e) the energy stored by the inductor
(f) the energy delivered by the current source.

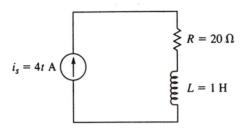

Figure 7.1

7.7: The current through a 240 mH inductor is given by $i = 4 \sin 400t$ A. Determine
(a) the instantaneous power at $t = \pi/200$ s and
(b) the energy stored during the period $0 \le t \le \pi/200$ s.

7.8: If the current through a 135 mH inductor is given by $i = 8t^2 + 4t + 2$ A, how much energy is stored or removed in the inductor during the period, $0 \le t \le 1.2$ s?

7.9: In Fig 7.2, the current source is supplying 288 W at $t = 2$ s. Determine the value of R

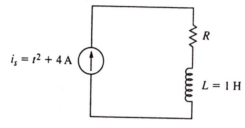

Figure 7.2

7.10: Two inductors of 0.080 H and 0.120 H are connected in parallel. If the current at a particular instant of time is 12.5 A, determine the currents in each inductor.

7.11: A pair of inductors, L_1 and L_2, are connected in parallel and the parallel combination is connected to an inductor with $L = 0.80$ H. Determine the value of L_1 and L_2 if the equivalent inductance for the combination is $L_{eq} = 1.60$ H and $L_1 = 4L_2$.

7.12: The current through a 400 mH inductor, over a period of 8 s, is described by the waveform

$$
i = \begin{cases}
2t^2 \text{ A}; & 0 < t < 2\,\text{s} \\
8 \text{ A}; & 2\,\text{s} < t < 4\,\text{s} \\
40 - 8t \text{ A}; & 4\,\text{s} < t < 8\,\text{s}
\end{cases}
$$

Determine the voltage across the inductor and sketch the waveform.

CAPACITORS

7.13: Determine the current through and the charge on a $50\,\mu\text{F}$ capacitor subjected to a voltage of

$$v = 120\sin 250t \text{ V}$$

7.14: An $80\,\mu\text{F}$ capacitor is charged so that it stores $0.40\,\text{J}$. If an uncharged $120\,\mu\text{F}$ capacitor is hooked up across the terminals of the $80\,\mu\text{F}$ capacitor, determine the final energy in the system.

7.15: If an $80\,\mu\text{F}$ capacitor is charged to $600\,\mu\text{C}$ and then connected across the terminals of an uncharged $160\,\mu\text{F}$ capacitor, determine the charge transferred from the $80\,\mu\text{F}$ capacitor to the $160\,\mu\text{F}$ capacitor.

7.16: Two capacitors of $100\,\mu\text{F}$ and $400\,\mu\text{F}$ are connected in series. Determine the equivalent capacitance.

7.17: If the voltage across a $200\,\mu\text{F}$ capacitor is

$$v = 120\sin 400t \text{ V}$$

determine the power provided to the capacitor.

7.18: A parallel combination of two capacitors ($C_1 = 80\,\mu\text{F}$ and $C_2 = 120\,\mu\text{F}$) is placed in parallel with a series combination of two more capacitors ($C_3 = 40\,\mu\text{F}$ and $C_4 = 160\,\mu\text{F}$). Determine the equivalent capacitance.

7.19: The simple network shown in Fig 7.3 is connected to a voltage source at $t = 0$. The voltage provided by the source is $v = 2t + 8\,\text{V}$. At $t = 4\,\text{s}$, find

(a) the instantaneous power drawn by the resistor,
(b) the instantaneous power drawn by the capacitor,
(c) the instantaneous power delivered by the source,

and between $0 \le t \le 4\,\text{s}$,

(d) the energy dissipated by the resistor
(e) the energy stored by the capacitor
(f) the energy delivered by the current source.

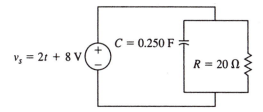

Figure 7.3

7.20: The voltage across a $40\,\mu F$ capacitor is given by $v = 4e^{-t} - 2^{-2t}\,V$. Determine how much energy will be stored or removed during the period $0 \le t \le 2\,s$.

7.21: How much energy is stored in an initially uncharged $125\,\mu F$ capacitor between 0 and $8\,s$ when a voltage of $18e^{-t/10}\,V$ is placed across its terminals.

7.22: Each pair of two pairs of capacitors are connected in parallel. The first pair has $C_1 = 60\,\mu F$ and $C_2 = 30\,\mu F$ and the second pair has $C_3 = 120\,\mu F$ with C_4 unknown. The two pairs of capacitors are connected in series and if the equivalent capacitance of the entire arrangement is $C_{eq} = 60\,\mu F$, determine the value of the unknown capacitor, C.

7.23: A $100\,\mu F$ capacitor is connected in series to a $300\,\mu F$ capacitor. At a certain instant of time, the total voltage across the two capacitors is $20\,V$. Determine the voltage distribution across the two capacitors.

7.24: The voltage across a $1\,F$ capacitor during a period of $10\,s$ is described by the waveform

$$v = \begin{cases} t\,V; & 0 < t < 2\,s \\ 3 - 0.5t\,V; & 2\,s < t < 4\,s \\ 1.5t - 5\,V; & 4\,s < t < 6\,s \\ 22 - 3t\,V; & 6\,s < t < 7\,s \\ 0.5t - 2.5\,V; & 7\,s < t < 9\,s \\ t - 7\,V; & 9\,s < t < 10\,s \end{cases}$$

Determine the current through the capacitor and sketch the waveform.

DUAL NETWORKS

7.25: Construct the dual network for the network in Fig 7.4 and then verify that the node equations for the dual network are in the same form as the mesh equations for the network in Fig 7.4.

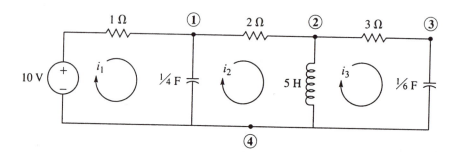

Figure 7.4

7.26: Construct the dual network for the network in Fig 7.5 and then verify that the node equations for the dual network are in the same form as the mesh equations for the network in Fig 7.5

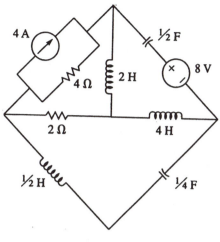

Figure 7.5

7.27: Construct the dual network for the network in Fig 7.6 and then verify that the node equations for the dual network are in the same form as the mesh equations for the network in Fig 7.6.

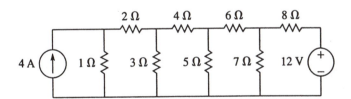

Figure 7.6

7.28: Construct the dual network for the network in Fig 7.7 and then verify that the node equations for the dual network are in the same form as the mesh equations for the network in Fig 7.7.

64

7.28: Construct the dual network for the network in Fig 7.7 and then verify that the node equations for the dual network are in the same form as the mesh equations for the network in Fig 7.7.

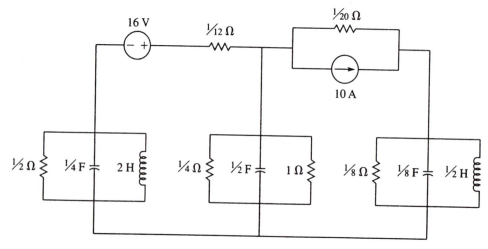

Figure 7.7

CHAPTER EIGHT

FIRST ORDER RL AND RC CIRCUITS

THE EXPONENTIAL FUNCTION

8.1: Measurements of the exponentially decaying charge on a sphere provide the following data:

Time, ms	Charge, μC
0.40	170.429
0.425	168.733

Determine the initial value of the charge, the time constant and the charge at $t = 0.625\,\text{ms}$.

8.2 Measurements of an exponentially decaying voltage provide the following data:

Time, s	Voltage, V
0.12	658.574
0.15	566.840

Determine the initial value of the voltage, the time constant and the voltage at $t = 32.5\,\text{ms}$.

8.3: If an exponentially decaying current has a time constant of 100 ms and if the current has a value of 12.42 A at $t = 165\,\text{ms}$, determine the value of the current at $t = 0$ and the time at which the current has a value of 4.571 A.

8.4: An exponentially decaying voltage has a time constant of 400 ms and an initial value of 1.2 kV. What is the value of the voltage at $t = 125\,\text{ms}$ and what is the value of the voltage five time constants later.

8.5 Measurements of an exponentially decaying current yield the following data:

Time, s	Current, A
0.250	294.30
0.275	266.30

Determine the initial value of the current and the time constant.

8.6 Measurements of an exponentially decaying voltage yield the following data:

Time, ms	Voltage, A
10	53.375
25	25.214

Determine the initial value of the voltage and the time constant.

EVALUATION OF INITIAL CONDITIONS

8.7: In Fig 8.1 the switch has been in position-a for a long period of time. At $t = 0,^+$ it moves instantaneously to position-b. Determine

$$
\begin{array}{cccc}
i_R(0^-) & i_L(0^-) & i_C(0^-) & v_C(0^-) \\
v_L(0^-) & i_R(0^+) & i_L(0^+) & i_C(0^+) \\
v_C(0^+) & v_L(0^+) & i'_L(0^+) &
\end{array}
$$

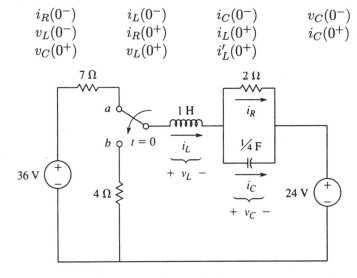

Figure 8.1

8.8: In Fig 8.2, the switch has been closed for a long period of time. At $t = 0,^+$ it opens instantaneously. Consider $t = 0^-$ and $t = 0^+$ as the instants just before and just after the switch opens and determine:

$$
\begin{array}{cccc}
i_{R1}(0^-) & i_L(0^-) & i_C(0^-) & i_{R2}(0^-) \\
v_L(0^-) & v_{R1}(0^-) & v_C(0^-) & v_{R2}(0^-) \\
i_{R1}(0^+) & i_L(0^+) & i_C(0^+) & i_{R2}(0^+) \\
v_L(0^+) & v_{R1}(0^+) & v_C(0^+) & v'_C(0^+)
\end{array}
$$

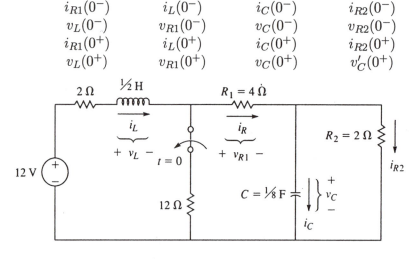

Figure 8.2

8.9: In Fig 8.3, the switch has been open for a long period of time and the voltage across the capacitor is 40 V. At $t = 0,^+$, the switch closes instantaneously. Consider $t = 0^-$ and $t = 0^+$ as the instants just before and just after the switch opens and determine:

$$i_{R1}(0^+) \qquad i_L(0^+) \qquad i_C(0^+) \qquad i_{R2}(0^+)$$
$$v_L(0^+) \qquad v_{R1}(0^+) \qquad v_C(0^+) \qquad i'_L(0^+)$$

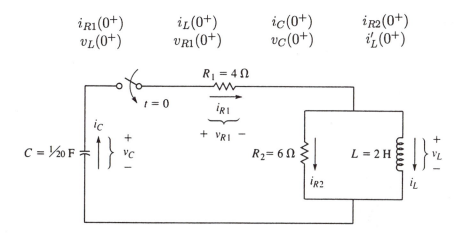

Figure 8.3

8.10: In Fig 8.4, the switch has been open for a long period of time and the voltage across the $1/100$ F capacitor is 80 V. At $t = 0,^+$, the switch closes instantaneously. Consider $t = 0^-$ and $t = 0^+$ as the instants just before and just after the switch closes and determine:

$$i_{R1}(0^+) \qquad v_{R1}(0^+) \qquad v'_C(0^+)$$
$$i_{R3}(0^+) \qquad i_{R2}(0^+) \qquad i_C(0^+)$$
$$v_{R3}(0^+) \qquad v_{R2}(0^+) \qquad v_C(0^+)$$

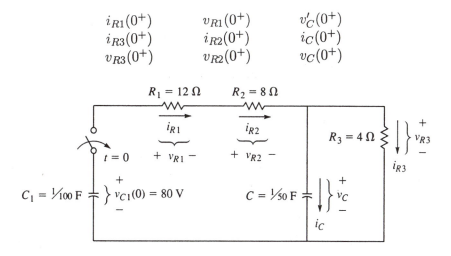

Figure 8.4

THE UNDRIVEN FIRST ORDER NETWORK

8.11: In the network of Fig 8.5, the switch opens instantaneously at $t = 0$. Determine the current response for all $t \geq 0$.

8.12: In the network of Fig 8.6, the switch opens instantaneously at $t = 0$. Determine the voltage response for all $t \geq 0$.

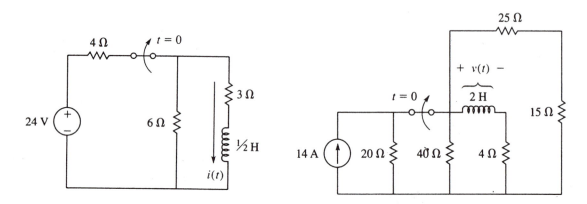

Figure 8.5   Figure 8.6

8.13: In the network of Fig 8.7, the switch opens instantaneously at $t = 0$. Determine the voltage across the $16\,\Omega$ resistor for all $t \geq 0$.

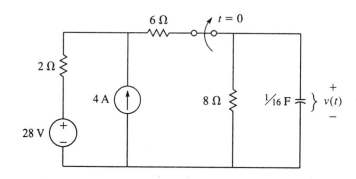

Figure 8.7

8.14: In the network of Fig 8.8, the switch opens instantaneously at $t = 0$. Determine the voltage across the $16\,\Omega$ resistor for all $t \geq 0$.

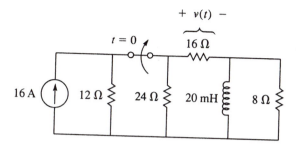

Figure 8.8

8.15: In the network of Fig 8.9, the switch moves from position-1 to position-2 instantaneously at $t = 0$. Determine the voltage response for all $t \geq 0$.

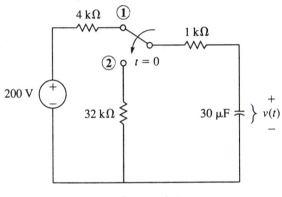

Figure 8.9

8.16: In the network of Fig 8.10, capacitor, C_1 is charged to 28 V and C_2 is uncharged. The switch closes instantaneously at $t = 0$. Determine the current $i(t)$ leaving C_1 for all $t \geq 0$.

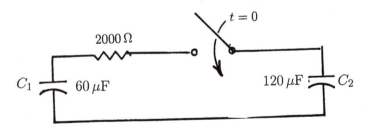

Figure 8.10

8.17: In the network of Fig 8.11, the switch closes instantaneously at $t = 0$. Determine the voltage across the capacitor for all $t \geq 0$.

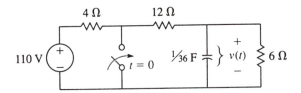

Figure 8.11

8.18: In the network of Fig 8.12, the switch closes instantaneously at $t = 0$. Determine the current through the inductor and the voltage across the $12\,\Omega$ resistor for all $t \geq 0$

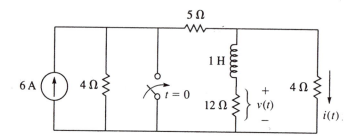

Figure 8.12

THE DRIVEN FIRST ORDER NETWORK

8.19: In the network of Fig 8.13, the switch closes instantaneously at $t = 0$. If the capacitors are initially uncharged, determine $i_1(t)$ and $i_2(t)$ for all $t \geq 0$.

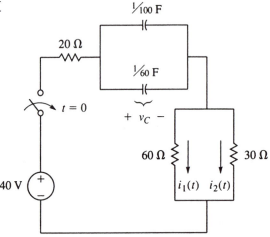

Figure 8.13

8.20: In the network of Fig 8.14, the switch closes instantaneously at $t = 0$. Determine $v_1(t)$ and $v_2(t)$ for all $t \geq 0$.

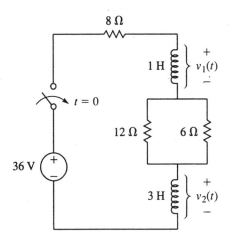

Figure 8.14

8.21: Figure 8.15 shows a network that can be used to control the current through the inductor to values between 1.10 and 1.30 A. Switch-1 (S_1) closes at $t = 0$ and Switch-2 (S_2) opens at $t = t_1$ when $i(t) = 1.3$ A. Switch-2 then remains open until $t = t_2$ when $i(t) = 1.1$ A at which time it closes and remains closed until $t = t_3$ when, once again, $i(t) = 1.3$ A. Determine the times, t_1, t_2 and t_3 and the steady-operation ratio of switch-2 *open time* to *closed time*.

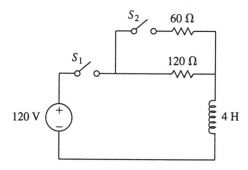

Figure 8.15

8.22: The purpose of the network in Fig 8.16 is to use switch-1 and R_1 to permit the capacitor to charge during a specified time interval and to use switch-2 and R_2 to hold the charge at the specified level. Select a value for R_1 to make the capacitor charge to 16 V in 20 ms and then select the value of R_2 to hold the capacitor voltage to 16 V.

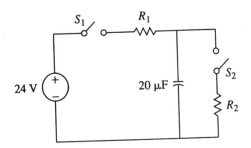

Figure 8.16

8.23: In the network of Fig 8.17, both switches close instantaneously at $t = 0$. Determine the voltage across the capacitor for all $t \geq 0$.

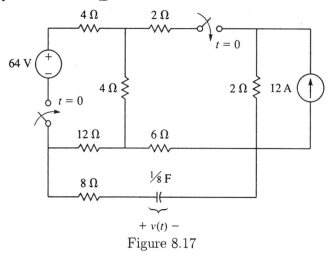

Figure 8.17

8.24: In the network of Fig 8.18, the switch moves instantaneously from position-1 to position-2 at $t = 0$. Determine the voltage across the inductor for all $t \geq 0$

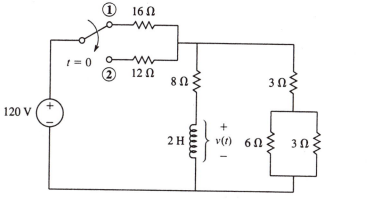

Figure 8.18

8.25: In the network of Fig 8.19, the switch closes instantaneously at $t = 0$. Determine the current through the inductor for all $t \geq 0$.

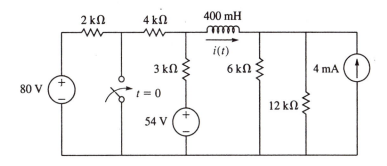

Figure 8.19

THE DRIVEN FIRST ORDER NETWORK - ALTERNATE SOLUTIONS

8.26: Provide an alternate solution to Problem 8.19 that is based on initial and final values of the response and the time constant.

8.27: Provide an alternate solution to Problem 8.20 that is based on initial and final values of the response and the time constant.

8.28: Provide an alternate solution to Problem 8.23 that is based on initial and final values of the response and the time constant.

8.29: Provide an alternate solution to Problem 8.24 that is based on initial and final values of the response and the time constant.

8.30: Provide an alternate solution to Problem 8.25 that is based on initial and final values of the response and the time constant.

OPERATIONAL AMPLIFIERS

8.31: In the ideal operational amplifier configuration shown in Fig 8.20, determine the value of the capacitor if the input voltage is $-12u(t)$, the initial charge on the capacitor is $0\,V$ and the output voltage is required to be $48r(t)$.

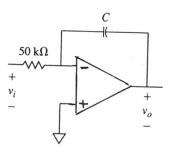

Figure 8.20

8.32: In the ideal operational amplifier configuration shown in Fig 8.21, the initial voltage across each capacitor is zero. If the input voltage, v_1, is a $4\,V$ step. determine the output voltage v_o.

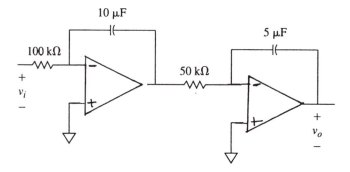

Figure 8.21

8.33: In the ideal operational amplifier configuration shown in Fig 8.22, the initial voltage across each capacitor is zero. If the input voltage, $v_{s1} = 12\sin 400t\,V$ and $v_{s2} = 12\sin 400t\,V$, determine the output voltage v_o.

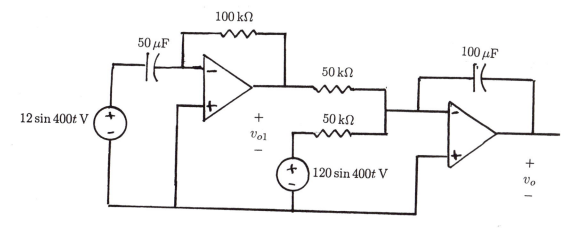

Figure 8.22

8.34: The set of simultaneous differential equations

$$\frac{dv_1}{dt} + 5v_1 - 2v_2 = 18$$

and

$$-2v_1 + 2\frac{dv_2}{dt} + 2v_2 = 0$$

is to be programmed for an analog computer. Draw a circuit diagram using ideal operational amplifiers and specifying values for all R's and C's and considering no initial voltages on any of the capacitors to accomplish this program.

8.35: The differential equation

$$\frac{d^2i}{dt^2} + 10\frac{di}{dt} + 16i = 40t$$

is to be programmed for an analog computer. Draw a circuit diagram using ideal operational amplifiers and specifying values for all R's and C's, a separate function generator for the $40t$ input and considering no initial voltages on any of the capacitors to accomplish this program.

CHAPTER NINE
SECOND ORDER RL AND RC CIRCUITS

THE SINUSOID

9.1: A sinusoidal voltage that has a frequency of 80 Hz and an amplitude of 120 V passes through 0 V with a positive slope at $t = 0.80$ ms. Find V, ω and ϕ in

$$v(t) = V \cos(\omega t + \phi)$$

9.2 A sinusoidal current has a period of 40μs and passes through a maximum of 40 ma at $t = 4\mu$ s. Find I, ω and ϕ in

$$i(t) = I \cos(\omega t + \phi)$$

9.3: A sinusoidally varying charge reaches a maximum of 20 μC at $t = 8$ ms and the next negative maximum of at $t = 16$ ms. Find Q, ω and ϕ in

$$q(t) = Q \cos(\omega t + \phi)$$

9.4: A sinusoidally varying voltage has a positive maximum of 208 V at $t = 0$ and decreases to a value of 120 V at $t = 0.125$ ms. Find V, ω and ϕ in

$$v(t) = V \cos(\omega t + \phi) \, \text{V}$$

9.5 Determine the frequency and the period of a sinusoidally varying voltage that has a value of 60 V at $t = 0$ and reaches its first maximum of 120 V at $t = 2.5$ ms.

9.6 A sinusoidally varying current passes through a negative maximum of 20 mA at $t = 2$ ms and then passes through the next positive maximum at $t = 12$ ms. Find I, ω and ϕ in

$$i(t) = I \cos(\omega t + \phi) \, \text{V}$$

FORMS OF RESPONSE

9.7: The current in an RLC series network is governed by the differential equation

$$L\frac{d^2 i}{dt^2} + R\frac{di}{dt} + \frac{1}{C}i = 0$$

If $R = 4\,\Omega$ and $C = 1/40$ F, determine the value of L to make the response critically damped?

9.8: The current in an RLC series network is governed by the differential equation

$$L\frac{d^2 i}{dt^2} + R\frac{di}{dt} + \frac{1}{C}i = 0$$

If $R = 8\,\Omega$ and $C = 1/100\,\mathrm{F}$, determine the value of L to make the response underdamped with a damped natural frequency of $6\,\mathrm{rad/s}$. Then, determine the natural frequency and the damping factor.

9.9: The voltage in an RLC parallel network is governed by the differential equation

$$C\frac{d^2v}{dt^2} + \frac{1}{R}\frac{dv}{dt} + \frac{1}{L}v = 0$$

If $R = 4\,\Omega$ and $L = 1/8\,\mathrm{H}$, determine the value of C to make the response overdamped with a damping factor of 2.0.

9.10: The voltage in an RLC parallel network is governed by the differential equation

$$C\frac{d^2v}{dt^2} + \frac{1}{R}\frac{dv}{dt} + \frac{1}{L}v = 0$$

If $R = 4\,\Omega$ and $L = 70.82\,\mathrm{mH}$, determine the value of C required to yield a damped natural frequency of $36\,\mathrm{rad/s}$. What is the damping factor?

9.11: The voltage in the network shown in Fig 9.1 is governed by the differential equation

$$C\frac{d^2v}{dt^2} + \frac{1}{R}\frac{dv}{dt} + \frac{1}{L}v = 0$$

Determine the value of R to make the response overdamped with a damping factor of $\zeta = 2.50$.

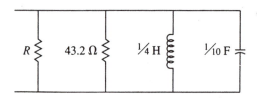

Figure 9.1

9.12: The current in the network shown in Fig 9.2 is governed by the differential equation

$$L\frac{d^2i}{dt^2} + R\frac{di}{dt} + \frac{1}{C}i = 0$$

Determine the value of C to make the response critically damped.

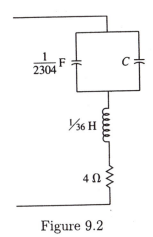

Figure 9.2

UNDRIVEN SECOND ORDER NETWORKS

9.13: In the network of Fig 9.3, the switch closes instantaneously at $t = 0$. Find the current response for $t \geq 0$.

9.14: In the network of Fig 9.4, the switch opens instantaneously at $t = 0$. Find the voltage response for $t \geq 0$.

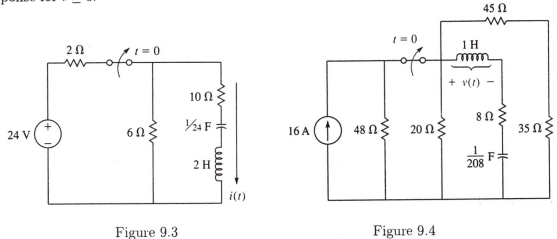

Figure 9.3   Figure 9.4

9.15: In the network of Fig 9.5, the switch moves instantaneously from position-1 to position-2 at $t = 0$. Find the voltage response for $t \geq 0$.

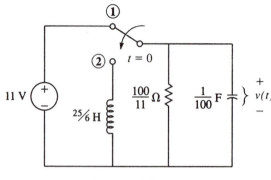

Figure 9.5

9.16: In the network of Fig 9.6, the switch opens instantaneously at $t = 0$. Find the current response for $t \geq 0$.

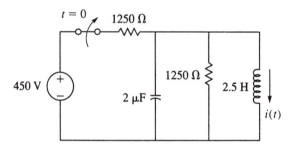

Figure 9.6

9.17: In the network of Fig 9.7 the switch goes from position-1 to position-2 instantaneously at $t = 0$. Determine the voltage across the capacitor for $t \geq 0$ if the capacitor is initially uncharged.

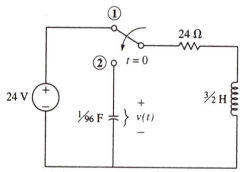

Figure 9.7

9.18: In the network of Fig 9.8 the switch goes from position-1 to position-2 instantaneously at $t = 0$. Determine the voltage across the capacitor for $t \geq 0$ if the capacitor is initially uncharged.

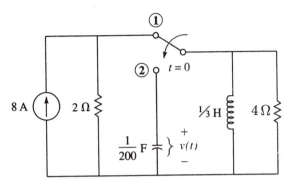

Figure 9.8

DRIVEN SECOND ORDER NETWORKS

9.19: In Fig 9.9, the current is applied to the network instantaneously at $t = 0$ when the voltage across the capacitor is 16 V and no current is flowing in the network. With $R = 4/5\,\Omega$, $L = 1/3\,$H, $C = 1/8\,$F and $i_s = 6\,$A, determine $v(t)$ for all $t \geq 0$.

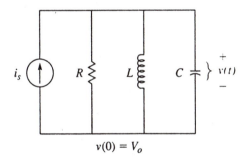

Figure 9.9

9.20: In Fig 9.10, the voltage is applied to the network instantaneously at $t = 0$ when there is no current flowing in the network. With the voltage across the capacitor at 2 V, $R = 2\,\Omega$, $L = 1\,$H, $C = 1/5\,$F and $v_s = 20\,$V, determine $i(t)$ for all $t \geq 0$.

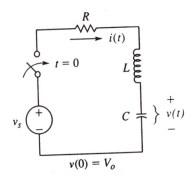

Figure 9.10

9.21: In Fig 9.9, the current is applied to the network instantaneously at $t = 0$ when the voltage across the capacitor is $12\,V$ and no current is flowing in the network. With $R = 5/8\,\Omega, L = 1/5\,H$, $C = 1/5\,F$ and $i_s = 20\,A$, determine $v(t)$ for all $t \geq 0$.

9.22: In Fig 9.10, the voltage is applied to the network instantaneously at $t = 0$ when there is no current flowing in the network. With the voltage across the capacitor at $8\,V$, $R = 2\,\Omega, L= 1/2\,H$, $C = 1/2\,F$ and $v_s = 12\,V$, determine $i(t)$ for all $t \geq 0$.

9.23: In Fig 9.9, the current is applied to the network instantaneously at $t = 0$ when the voltage across the capacitor is $8\,V$ and no current is flowing in the network. With $R = 1\,\Omega, L = 1\,H$, $C = 1/4\,F$ and $i_s = 36\,A$, determine $v(t)$ for all $t \geq 0$.

9.24: In Fig 9.10, the voltage is applied to the network instantaneously at $t = 0$ when there is no current flowing in the network. With the voltage across the capacitor at $48\,V$, $R = 7\,\Omega, L= 1\,H$, $C = 1/12\,F$ and $v_s = 24\,V$, determine $i(t)$ for all $t \geq 0$.

CHAPTER TEN
SINUSOIDAL STEADY STATE ANALYSIS BY PHASOR METHODS

COMPLEX NUMBER ALGEBRA

10.1: If $N_1 = 4 - j3$, $N_2 = 2\sqrt{2}\underline{/45°}$ and $N_3 = 5e^{90°}$, determine

$$N = N_1 + N_2 + N_3$$

10.2: If $N_1 = 4 - j3$, $N_2 = 2\sqrt{2}\underline{/45°}$ and $N_3 = 5e^{90°}$, determine

$$N = \frac{N_1 N_2}{N_3}$$

10.3: If $N_1 = 4 - j3$, $N_2 = 2\sqrt{2}\underline{/45°}$ and $N_3 = 5e^{90°}$, determine

$$N = (N_1 - N_2)^2 N_3$$

10.4: If $N_1 = 4 - j3$, $N_2 = 2\sqrt{2}\underline{/45°}$ and $N_3 = 5e^{90°}$, determine

$$N = 3N_1 N_2 N_3$$

10.5: If $N_1 = 4 - j3$, $N_2 = 2\sqrt{2}\underline{/45°}$ and $N_3 = 5e^{90°}$, determine

$$N = \frac{N_1}{N_2} + \frac{N_1}{N_3}$$

10.6: If $N_1 = 4 - j3$, $N_2 = 2\sqrt{2}\underline{/45°}$ and $N_3 = 5e^{90°}$, determine

$$N = (\sqrt{N_1 N_2} - N_3)(N_1)$$

10.7: If $N_1 = 4 - j3$, $N_2 = 2\sqrt{2}\underline{/45°}$ and $N_3 = 5e^{90°}$, determine

$$N = \frac{N_1 + N_2}{N_1 - N_2}$$

10.8: If $N_1 = 4 - j3$, $N_2 = 2\sqrt{2}\underline{/45°}$ and $N_3 = 5e^{90°}$, determine

$$N = (N_1 + N_2 + N_3)(N_1 N_2 N_3)^2$$

10.9: If $N_1 = 4 - \jmath 3$, $N_2 = 2\sqrt{2}\underline{/45°}$ and $N_3 = 5e^{90°}$, determine

$$N = N_1^3 + N_2^2 + N_3$$

10.10: If $N_1 = 4 - \jmath 3$, $N_2 = 2\sqrt{2}\underline{/45°}$ and $N_3 = 5e^{90°}$, determine

$$N = N_1 N_2 + N_2 N_3 + N_1 N_3$$

IMPEDANCE AND ADMITTANCE

10.11: If $\omega = 400\,\mathrm{rad/s}$, find the driving point impedance at terminal-a and -b in the network of Fig 10.1.

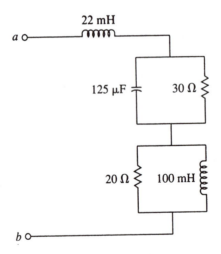

Figure 10.1

10.12: If $\omega = 200\,\mathrm{rad/s}$, find the driving point impedance at terminal-a and -b in the network of Fig 10.2.

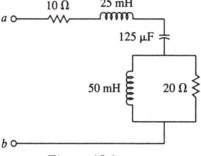

Figure 10.2

10.13: If $\omega = 100\,\mathrm{rad/s}$, find the driving point admittance at terminal-a and -b in the network of Fig 10.3.

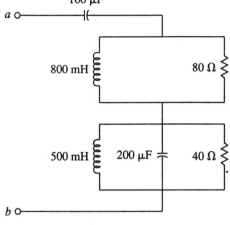

Figure 10.3

10.14: If $\omega = 500\,\mathrm{rad/s}$, find the driving point admittance at terminal-a and -b in the network of Fig 10.4.

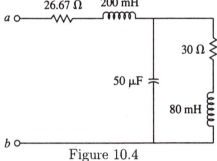

Figure 10.4

10.15: If $\omega = 1000\,\mathrm{rad/s}$, find the driving point admittance at terminal-a and -b in the network of Fig 10.5.

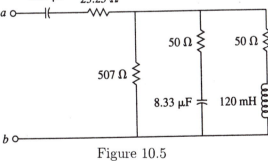

Figure 10.5

PHASORS

10.16: If, in the form

$$f(t) = A\cos\omega t + B\sin\omega t$$

$$A = -4 \qquad \text{and} \qquad B = 4$$

determine F and ϕ in

$$f(t) = F\cos(\omega t + \phi)$$

10.17: If, in the form

$$f(t) = A\cos\omega t + B\sin\omega t$$

$$A = -64 \qquad \text{and} \qquad B = 120$$

determine F and ϕ in

$$f(t) = F\cos(\omega t + \phi)$$

10.18: If, in the form

$$f(t) = A\cos\omega t + B\sin\omega t$$

$$A = -60 \qquad \text{and} \qquad B = -45$$

determine F and ϕ in

$$f(t) = F\cos(\omega t + \phi)$$

10.19: If, in the form

$$f(t) = A\cos\omega t + B\sin\omega t$$

$$A = -70 \qquad \text{and} \qquad B = -168$$

determine F and ϕ in

$$f(t) = F\cos(\omega t + \phi)$$

10.20: If, in the form

$$f(t) = A\cos\omega t + B\sin\omega t$$

$$A = 28 \qquad \text{and} \qquad B = -96$$

determine F and ϕ in

$$f(t) = F\cos(\omega t + \phi)$$

10.21: If $F = 275$ and $\phi = -143.13°$ in the instantaneous form

$$f(t) = F\cos(\omega t + \phi)$$

determine A and B in
$$f(t) = A\cos\omega t + B\sin\omega t$$

10.22: If $F = 390$ and $\phi = 112.62°$ in the instantaneous form
$$f(t) = F\cos(\omega t + \phi)$$

determine A and B in
$$f(t) = A\cos\omega t + B\sin\omega t \Leftarrow$$

10.23: If $F = 800$ and $\phi = -196.26°$ in the instantaneous form
$$f(t) = F\cos(\omega t + \phi)$$

determine A and B in
$$f(t) = A\cos\omega t + B\sin\omega t \Leftarrow$$

10.24: If $F = 120$ and $\phi = -150°$ in the instantaneous form
$$f(t) = F\cos(\omega t + \phi)$$

determine A and B in
$$f(t) = A\cos\omega t + B\sin\omega t \Leftarrow$$

10.25: If $F = 1000$ and $\phi = -135°$ in the instantaneous form
$$f(t) = F\cos(\omega t + \phi)$$

determine A and B in
$$f(t) = A\cos\omega t + B\sin\omega t \Leftarrow$$

10.26: If a voltage is given by
$$v(t) = 208\cos(377t + 60°)$$

determine the voltage phasor in rectangular, polar and exponential form.

10.27: If a current is given by
$$i(t) = 4\sqrt{2}\cos(400t - 135°)$$

determine the current phasor in rectangular, polar and exponential form.

10.28: If a current is given by
$$i(t) = 120\sqrt{3}\cos(600t + 15°)$$

determine the current phasor in rectangular, polar and exponential form.

88

10.29: If a voltage is given by

$$v(t) = 1750\cos(200t - 163.74°)$$

determine the voltage phasor in rectangular, polar and exponential form.

10.30: If a voltage is given by

$$v(t) = 400\cos(1200t + 90°)$$

determine the voltage phasor in rectangular, polar and exponential form.

10.31: Two elements are connected in parallel as indicated in Fig 10.6. The line current is $i(t) = 10\sqrt{2}\cos(377t + 135)°$ A and $i_1(t) = 6\cos(377t + 30)°$ Express $i_2(t)$ instantaneous form.

10.32: Three elements are connected in series as indicated in Fig 10.7. The line voltage is $v(t) = 120\cos(377t+90)°$ V, $v_1(t) = 40\sqrt{2}\cos(377t+45)°$ V and $v_3(t) = 60\cos(377t-53.13)°$ V. Express $v_2(t)$ instantaneous form.

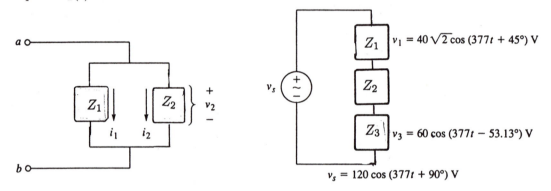

Figure 10.6   Figure 10.7

10.33: In Fig 10.8, $\mathbf{I}_1 = 2\sqrt{2}\underline{/45°}$ and $\mathbf{V}_2 = 120\underline{/53.13°}$ V. If the line current is $i = 4\cos(\omega t + 90°)$ A and the frequency is $f = 250/2\pi$ Hz, Determine the four components that comprise the parallel combination.

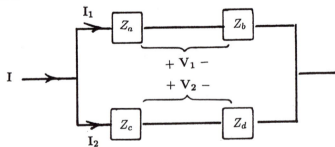

Figure 10.8

10.34: In Fig 10.9, $i_s(t) = 8\cos(400t + 90°)$ A, $v_{s1}(t) = 200\cos(400t + 36.87°)$ V and $v_{s2}(t) = 200\sqrt{2}\cos(400t+135°)$ A. Combine the three sources to a single voltage source connected across terminals a-b.

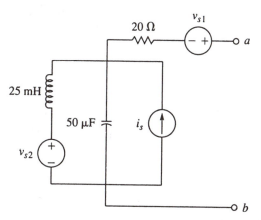

Figure 10.9

LADDER NETWORKS AND PHASOR DIAGRAMS

10.35: Determine all branch currents and branch voltages in the network of Fig 10.10 and then draw a phasor diagram.

10.36: Determine all branch currents and branch voltages in the network of Fig 10.11 and then draw a phasor diagram.

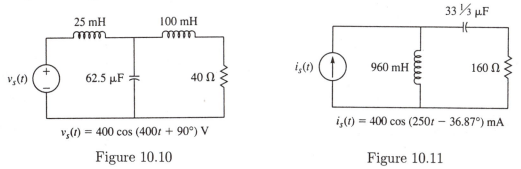

$v_s(t) = 400\cos(400t + 90°)$ V

Figure 10.10

$i_s(t) = 400\cos(250t - 36.87°)$ mA

Figure 10.11

10.37: Determine all branch currents and branch voltages in the network of Fig 10.12 and then draw a phasor diagram.
10.38: Determine all branch currents and branch voltages in the network of Fig 10.13 and then draw a phasor diagram.

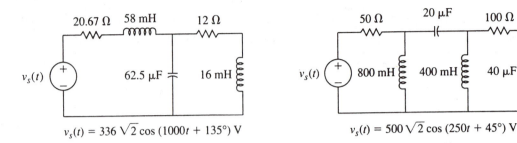

$v_s(t) = 336\sqrt{2}\cos(1000t + 135°)$ V

$v_s(t) = 500\sqrt{2}\cos(250t + 45°)$ V

Figure 10.12   Figure 10.13

NETWORK ANALYSIS

10.39: Rework Problem 10.35 (Fig 10.10) using a nodal analysis to determine the current through the 25 mH inductor and the voltage across the 62.5 μF capacitor.

10.40: Rework Problem 10.36 (Fig 10.11) using a nodal analysis to determine the current through the 960 mH inductor and the voltage across the 160 Ω resistor.

10.41: Rework Problem 10.38 (Fig 10.13) using a nodal analysis to determine the current through the 800 mH inductor and the voltage across the 40 μF capacitor.

10.42: Use nodal analysis to find the current through the 50 mH inductor and the voltage across the 40 Ω resistor in Fig 10.14.

10.43: Use nodal analysis to find the current through the 10 Ω resistor in Fig 10.15.

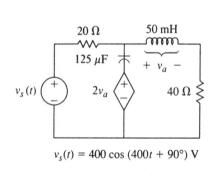

$v_s(t) = 400\cos(400t + 90°)$ V

Figure 10.14

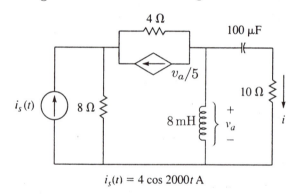

$i_s(t) = 4\cos 2000t$ A

Figure 10.15

10.44: Rework Problem 10.35 (Fig 10.10) using a mesh analysis to determine the current through the 25 mH inductor and the voltage across the 62.5 μF capacitor.

10.45: Rework Problem 10.36 (Fig 10.11) using a mesh analysis to determine the current through the 960 mH inductor and the voltage across the 160 Ω resistor.

10.46: Rework Problem 10.42 (Fig 10.14) using a mesh analysis to find the current through the 50 mH inductor and the voltage across the 40 Ω resistor.

10.47: Rework Problem 10.43 (Fig 10.15) using mesh analysis to find the current through the 10 Ω resistor in Fig 10.15.

10.48: Use mesh analysis to determine the current through the 20 mH inductor in Fig 10.16.

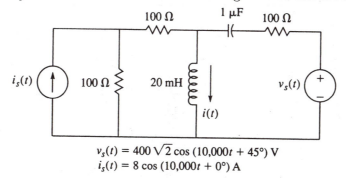

$$v_s(t) = 400\sqrt{2}\cos(10{,}000t + 45°) \text{ V}$$
$$i_s(t) = 8\cos(10{,}000t + 0°) \text{ A}$$

Figure 10.16

10.49: Rework Problem 10.48 (Fig 10.16) using superposition to determine the current through the 20 mH inductor.

10.50: Use superposition to determine the voltage across the 58 mH inductor in Fig 10.17.

10.51: Use superposition to determine the voltage across the 50 μF capacitor in Fig 10.18.

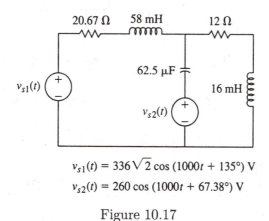

$$v_{s1}(t) = 336\sqrt{2}\cos(1000t + 135°) \text{ V}$$
$$v_{s2}(t) = 260\cos(1000t + 67.38°) \text{ V}$$

Figure 10.17

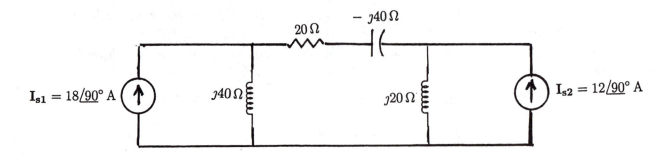

Figure 10.18

10.52: Use superposition to determine the current through the 20 mH inductor in Fig 10.19.

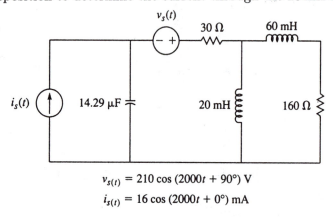

$$v_{s(t)} = 210 \cos (2000t + 90°) \text{ V}$$
$$i_{s(t)} = 16 \cos (2000t + 0°) \text{ mA}$$

Figure 10.19

10.53: Rework Problem 10.35 (Fig 10.10) using Thevenin's theorem to determine the current through the 40 Ω resistor.

10.54: Rework Problem 10.51 (Fig 10.18) using Thevenin's theorem to determine the current through the 50 μF capacitor.

10.55: Rework Problem 10.43 (Fig 10.15) using Thevenin's theorem to determine the current through the 10 Ω resistor.

10.56: Rework Problem 10.37 (Fig 10.12) using Thevenin's theorem to determine the current through the 16 mH inductor.

10.57: Rework Problem 10.38 (Fig 10.13) using Thevenin's theorem to determine the current through the 40 μF capacitor.

CHAPTER ELEVEN
SINUSOIDAL STEADY STATE POWER CALCULATIONS

AVERAGE AND EFFECTIVE VALUES

11.1: If the period, $T = 10\,$s, determine the average and effective values for the voltage wave shown in Fig 11.1.

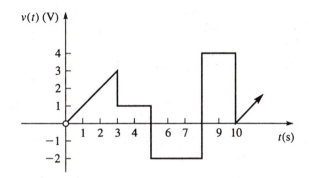

Figure 11.1

11.2: If the period, $T = 10\,$s, determine the average and effective values for the current wave shown in Fig 11.2.

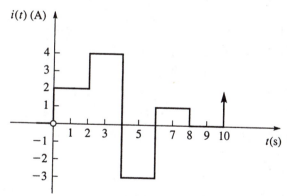

Figure 11.2

11.3: If the period, $T = 10\,\text{s}$, determine the average and effective values for the voltage wave shown in Fig 11.3.

11.4: If the period, $T = 10\,\text{s}$, determine the average and effective values for the current wave shown in Fig 11.4.

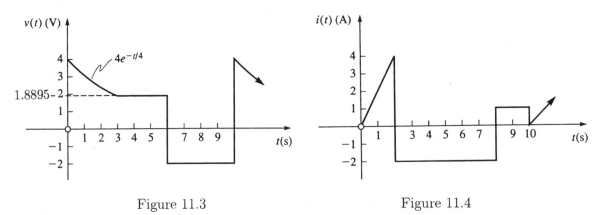

Figure 11.3   Figure 11.4

11.5: If the period, $T = 40\,\text{ms}$, determine the average and effective values for the voltage wave shown in Fig 11.5.

11.6: If the period, $T = 6\,\text{s}$, determine the average and effective values for the voltage wave shown in Fig 11.6.

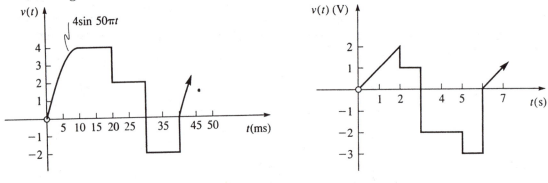

Figure 11.5   Figure 11.6

SINUSOIDAL STEADY STATE

11.7: Determine the average power and the power factor for an RLC series network operating at $400\,\text{Hz}$ with $R = 100\,\Omega, L = 39.79\,\text{mH}$ and $C = 15.92\,\mu\text{F}$ if the network is subjected to a sinusoidal voltage with a maximum amplitude of $180\,\text{V}$.

11.8: Determine the average power and the power factor for an RLC parallel network operating at $250\,\text{Hz}$ with $R = 50\,\Omega, L = 50.93\,\text{mH}$ and $C = 21.22\mu\text{F}$ if the network is subjected to a sinusoidal current with a maximum amplitude of $16\,\text{A}$.

11.9 Two impedances, $Z_1 = 40\sqrt{2}\underline{/45°}\,\Omega$ and $Z_2 = 80 - j60\,\Omega$ are connected in parallel and the combination is then connected in series with a $24\,\Omega$ resistor. If the combination is connected across a source having an rms value of $120\,\text{V}$, how much power is drawn and what is the power factor?

11.10: Two impedances, $Z_1 = 25\underline{/53.13°}\,\Omega$ and $Z_2 = 39\underline{/67.38°}\,\Omega$ are connected in parallel and the combination is then connected in series with an impedance of $17\underline{/28.07°}\,\Omega$. If the combination is connected across a source having an rms value of $20\,\text{A}$, how much power is drawn and what is the power factor?

11.11: An RLC series network is subjected to an rms voltage of

$$v(t) = 200\cos 400t\,\text{V}$$

An impedance meter has measured the impedance angle as $36.87°$. If, $R = 40\,\Omega, L = 387.5\,\text{mH}$ and the power dissipated is $640\,\text{W}$, determine the value of C.

11.12: A series RC circuit absorbs $22.15\,\text{W}$ at a power factor of 0.9231 when it is connected to an rms voltage source of

$$v(t) = 120\cos 400t\,\text{V}$$

Determine the values of R and C.

11.13: Determine the total power dissipated by all of the resistors in the phasor domain network of Fig 11.7.

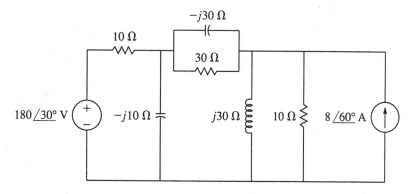

Figure 11.7

11.14: Determine the complex power delivered to the phasor domain network in Fig 11.8.

94

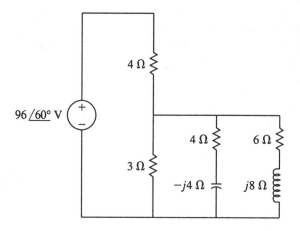

Figure 11.8

11.15: Determine the complex power delivered to the network in Fig 11.9.

11.16: Determine the complex power delivered to the network in Fig 11.10.

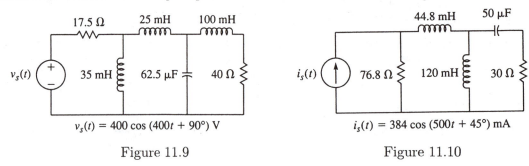

$v_s(t) = 400 \cos (400t + 90°)$ V

Figure 11.9

$i_s(t) = 384 \cos (500t + 45°)$ mA

Figure 11.10

11.17: Determine the complex power delivered to the network in Fig 11.11.

11.18: For the phasor domain network shown in Fig 11.12, Determine the real power, the apparent power, the magnitude of the apparent power and the power factor.

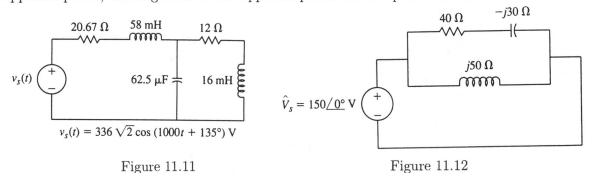

$v_s(t) = 336 \sqrt{2} \cos (1000t + 135°)$ V

Figure 11.11

$\hat{V}_s = 150\underline{/0°}$ V

Figure 11.12

POWER FACTOR CORRECTION

11.19: In the power distribution system shown in Fig 11.13, the frequency is 60 Hz. Determine:

- (a)   The real power
- (b)   The reactive power
- (c)   The magnitude of the apparent power
- (d)   The power factor
- (e)   The correction necessary to make the power factor 0.950
- (f)   the component necessary to achieve the correction in part (e)

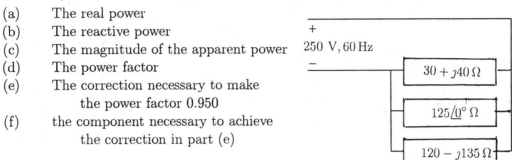

Figure 11.13

11.20: In the power distribution system shown in Fig 11.14, the frequency is 60 Hz. Determine:

- (a)   The real power
- (b)   The reactive power
- (c)   The magnitude of the apparent power
- (d)   The power factor
- (e)   The correction necessary to make the power factor 0.935
- (f)   the component necessary to achieve the correction in part (e)

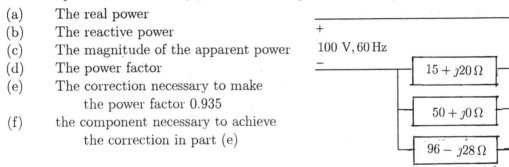

Figure 11.14

11.21: In the power distribution system shown in Fig 11.15, the frequency is 60 Hz. Determine:

- (a)   The real power
- (b)   The reactive power
- (c)   The magnitude of the apparent power
- (d)   The power factor
- (e)   The correction necessary to make the power factor 0.920.
- (f)   the component necessary to achieve the correction in part (e)

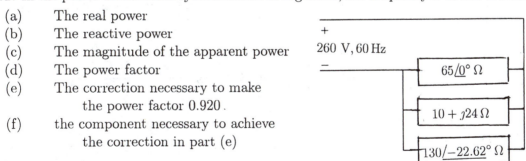

Figure 11.15

11.22: In the power distribution system shown in Fig 11.16, the frequency is 100 Hz. Determine:

(a)   The real power
(b)   The reactive power
(c)   The magnitude of the apparent power
(d)   The power factor
(e)   The correction necessary to make
         the power factor 0.945
(f)   the component necessary to achieve
         the correction in part (e)

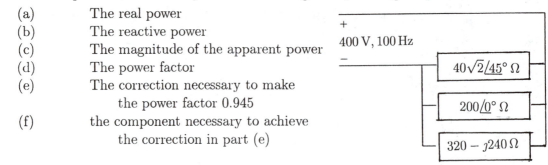

Figure 11.16

MAXIMUM POWER TRANSFER

11.23: In the network of Fig 11.17, determine what elements should be placed across the terminals a-b to make the power factor unity.

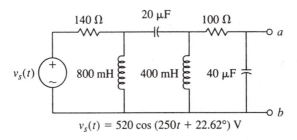

$$v_s(t) = 520 \cos (250t + 22.62°) \text{ V}$$

Figure 11.17

11.24: In the phasor domain network of Fig 11.18, determine the load to be placed across terminals a-b to make the power drawn by the load a maximum and then determine the value of this maximum power.

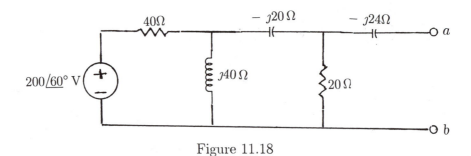

Figure 11.18

11.25: In the phasor domain network of Fig 11.19, determine the load to be placed across terminals *a-b* to make the power drawn by the load a maximum and then determine the value of this maximum power.

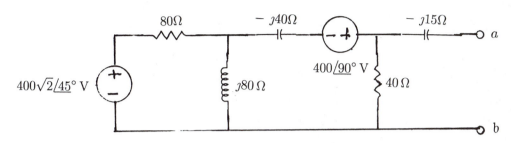

Figure 11.19

11.26: In the phasor domain network of Fig 11.20, determine the load to be placed across terminals *a-b* to make the power drawn by the load a maximum and then determine the value of this maximum power.

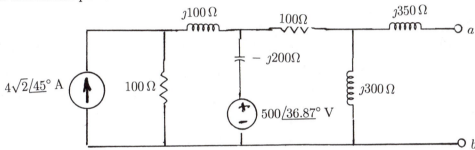

Figure 11.20

CHAPTER TWELVE

THREE PHASE POWER SYSTEMS

BALANCED SYSTEMS

12.1: In Fig 12.1, $\mathbf{V}_{AN} = \mathbf{V}_{NB} = 120\underline{/0^\circ}$ V. If $Z_1 = Z_2 = Z_3 = 40\,\Omega$, determine the total power delivered to the system of loads.

12.2: In Fig 12.1, $\mathbf{V}_{AN} = \mathbf{V}_{NB} = 120\underline{/0^\circ}$ V. If $Z_1 = Z_2 = Z_3 = 30\underline{/16.26^\circ}\,\Omega$, determine the total power delivered to the system of loads.

12.3: For the balanced three phase generator connected as shown in Fig 12.2, the phase voltages have an rms amplitude of 440 V and are connected in the positive sequence. Determine the currents, $\mathbf{I}_a, \mathbf{I}_b$ and $\mathbf{I}_c$ if the load is balanced with all $Z = 44\underline{/0^\circ}\,\Omega$.

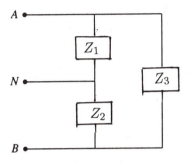

Figure 12.1

12.4: The line voltages in the wye connected generator that supplies the loads shown in Fig 12.3 have a magnititude of 208 V and are in the positive sequence at 60 Hz. Determine all phase and line voltages in instantaneous form.

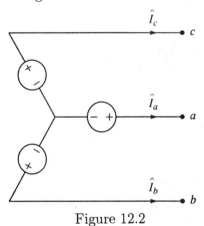

Figure 12.2

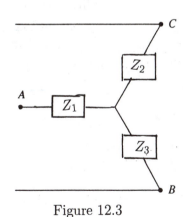

Figure 12.3

100

12.5: The line voltages in the wye connected generator that supplies the loads shown in Problem 12.4 (Fig 12.3) have a magnititude of 208 V and are in the *negative sequence* at 60 Hz. Determine all phase and line voltages in instantaneous form.

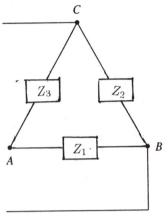

Figure 12.4

12.6: The line voltages in the delta connected generator that supplies the loads shown in Fig 12.4 have a magnititude of 208 V and are in the positive sequence at 60 Hz. Determine all phase and line voltages in instantaneous form.

12.7: The line voltages in the delta connected generator that supplies the loads shown in Problem 12.6 (Fig 12.4) have a magnititude of 208 V and are in the *negative sequence* at 60 Hz. Determine all phase and line voltages in instantaneous form.

12.8: The line-to-line voltages in Fig 12.5 have a magnitude of 440 V and are in the positive sequence at 60 Hz. The loads are balanced with $Z = Z_1 = Z_2 = Z_3 = 25\underline{/36.87°}\ \Omega$. Determine all phase and line and phase voltages and load currents in instantaneous form.

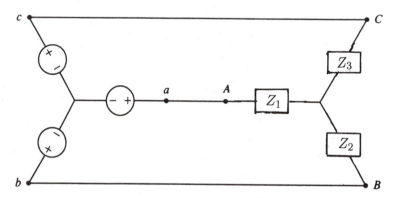

Figure 12.5

12.9 In Problem 12.8, determine the power drawn by the load.

12.10: In Fig 12.5, the rms amplitude of the line-to-line voltages in the three phase source is 240 V. The source is connected in the positive sequence and operates at 60 Hz. If the balanced load consists of $Z = Z_1 = Z_2 = Z_3 = 30\underline{/36.87°}\,\Omega$, determine the phase voltages and currents and the power delivered to the load.

12.11: In Fig 12.5, the rms amplitude of the line-to-line voltages in the three phase source is 240 V. The source is connected in the positive sequence and operates at 60 Hz. If the balanced load consists of $Z = Z_1 = Z_2 = Z_3 = 40\underline{/53.13°}\,\Omega$, determine the phase voltages and currents and the power delivered to the load.

12.12: In Fig 12.6, the rms amplitude of the line-to-line voltages in the three phase source is 240 V. The source is connected in the positive sequence and operates at 60 Hz. If the balanced load consists of $Z = Z_1 = Z_2 = Z_3 = 30\underline{/36.87°}\,\Omega$, determine the line and phase voltages, the line currents and the power delivered to the load.

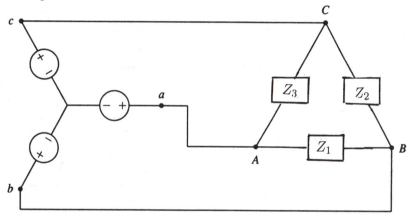

Figure 12.6

12.13: In Problem 12.12 (Fig 12.6), determine the phase currents and the power delivered to the balanced load if $Z = Z_1 = Z_2 = Z_3 = 50\underline{/16.26°}\,\Omega\,V$.

·**12.14:** In Fig 12.7, the rms amplitude of the line-to-line voltages in the three phase source is 240 V. The source is connected in the positive sequence and operates at 60 Hz. If the balanced load consists of $Z = Z_1 = Z_2 = Z_3 = 30\underline{/36.87°}\,\Omega$, determine the line voltages and currents and the power delivered to the load.

12.15: Rework Problem 12.14 (Fig 12.7) to determine the power delivered to the load with the balanced load consisting of $Z = Z_1 = Z_2 = Z_3 = 40\underline{/53.13°}\,\Omega$.

102

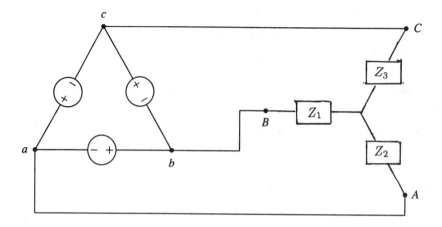

Figure 12.7

12.16: In Fig 12.8, the rms amplitude of the line-to-line voltages in the three phase source is 240 V. The source is connected in the positive sequence and operates at 60 Hz. If the balanced load consists of $Z = Z_1 = Z_2 = Z_3 = 30\underline{/36.87°}\ \Omega$, determine the phase voltages and currents and the power delivered to the load.

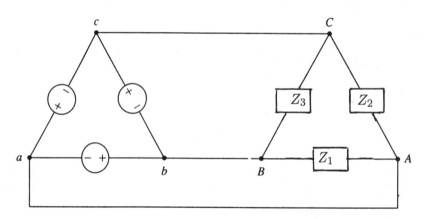

Figure 12.8

12.17: Rework Problem 12.16 (Fig 12.8) to determine the power delivered to the load with the balanced load consisting of $Z = Z_1 = Z_2 = Z_3 = 50\underline{/16.26°}\ \Omega$.

UNBALANCED SYSTEMS

12.18: In Fig 12.1, $\mathbf{V}_{AN} = \mathbf{V}_{NB} = 120\underline{/0°}$ V. If $Z_1 = 8 + j6\,\Omega$, $Z_2 = 8 - j6\,\Omega$ and $Z_3 = 4 - j20\,\Omega$, determine the total power delivered to the system of loads.

12.19: In Fig 12.1, $\mathbf{V}_{AN} = \mathbf{V}_{NB} = 120\underline{/0°}$ V. If $Z_1 = 20\underline{/30°}$, $Z_2 = 30\underline{/36.87°}$ and $Z_3 = 30\underline{/45°}\,\Omega$, determine the total power delivered to the system of loads.

12.20: In the system of Fig 12.9, the line voltages are 240 V and are in the positve sequence. Determine $\mathbf{I}_b$.

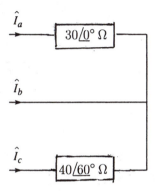

Figure 12.9

12.21: In Fig 12.10, determine the line currents. $\mathbf{I}_1, \mathbf{I}_2$ and $\mathbf{I}_3$ if the line-to-line voltages all have a magnitude of 300 V and are connected in the positive sequence.

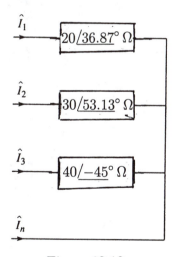

Figure 12.10

104

12.22: In Fig 12.10, determine the line currents. $\mathbf{I_1}, \mathbf{I_2}$ and $\mathbf{I_3}$ if the line-to-line voltages all have a magnitude of 300 V and are connected in the *negative sequence.*

12.23: In Fig 12.11, the rms amplitude of the line-to-line voltages in the three phase source is 440 V. The source is connected in the positive sequence and operates at 60 Hz. If the unbalanced load consists of $Z_1 = 88\underline{/0°}\,\Omega, Z_2 = 44\sqrt{2}\underline{/45°}\,\Omega$ and $Z_3 = 44\underline{/36.87°}\,\Omega$, determine the power delivered to the load.

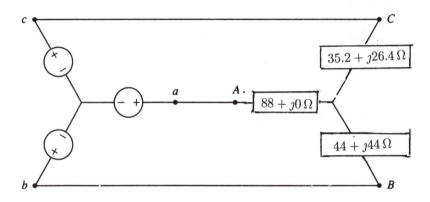

Figure 12.11

12.24: Determine the overall power factor for the system treated in Problem 12.23.

12.25: In Fig 12.12, the rms amplitude of the line-to-line voltages in the three phase source is 300 V. The source is connected in the positive sequence and operates at 60 Hz. If the unbalanced load consists of $Z_1 = 60\underline{/0°}\,\Omega, Z_2 = 75\sqrt{2}\underline{/-45°}\,\Omega$, and $Z_3 = 50\underline{/36.87°}\,\Omega$, determine the power delivered to the load.

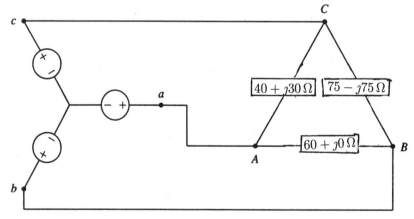

Figure 12.12

12.26: Determine the overall power factor for the system treated in Problem 12.25.

12.27: Figure 12.13 shows a wye-connected generator delivering power to a wye-connected load in a four wire system. The rms amplitude of the line-to-line voltages in the three phase source is 240 V. The source is connected in the positive sequence and operates at 60 Hz. If the unbalanced load consists of three impedances, $Z_1 = 48\underline{/0°}\,\Omega$, $Z_2 = 24\sqrt{2}\underline{/-135°}\,\Omega$ and $Z_3 = 60\underline{/36.87°}\,\Omega$, determine the power delivered to the load.

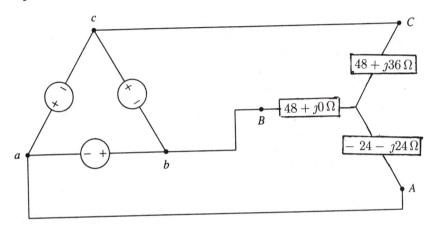

Figure 12.13

PROBLEM ANSWERS

CHAPTER ONE
INTRODUCTION AND BASIC CONCEPTS

CHARGE AND CURRENT

1.1: 1.248×10^{21} electrons

1.2: $I = 20\,\text{A}$

1.3: $I = 32.04\,\text{A}$

1.4: $Q = 164\,\text{C}$

ENERGY, POWER, VOLTAGE, CURRENT AND CHARGE

1.5: $W = 5.44 \times 10^{-3}\,\text{J}$

1.6: $V = 29\,\text{V}$

1.7: $W = 400\,\text{J}$

1.8: $Q = 1.914\,\text{C}$

1.9: $W = 17.904\,\text{MJ}$

1.10: $4500\,\text{C}$, 28.09×10^{21} electrons

1.11: \$7.26

1.12: $I = 14\,\text{A}$

1.13: $W = 44\,\text{J}$

1.14: $W = 0\,\text{J}$

RESISTANCE

1.15: $V = 25.98\,\text{V}$

1.16: $R = 3.48\,\Omega$

1.17: $R = 8.105 \times 10^{-4}\,\Omega$

1.18: $L = 0.2031\,\text{m}$

1.19: $L = 123.48\,\text{m}$

1.20: $T = 81.99°\text{C}$, $T = -31.42°\text{C}$

1.21: $\alpha = 0.00360°\text{C}^{-1}$

1.22: $R(T) = 250\,\Omega\text{-}°\text{C}^{-1}$, $T = 1895°\text{C}$

1.23: $I = 15\,\text{A}$, $R = 8\,\Omega$

1.24: $R = 5\,\Omega$

APPLICATIONS OF OHM'S LAW

1.25: $I_1/I_2 = 0.711$, $P_1/P_2 = 0.506$

1.26: $V_1/V_2 = 0.711$, $P_1/P_2 = 0.360$

1.27: $I = 12\,\text{A}$, $P_R = 72\,\text{W}$, $P_{\text{source}} = 72\,\text{W}$

1.28: $L = 39.35\,\text{m}$

1.29: $R = 48\,\Omega$, $P_{\text{source}} = 60\,\text{W}$

1.30: $R = 4\,\Omega$

1.31: $V_{30} = 150\,\text{V}$, $V_3 = 75\,\text{V}$, $I_4 = 18.75\,\text{A}$, $I_{12} = 6.25\,\text{A}$

1.32: $V_5 = 15\,\text{V}$, $V_1 = 3\,\text{V}$, $I_3 = 2\,\text{A}$, $I_6 = 1\,\text{A}$

CONTROLLED SOURCES

1.33: $\beta = 4$

1.34: $|V_{\text{out}}/V_{\text{in}}| = 8$, $|P_{\text{out}}/P_{\text{in}}| = 32$

1.35: $R_1 = 8\,\Omega$

1.36: $|I_2/I_1| = 5$

1.37: $|V_{\text{out}}/V_{\text{in}}| = 16$, $|P_{\text{out}}/P_{\text{in}}| = 128$

1.38: $|I_3/I_{\text{in}}| = 12$

CHAPTER TWO
KIRCHHOFF'S CURRENT AND VOLTAGE LAWS AND SERIES-PARALLEL RESISTIVE CIRCUITS

KCL PROBLEMS

2.1: $i_1 = -7\,\text{A}$, $i_2 = 6\,\text{A}$, $i_3 = -3\,\text{A}$

2.2: $i_1 = 5\,\text{A}$, $i_2 = 8\,\text{A}$, $i_3 = 3\,\text{A}$, $i_4 = 1\,\text{A}$, $i_5 = 3\,\text{A}$

2.3: $i_1 = 2\,\text{A}$, $i_2 = 0\,\text{A}$, $i_3 = 2\,\text{A}$, $i_4 = -5\,\text{A}$

KVL PROBLEMS

2.4: $v_1 = 6\,\text{V}$, $v_2 = -4\,\text{V}$, $v_3 = 12\,\text{V}$, $v_4 = 2\,\text{V}$, $v_5 = 13\,\text{V}$

2.5: $v_1 = 10\,\text{V}$, $v_2 = 20\,\text{V}$, $v_3 = 20\,\text{V}$, $v_4 = 18\,\text{V}$,
$v_5 = 18\,\text{V}$, $v_6 = 26\,\text{V}$, $v_7 = 12\,\text{V}$

2.6: $v_1 = -30\,\text{V}$, $v_2 = -20\,\text{V}$, $v_3 = -36\,\text{V}$, $v_4 = 32\,\text{V}$,
$v_5 = 36\,\text{V}$, $v_6 = -12\,\text{V}$

EQUIVALENT RESITANCE PROBLEMS

2.7: (a) $R_{\text{eq}} = 10\,\Omega$, (b) $R_{\text{eq}} = 16\,\Omega$, (c) $R_{\text{eq}} = 20\,\Omega$
(d) $R_{\text{eq}} = 0\,\Omega$, (e) $R_{\text{eq}} = 14.29\,\Omega$

2.8: $R = 21\,\Omega$

2.9: $R_{\text{eq}} = 48\,\Omega$

2.10: $R_{\text{eq}} = 19\,\Omega$

2.11: $R_{\text{eq}} = 25\,\Omega$

2.12: $R_{\text{eq}} = 110\,\Omega$

OHM'S LAW PROBLEMS

2.13: (a) $I_4 = I_{12} = 3/2\,\text{A}$, (b) $V_4 = 6\,\text{V}$, $V_{12} = 18\,\text{V}$

2.14: $P_4 = 9\,\text{W}$, $P_{12} = 27\,\text{W}$

2.15: $V_4 = 6\,\text{V}$, $V_{12} = 18\,\text{V}$

2.16: (a) $V = 36\,\text{V}$, (b) $I_4 = 9\,\text{A}$, $I_{12} = 3\,\text{A}$

2.17: $P_4 = 324\,\text{W}$, $P_{12} = 108\,\text{W}$

2.18: $I_4 = 9\,\text{A}$, $I_{12} = 3\,\text{A}$

2.19: $I_3 = I_9 = 2\,\text{A}$, $I_4 = 6\,\text{A}$, $I_6 = 4\,\text{A}$,

2.20: $P_3 = 12\,\text{W}$, $P_4 = 144\,\text{A}$, $P_6 = 96\,\text{W}$, $P_9 = 36\,\text{W}$

2.21: $R_1 = 3\,\Omega, R_2 = 6\,\Omega, R_3 = 1\,\Omega$

OHM'S LAW PROBLEMS (Cont'd)

2.22: $I_1 = 4\,\text{A}, I_3 = 4/3\,\text{A}, I_2 = I_4 = 2/3\,\text{A}$
$V_1 = 4\,\text{V}, V_2 = 4/3\,\text{V}, V_3 = 4\,\text{V}, V_4 = 8/3\,\text{V}$

2.23: $I_4 = 3\,\text{A}, I_5 = 4\,\text{A}, I_6 = 8\,A, I_8 = 4\,\text{A}, I_{12} = 1\,\text{A}$
$V_4 = 12\,\text{V}, V_5 = 20\,\text{V}, V_6 = 48\,\text{V}, V_8 = 32\,\text{V}, V_{12} = 12\,\text{V}$

2.24: $I_1 = 6\,\text{A}, I_2 = 2\,\text{A}, V_6 = 18\,\text{V}$

2.25: $I_1 = 6\,\text{A}, I_2 = 2\,\text{A}, V_6 = 18\,\text{V}$

2.26: $I_4 = 9/2\,\text{A}, I_9 = 12\,\text{A}, V_7 = 21/2\,\text{V}$

2.27: $I_4 = 9/2\,\text{A}, I_9 = 12\,\text{A}, V_7 = 21/2\,\text{V}$

2.28: $I_3 = 18\,\text{A}, I_4 = 3\,\text{A}, V_{12} = 24\,\text{V}, V_{14} = 14\,\text{V}$

2.29: $I_6 = 2\,\text{A}, I_{10} = 2\,\text{A}, V_2 = 8/3\,\text{V}$

2.30: $I_9 = 9\,\text{A}, I_{40} = 9/20\,\text{A}, V_7 = 63/5\,\text{V}$

2.31: $I_{13} = 1\,\text{A}, I_{14} = 1\,\text{A}, V_6 = 6\,\text{V}, V_7 = 21\,\text{V}$

2.32: $I_5 = 5/4\,\text{A}, V_{20} = 100\,\text{V}, P_{24} = 9.375\,\text{W}$

CHAPTER THREE
NODAL AND LOOP ANALYSES

SINGLE NODE PROBLEMS

3.1: $V_1 = 24\,\text{V}$

3.2: $V_1 = 64/3\,\text{V}$

3.3: $V_1 = 400\,\text{V}$

3.4: $I_2 = -5/2\,\text{A}$

3.5: $I_5 = -20/7\,\text{A}$

3.6: $V_1 = 3\,\text{V}, V_2 = 6\,\text{V}$

3.7: $I_4 = 0.590\,\text{A}$

3.8:
$$\begin{bmatrix} 14 & -4 & -2 & 0 & 0 \\ -4 & 16 & -6 & 0 & -5 \\ -2 & -6 & 12 & -4 & 0 \\ 0 & 0 & -4 & 20 & -16 \\ 0 & -5 & 0 & -16 & 33 \end{bmatrix} \begin{bmatrix} V_1 \\ V_2 \\ V_3 \\ V_4 \\ V_5 \end{bmatrix} = \begin{bmatrix} -6 \\ -2 \\ 14 \\ -11 \\ 8 \end{bmatrix}$$

3.9: $I_s = -576\,\text{A}$

3.10: $I_{40} = -0.81\,\text{A}$

3.11: $V_1 = 224/13\,\text{V}, V_2 = 96/13\,\text{V}$

3.12: $I_{1/8} = 18.49\,\text{A}$

3.13: $V_1 = -19.733\,\text{V}, V_2 = -10.667\,\text{V}, V_3 = -21.867\,\text{V}$

SUPERNODE PROBLEMS

3.14: $V_1 = 2.958\,\text{V}, V_2 = -17.917\,\text{V}, V_3 = 10.083\,\text{V}$

3.15: $V_1 = 76.952\,\text{V}, V_2 = 102.095\,\text{V}, V_3 = 42.667\,\text{V}, V_4 = 106.667\,\text{V}$

3.16: $V_1 = 5.551\,\text{V}, V_2 = -21.279\,\text{V},$

3.17: $V_1 = 12\,\text{V}, V_2 = 2\,\text{V}, V_3 = -8\,\text{V}$

A LINK TO CHAPTER TWO

3.18: $I_1 = 4\,\text{A}, I_3 = 4/3\,\text{A}, I_2 = I_4 = 2/3\,\text{A}$
$V_1 = 4\,\text{V}, V_2 = 4/3\,\text{V}, V_3 = 4\,\text{V}, V_4 = 8/3\,\text{V}$

3.19: $I_1 = 6\,\text{A}, I_2 = 2\,\text{A}, V_6 = 18\,\text{V}$

3.20: $I_3 = 18\,\text{A}, I_4 = 3\,\text{A}, V_{12} = 24\,\text{V}, V_{14} = 14\,\text{V}$

3.21: $I_5 = 5/4\,\text{A}, V_{20} = 100\,\text{V}, P_{24} = 9.375\,\text{W}$

SINGLE LOOP ANALYSIS

3.22: $\qquad$ $I = 400\,\text{mA}$

3.23: $\qquad$ $I = 40\,\text{mA}$

3.24: $\qquad$ $I = 37.5\,\text{mA}$

3.25: $\qquad$ $I = 1.50\,\text{A}, R_{\text{eq}} = 68\,\Omega$

3.26: $\qquad$ $I_5 = -20/7\,\text{A}$

3.27: $\qquad$ $I = 50\,\text{mA}$

MULTIPLE LOOP ANALYSIS

3.28: $\qquad$ $I_4 = 0.590\,\text{A}$

3.29:
$$\begin{bmatrix} 32 & -10 & -8 & -2 & 0 \\ -10 & 42 & 0 & -6 & -2 \\ -8 & 0 & 24 & -4 & 0 \\ -2 & -6 & -4 & 22 & -10 \\ 0 & -2 & 0 & -10 & 16 \end{bmatrix} \begin{bmatrix} I_1 \\ I_2 \\ I_3 \\ I_4 \\ I_5 \end{bmatrix} = \begin{bmatrix} 8 \\ -40 \\ 64 \\ 0 \\ 36 \end{bmatrix}$$

3.30: $\qquad$ $V_s = 56\,\text{V}$

3.31: $\qquad$ $I_{16} = 1.808\,\text{A}, V_8 = -1.496\,\text{V}$

3.32: $\qquad$ $I_6 = 1.819\,\text{A}, I_4 = 5.455\,\text{A}, V_{12} = 32.724\,\text{V}, P_{24} = 401.7\,\text{W}$

3.33: $\qquad$ $I_6 = 2.208\,\text{A}, V_6 = 13.248\,\text{V}, P_6 = 29.25\,\text{W}$

3.34: $\qquad$ $P = 49.96\,\text{W}$

3.35: $\qquad$ $I_1 = 4\,\text{A}, I_3 = 4/3\,\text{A}, I_2 = I_4 = 2/3\,\text{A}$
$\qquad$ $V_1 = 4\,\text{V}, V_2 = 4/3\,\text{V}, V_3 = 4\,\text{V}, V_4 = 8/3\,\text{V}$

3.36: $\qquad$ $I_1 = 6\,\text{A}, I_2 = 2\,\text{A}, V_6 = 18\,\text{V}$

3.37: $\qquad$ $I_3 = 18\,\text{A}, I_4 = 3\,\text{A}, V_{12} = 24\,\text{V}, V_{14} = 14\,\text{V}$

3.38: $\qquad$ $I_5 = 5/4\,\text{A}, V_{20} = 100\,\text{V}, P_{24} = 9.375\,\text{W}$

CHAPTER FOUR
THE OPERATIONAL AMPLIFIER

ANALYSIS USING THE IDEAL OP-AMP MODEL

4.1: (a) $G = -5, R_{\text{in}} = 20\,\text{k}\Omega$, (b) $G = -2, R_{\text{in}} = 50\,\text{k}\Omega$,
(c) $G = -0.25, R_{\text{in}} = 40\,\text{k}\Omega$

4.2: (a) $G = -1, R_{\text{in}} = 50\,\text{k}\Omega$, (b) $G = -3, R_{\text{in}} = 25\,\text{k}\Omega$,
(c) $G = -8, R_{\text{in}} = 11\,\text{k}\Omega$

4.3: $v_o/v_i = 50$

4.4: $\dfrac{v_o}{v_i} = -\dfrac{R_2 R_3 (R_4 + R_5)}{R_1 (R_3 R_4 + R_3 R_5 + R_2 R_4)}$

4.5: (a) $G = 10$, (b) $G = 16$, (c) $G = 12$

4.6: $v_o = 24\,\text{V}$

4.7: $v_o = \left(1 + \dfrac{R_2}{R_1}\right)\left(\dfrac{R_2}{R_1 + R_2}\right) v_{i2} - \dfrac{R_2}{R_1} v_{i1},\ v_o = v_{i2} - v_{i1}$

4.8: $\Delta R = -(R_a + R_b)\dfrac{v_o}{v_i}$

4.9: $v_0 = -\dfrac{R_a}{R_1}\left[1 + \left(\dfrac{R_a + R_b}{R_b}\right)\right] v_i,\ R_a = 10,900\,\Omega$

4.10: $R_{\text{in}} = -\dfrac{R}{R_f} R_1$

4.11: $R_f = 5\,\text{M}\Omega, R_1 = 40,000\,\Omega$

4.12: $R == 80\,\text{k}\Omega$

GENERAL SUMMING AMPLIFIER PROBLEMS
(THE IDEAL OP-AMP MODEL)

4.13: $v_o = 5\,\text{V}$

4.14: $v_3 = 3\,\text{V}$

4.15: $v_o = 20\,\text{V}$

4.16: $R_f = 20\,\text{k}\Omega,\ R_{1,2} = 2.5\,\text{k}\Omega,\ R_{1,3} = 2\,\text{k}\Omega$

4.17: $i_o = -1.06\,\text{mA}\ P = 2.592\,\text{mW}$

4.18: $v_o = 34.29\,\text{V}$

NON IDEAL OP-AMPS: FINITE OPEN LOOP GAIN

4.19: $v_o = -19.76\,\text{V}$

4.20: $A = 74.00$

4.21: $R_f \approx 202,500\,\Omega$

4.22: $R_1 \approx 10,560\,\Omega$

SATURATION EFFECTS

4.23:   see sketch in example.

4.24:   for negative saturation, 7.85 ms to 23.56 ms

               for positive saturation region, 39.27 ms to 54.98 ms

4.25: $-20\text{mV} \le v_d \le 20\,\text{mV}\ v_o = -2.4\,\text{V}$

4.26:   (a) $v_o = -2.5\,\text{V}, v_d = 0\,\text{V}$, (b) $v_o = 8\,\text{V}, v_d = -0.80\,\text{V}$

               (c) $v_o = -8\,\text{V}, v_d = -1.6\,\text{V}$

SELECTION OF OPERATIONAL AMPLIFIER COMPONENTS

4.27: $R_f = 100\text{k}\Omega, \Delta R = 16.67\text{k}\Omega, R_g = 100\text{k}\Omega$

               $R_{a1} = 12.5\text{k}\Omega, R_{a2} = 25\text{k}\Omega, R_{a3} = 50\text{k}\Omega$

               $R_{b1} = 33.33\text{k}\Omega, R_{b2} = 16.67\text{k}\Omega, R_{b3} = 11.11\text{k}\Omega$

4.28: $v_o = 5\,\text{V}$

4.29: $R_f = 100\text{k}\Omega, R_g = 12.5\text{k}\Omega, R_{a1} = 25\text{k}\Omega$

               $R_{a2} = 50\text{k}\Omega, R_{a3} = 100\text{k}\Omega, R_{a4} = 12.5\text{k}\Omega$

               $R_{b1} = 20\text{k}\Omega, R_{b2} = 50\text{k}\Omega$

4.30: $v_o = -8\,\text{V}$

CHAPTER FIVE
SUPERPOSITION AND SOURCE TRANSFORMATIONS

LINEARITY
5.1: (a) linear, (b) nonlinear
5.2: linear
5.3: linear
5.4: nonlinear
5.5: nonlinear

PROPORTIONALITY
5.6: $I_3 = 18\,\text{A}, I_4 = 3\,\text{A}$ $V_{12} = 24\,\text{V}$ $V_{14} = 14\,\text{V}$
5.7: $I_6 = 3\,\text{A}, V_3 = 12\,\text{V}$
5.8: $I_4 = 4\,\text{A}, I_6 = 4/3\,\text{A}$ $V_2 = 16/3\,\text{V}$
5.9: $I_2 = 1\,\text{A}, I_4 = 3\,\text{A}$ $V_8 = 24\,\text{V}$
5.10: $I_9 = 9\,\text{A}, I_{40} = 9/20\,\text{A}$ $V_7 = 63/5\,\text{V}$

SOURCE TRANSFORMATION PROBLEMS
5.11: (a) $I = 4\,\text{A}, R = 6\,\Omega$, (b) $I = 2\,\text{A}, R = 6\,\Omega$
(c) $I = 2I_a\,\text{A}$ $R = 2\,\Omega$
5.12: (a) $V = 28\,\text{V}, R = 7\,\Omega$, (b) $V = 32\,\text{V}, R = 4\,\Omega$
(c) $V = 20I_a\,\text{V}$ $R = 4\,\Omega$
5.13: $V = 18\,\text{V}, R = 18\,\Omega$
5.14: $V = 24\,\text{V}, R = 8\,\Omega$
5.15: $V = 33\,\text{V}, R = 22\,\Omega$

SUPERPOSITION PROBLEMS
5.16: $I = 3\,\text{A}$
5.17: $I = 3/4\,\text{A}$
5.18: $I = 3.420\,\text{A}$
5.19: $I = -0.323\,\text{A}$
5.20: $V = -55\,\text{V}$
5.21: $I = 5/14\,\text{A}$
5.22: $V = -8/11\,\text{V}$
5.23: $I = -0.261\,\text{A}$
5.24: $V_s = 22\,\text{V}$
5.25: $P = 18.14\,\text{mW}$

SUPERPOSITION AND OPERATIONAL AMPLIFIERS

5.26: $\qquad v_o = \dfrac{R_2}{R_1}(v_{i2} - v_{i1})$

5.27: $\qquad R = 25\,\text{k}\Omega$

5.28: $\qquad v_0 = 6.5\,\text{V}$

5.29: $\qquad v_o = \left(\dfrac{1 + \dfrac{R_2}{R_1}}{1 + \dfrac{R_3}{R_4}}\right) v_{i2} - \dfrac{R_2}{R_1} v_{i1}$

5.30: $\qquad R_1 = 50\,\text{k}\Omega,\ R_3 = 40\,\text{k}\Omega$

CHAPTER SIX

THEVENIN, NORTON AND
MAXIMUM POWER TRANSFER THEOREMS

THEVENIN AND NORTON THEOREMS FOR NETWORKS WITHOUT CONTROLLED SOURCES

6.1: $R = 12\,\Omega$

6.2: $V_T = 19.51\,\text{V}\ R_T = 7.63\,\Omega$

6.3: $I = 0.718\,\text{A}$

6.4: $I = 0.352\,\text{A}$

6.5: $I_N\ 0.0965\,\text{A}\ \ R_N = 600\,\Omega$

6.6: $V_{320} = 755.53\,\text{V}$

6.7: $R = 8\,\Omega$

6.8: $I_{10} = 3/4\,\text{A}$

THEVENIN AND NORTON THEOREMS FOR NETWORKS WITH CONTROLLED SOURCES

6.9: $P_{12} = 117.6\,\text{W}$

6.10: $I_N = 5.00\,\text{mA}, R_N = 1800\,\Omega$

6.11: $I_{1.6} = 3\,\text{A}$

6.12: $I_4 = 22/7\,\text{A}$

6.13: $I_{16} = 3.71\,\text{A}$

6.14: $P_{300} = 7.68\,\text{W}$

6.15: $I_6 = 5.76\,\text{A}$

6.16: $I_3 = -9.6\,\text{A}$

MAXIMUM POWER TRANSFER

6.17: $R = 3\,\Omega,\ P = 12\,\text{W}$

6.18: $R = 16\,\Omega,\ P = 324\,\text{W}$

6.19: $R = 30/7\,\Omega,\ P = 33.6\,\text{W}$

6.20: $R = 1100\,\Omega,\ P = 16.57\,\text{W}$

6.21: $R = 160\,\Omega,\ P \approx 2007\,\text{W}$

6.22: $R = 3.619\,\Omega,\ P \approx 2763\,\text{W}$

CHAPTER SEVEN
INDUCTORS, CAPACITORS AND DUALITY

INDUCTORS

7.1: $v = 12.5\,\text{V}$

7.2: $L_{\text{eq}} = = 0.200\,\text{H}$

7.3: $p = 12,800\sin 800t\,\text{W}$

7.4:   (a) $w_L(t = 0\,\text{s}) = 0\,\text{J}$, (b) $w_L(t = 0.5\,\text{s}) = 0.1823\,\text{J}$
        (c) $v_L(t = 0.5\,\text{s}) = 0.2068\,\text{V}$, (d) $p(t = 0.5\,\text{s}) = 0.1974\,\text{J}$

7.5: $L_{\text{eq}} = = 2.20\,\text{H}$

7.6:   (a) $p_R(t = 5\,\text{s}) = 8000\,\text{W}$ (b) $w_L(t = 5\,\text{s}) = 80\,\text{W}$
        (c) $p_s(t = 5\,\text{s}) = 8080\,\text{W}$ (d) $W_R(0 \rightarrow 4\,\text{s}) = 6826.67\,\text{J}$
        (e) $W_L(0 \rightarrow 4\,\text{s}) = 128\,\text{J}$ (f) $W_s(0 \rightarrow 4\,\text{s}) = 6954.67\,\text{J}$

7.7:   (a) $p(t = \pi/200\,\text{s}) = 0\,\text{W}$ (b) $p(0 \rightarrow \pi/200\,\text{s}) = 0\,\text{J}$

7.8: $W(0 \rightarrow 1.2\,\text{s}) = 22.38\,\text{J}$

7.9: $R = 4\,\Omega$

7.10: $i_1 = 7.5\,\text{A}, i_2 = 5\,\text{A}$

7.11: $L_1 = 4.0\,\text{H}, L_2 = 1.0\,\text{H}$

7.12:   see sketch

CAPACITORS

7.13: $i = 1.50\cos 250t\,\text{A}, q = 6\sin 250t\,\text{mC}$

7.14: $W = 0.160\,\text{J}$

7.15: $Q_1 = 200\,\mu\text{C}, Q_2 = 400\,\mu\text{C}$

7.16: $C_{\text{eq}} = 80\,\mu\text{F}$

7.17: $p = 576\sin 800t\,\text{W}$

7.18: $C_{\text{eq}} = 232\,\mu\text{F}$

7.19:   (a) $p_R(t = 4\,\text{s}) = 12.80\,\text{W}$ (b) $p_C(t = 4\,\text{s}) = 8\,\text{W}$
        (c) $p_s(t = 4\,\text{s}) = 20.80\,\text{W}$, (d) $W_R(0 \rightarrow 4\,\text{s}) = 29.87\,\text{J}$
        (e) $W_C(0 \rightarrow 4\,\text{s}) = 24\,\text{J}$, (f) $W_s(0 \rightarrow 4\,\text{s}) = 53.87\,\text{J}$

7.20: $W_C = -74.91\,\mu\text{J}$ removed.

7.21: $W_C = 4.088\,\text{mJ}$

7.22: $C = 60\,\mu\text{F}$

7.23: $v_1 = 15\,\text{V}, v_2 = 5\,\text{V}$

7.24:   see sketch

CHAPTER EIGHT
FIRST ORDER RL AND RC CIRCUITS

THE EXPONENTIAL FUNCTION

8.1: $q(t=0) = 200\,\mu\text{C}, T = 2.50\,\text{ms}, \ q(t=0.625\,\text{ms}) = 155.76\,\mu\text{C}$

8.2: $v(t=0) = 1200\,\text{V}, T = 0.20\,\text{s}, v(t=32.5\,\text{ms}) = 1020\,\text{V}$

8.3: $i(t=0) = 64.67\,\text{A}, t = 265\,\text{ms}$

8.4: $v(\text{t}=125\,\text{ms}) = 877.9\,\text{V}, v(\text{t}=2.125\,\text{s}) = 5.92\,\text{V}$

8.5: $i(t=0) = 800\,\text{A}, T = 0.25\,\text{s}$

8.6: $v(t=0) = 88\,\text{V}, T = 20\,\text{ms}$

EVALUATION OF INITIAL CONDITIONS

8.7:
$i_R(0^-) = 4/3\,\text{A}$ $i_L(0^-) = 4/3\ \text{A}$ $i_C(0^-) = 0\ \text{A}$ $v_C(0^-) = 8/3\,\text{V}$
$v_L(0^-) = 0\,\text{V}$ $i_R(0^+) = 4/3\,\text{A}$ $i_L(0^+) = 4/3\,\text{A}$ $i_C(0^+) = 0\,\text{A}$
$v_C(0^+) = 8/3\ \text{V}$ $v_L(0^+) = -32\,\text{V}$ $i_L'(0^+) = -32\,\text{A/s}$

8.8:
$i_{R1}(0^-) = 4/3\,\text{A}$ $i_L(0^-) = 2\,\text{A}$ $i_C(0^-) = 0\,\text{A}$ $i_{R2}(0^-) = 4/3\,\text{A}$
$v_L(0^-) = 0\,\text{V}$ $v_{R1}(0^-) = 8\,\text{V}$ $v_C(0^-) = 8/3\,\text{V}$ $v_{R2}(0^-) = 8/3\,\text{V}$
$i_{R1}(0^+) = 2\,\text{A}$ $i_L(0^+) = 2\,\text{A}$ $i_C(0^+) = 2/3\,\text{A}$ $i_{R2}(0^+) = 4/3\,\text{A}$
$v_L(0^+) = -8/3\,\text{V}$ $v_{R1}(0^+) = 8\,\text{V}$ $v_C(0^+) = 8/3\,\text{V}$ $v_C'(0^+) = 16/3\,\text{V/s}$

8.9:
$i_{R1}(0^+) = 4\,\text{A}$ $i_L(0^+) = 0\,\text{A}$ $i_C(0^+) = 4\,\text{A}$ $i_{R2}(0^+) = 4\,\text{A}$
$v_L(0^+) = 24\,\text{V}$ $v_{R1}(0^+) = 16\,\text{V}$ $v_C(0^+) = 40\,\text{V}$ $i_L'(0^+) = 12\,\text{A/s}$

8.10:
$i_{R1}(0^+) = 4\,\text{A}$ $v_{R1}(0^+) = 48\,\text{V}$ $v_C'(0^+) = 200\,\text{V/s}$ $i_{R3}(0^+) = 0\,\text{A}$
$i_{R2}(0^+) = 4\,\text{A}$ $i_C(0^+) = 4\ \text{A}$ $v_{R3}(0^+) = 0\,\text{V}$ $v_{R2}(0^+) = 32\,\text{V}$
$v_C(0^+) = 0\,\text{V}$

THE UNDRIVEN FIRST ORDER NETWORK

8.11: $i(t) = \dfrac{8}{3}e^{-18t}\,\text{A}$

8.12: $v(t) = -240e^{-12t}\,\text{V}$

8.13: $v(t) = 18e^{-2t}\,\text{V}$

8.14: $v_{16} = \dfrac{128}{9}e^{-1000t/3}\,\text{V}$

8.15: $v(t) = 200e^{-1.0101t}\,\text{V}$

8.16: $i(t) = 0.014e^{-12.5t}\,\text{A}$

8.17: $v(t) = 30e^{-9t}\,\text{V}$

8.18: $i_L = \dfrac{1}{2}e^{-128t/9}\,\text{A}, \ v_{12}(t) = 6e^{-128t/9}\,\text{V}$

THE DRIVEN FIRST ORDER NETWORK

8.19: $i_1(t) = \frac{1}{3}e^{-15t/16}$ A, $i_2(t) = \frac{2}{3}e^{-15t/16}$ A

8.20: $v_1(t) = 9e^{-3t}$ V, $v_2(t) = 27e^{-3t}$ V

8.21: $t_1 = 56.8\,\text{ms}, t_2 = 36.6\,\text{ms}, t_3 = 11.1\,\text{ms}, \text{ratio} = 3.29$

8.22: $R_1 = 910.2\,\Omega, R_2 = 1820.4\,\Omega$

8.23: $v(t) = 31.272(e^{-0.595t} - 1)$ V

8.24: $i(t) = 3.061 - 0.642e^{-5.765t}$ A

8.25: $i(t) = 2.600 + 5.178e^{-14,285t}$ A

THE DRIVEN FIRST ORDER NETWORK - ALTERNATE SOLUTIONS

8.26: $i_1(t) = \frac{1}{3}e^{-15t/16}$ A, $i_2(t) = \frac{2}{3}e^{-15t/16}$ A

8.27: $i(t) = 3(1 - e^{-3t})$ A

8.28: $v(t) = 31.272(e^{-0.595t} - 1)$ V

8.29: $i(t) = 3.061 - 0.642e^{-5.765t}$ A

8.30: $i(t) = 2.600 + 5.178e^{-14,285t}$ A

OPERATIONAL AMPLIFIERS

8.31: $C = 5\,\mu\text{F}$

8.32: $v_o = 8t^2$ V

8.33: $v_o = 12\sin 400t - 6(\cos 400t - 1)$ V

8.34: see circuit diagram

8.35: see circuit diagram

CHAPTER NINE
SECOND ORDER RL AND RC CIRCUITS

THE SINUSOID

9.1: $V = 120\,\text{V}, \omega = 160\pi\,\text{rad/s}, \phi = -1.973\,\text{rad}$

9.2: $I = 40\,\text{ma}, \omega = 50,000\pi\,\text{rad/s}, \phi = -0.20\,\text{rad}$

9.3: $Q = 20\,\mu\text{C}, \omega = 125\pi\,\text{rad/s}, \phi = \pm\pi\,\text{rad}$

9.4: $V = 208\,\text{V}, \omega = 7646.7\pi\,\text{rad/s}, \phi = 0\,\text{rad}$

9.5: $f = 66.67\,\text{Hz}, T = 15\,\text{ms}$

9.6: $I = 20\,\text{mA}, \omega = 100\pi\,\text{rad/s}, \phi = -1.20\,\text{rad}$

FORMS OF RESPONSE

9.7: $L = 0.10\,\text{H}$

9.8: $L = 0.171\,\text{H}, \zeta = 0.967, \omega_n = 24.18\,\text{rad/s}$

$L = 2.067\,\text{H}, \zeta = 0.248, \omega_n = 6.19\,\text{rad/s}$

9.9: $C = 488.3\,\mu\text{F}$

9.10: $C = 0.00125\,\text{F}, \zeta = 0.941$

$C = 0.009645\,\text{F}, \zeta = 0.339$

9.11: $R = 0.319\,\Omega$

9.12: $C = 0.00651\,\text{F}$

UNDRIVEN SECOND ORDER NETWORKS

9.13: $i(t) = \dfrac{9}{4}\left(e^{-6t} - e^{-2t}\right)\,\text{A}$

9.14: $v_L(t) = 96e^{-12t}(3\sin 8t - 2\cos 8t)\,\text{V}$

9.15: $v(t) = \dfrac{11}{5}\left(8e^{-8t} - 3e^{-3t}\right),\text{V}$

9.16: $i_L(t) = e^{-200t}(0.180\sin 400t + 0.360\cos 400t)\,\text{A}$

9.17: $v_C(t) = -96te^{-8t}\,\text{V}$

9.18: $v_C(t) = 160(e^{-30t} - e^{-20t})\,\text{V}$

DRIVEN SECOND ORDER NETWORKS

9.19: $v(t) = 24e^{-6t} - 8e^{-4t}\,\text{V}$

9.20: $i(t) = 9e^{-t}\sin 2t\,\text{A}$

9.21: $v(t) = e^{-4t}\left(12\cos 3t + \dfrac{52}{3}\sin 3t\right)\,\text{V}$

9.22: $i(t) = 8te^{-2t}\,\text{A}$

9.23: $v(t) = 8e^{-2t} + 128te^{-2t}\,\text{V}$

9.24: $i(t) = 24(e^{-4t} - e^{-3t})\,\text{A}$

CHAPTER TEN

SINUSOIDAL STEADY STATE
ANALYSIS BY PHASOR METHODS

COMPLEX NUMBER ALGEBRA

10.1: $6 + j4, 7.211\underline{/33.69°}, 7.211e^{33.69°}$

10.2: $2.828\underline{/-81.87°}, 0.40 - j2.80, 2.828e^{-81.87°}$

10.3: $145.00\underline{/-46.40°}, 100 - j105, 145e^{-46.40°}$

10.4: $100\sqrt{2}\underline{/98.13°}, -20 + j140, 100\sqrt{2}e^{98.13°}$

10.5: $-0.35 - j2.55, 2.574\underline{/-97.82°}, 2.574e^{-97.82°}$

10.6: $30.198\underline{/-88.47°}, 0.804 - j30.187, 30.198e^{-88.47°}$
$32.330\underline{/-162.33°}, -30.804 - j9.813, 32.330e^{-162.33°}$

10.7: $1.130\underline{/58.74°}, 0.586 + j0.966, 1.130e^{58.74°}$

10.8: $36,055\underline{/-130.05°}, -23,200 - j27,600, 36,055e^{-130.05°}$

10.9: $-44 - j104, 112.93\underline{/-112.93°}, 112.93e^{-122.93°}$

10.10: $19 + j32, 37.22\underline{/59.30°}, 37.22e^{59.30°}$

IMPEDANCE AND ADMITTANCE

10.11: $Z(\omega = 400\,\text{rad/s}) = 20\,\Omega$

10.12: $Z(j200\,\text{rad/s}) = 29.2\,\Omega$

10.13: $Y(\omega = j100\,\text{rad/s}) = 0.01\underline{/36.87°}\,\mho = 0.008 + j0.006\,\mho$

10.14: $Y(s = j500\,\text{rad/s}) = 0.01\underline{/-36.87°}\,\mho = 0.008 - j0.006\,\mho$

10.15: $Z(\omega = 1000\,\text{rad/s}) = 150 - j80\,\Omega = 170\underline{/-28.07°}\,\Omega$

PHASORS

10.16: $f(t) = 4\sqrt{2}\cos(\omega t - 135°)$

10.17: $f(t) = 136\cos(\omega t - 118.07°)$

10.18: $f(t) = 75\cos(\omega t + 143.13°)$

10.19: $f(t) = 182\cos(\omega t + 112.62°)$

10.20: $f(t) = 100\cos(\omega t + 73.74°)$

10.21: $f(t) = -220\cos\omega t + 165\sin\omega t$

10.22: $f(t) = -150\cos\omega t - 360\sin\omega t$

10.23: $f(t) = 768\cos\omega t - 224\sin\omega t$

10.24: $f(t) = -103.92\cos\omega t - 60\sin\omega t$

10.25: $f(t) = -500\sqrt{2}\cos\omega t + 500\sqrt{2}\sin\omega t$

10.26: $\quad \mathbf{V} = 208\underline{/60°}\,\mathrm{V},\ 104 + \jmath180.13\,\mathrm{V},\ 208e^{60°}\,\mathrm{V}$

10.27: $\quad \mathbf{i} = 4\sqrt{2}\underline{/-135°}\,\mathrm{A},\ -4 - \jmath4\,\mathrm{A},\ 4\sqrt{2}e^{-135°}\,\mathrm{A}$

10.28: $\quad \mathbf{i} = 120\sqrt{3}\underline{/15°}\,\mathrm{A},\ 200.76 + \jmath53.79\,\mathrm{A},\ 120\sqrt{3}e^{15°}\,\mathrm{A}$

10.29: $\quad \mathbf{V} = 1750\underline{/-163.74°}\,\mathrm{V},\ 1680 - \jmath490\,\mathrm{V},\ 1750e^{-163.74°}\,\mathrm{V}$

10.30: $\quad \mathbf{V} = 400\underline{/90°}\,\mathrm{V},\ 0 + \jmath400\,\mathrm{V},\ 400e^{90°}\,\mathrm{V}$

10.31: $\quad i_2(t) = 16.731\cos(377t + 155.27°)\,\mathrm{A}$

10.32: $\quad v_2(t) = 148.86\cos(377t + 120.70°)\,\mathrm{V}$

10.33: $\quad R_a = 42\,\Omega,\ L_b = 0.024\,\mathrm{H},\ R_c = 6\,\Omega,\ C_d = 95.24\,\mu\mathrm{F}$

10.34: $\quad \mathbf{V} = 415.93\underline{/117.18°},\ Z_{\mathrm{eq}} = 20 + \jmath12.5\,\Omega$

LADDER NETWORKS AND PHASOR DIAGRAMS

10.35: See solution for phasor diagram.

10.36: See solution for phasor diagram.

10.37: See solution for phasor diagram.

10.38: See solution for phasor diagram.

10.39: See solution for phasor diagram.

NETWORK ANALYSIS

10.39: $\quad \mathbf{i}_L = 8\underline{/126.87°}\,\mathrm{A},\ \mathbf{V}_C = 320\sqrt{2}\underline{/81.87°}\,\mathrm{V}$

10.40: $\quad \mathbf{V}_R = 76.80\underline{/16.26°}\,\mathrm{V},\ I_L = 400\underline{/-110.61°}\,\mathrm{mA}$

10.41: $\quad \mathbf{i}_L = 2\sqrt{2}\underline{/-45°}\,\mathrm{A},\ \mathbf{V}_C = 400\underline{/90°}\,\mathrm{A}$

10.42: $\quad \mathbf{V}_R = 160\underline{/53.13°}\,\mathrm{A},\ \mathbf{i}_L = 4\underline{/53.13°}\,\mathrm{V}$

10.43: $\quad \mathbf{i}_R = 0.973\underline{/46.85°}\,\mathrm{A}$

10.44: $\quad \mathbf{i}_L = 8\underline{/126.87°}\,\mathrm{A},\ \mathbf{V}_C = 452.54\underline{/81.87°}\,\mathrm{V}$

10.45: $\quad \mathbf{i}_L = 400\underline{/-110.61°}\,\mathrm{mA},\ V_R = 76.80\underline{/16.26°}\,\mathrm{V}$

10.46: $\quad \mathbf{i}_L = 4\underline{/53.13°}\,\mathrm{mA},\ \mathbf{V}_R = 76.80\underline{/16.26°}\,\mathrm{V}$

10.47: $\quad \mathbf{i}_R = 0.973\underline{/46.85°}\,\mathrm{A}$

10.48: $\quad 2\sqrt{2}\underline{/-45°}\,\mathrm{A}$

10.49: $\quad 2\sqrt{2}\underline{/-45°}\,\mathrm{A}$

10.50: $\quad \mathbf{V}_L = 118.89\underline{/-117.16°}\,\mathrm{V}$

10.51: $\quad \mathbf{V}_C = 480\sqrt{2}\underline{/45°}\,\mathrm{V}$

10.52: $\quad \mathbf{i}_L = 8.34\underline{/-98.13°}\,\mathrm{A}$

10.53: $\quad \mathbf{i}_R = 8\underline{/36.87°}\,\mathrm{A}$

10.54: $\quad \mathbf{V}_C = 480\sqrt{2}\underline{/45°}\,\mathrm{V}$

10.55: $\quad \mathbf{i}_R = 0.973\underline{/46.84°}\,\mathrm{A}$

10.56: $\quad \mathbf{i}_L = 10.67\underline{/0°}\,\mathrm{A}$

10.57: $\quad \mathbf{i}_C = 4\underline{/180°}\,\mathrm{A}$

CHAPTER ELEVEN
SINUSOIDAL STEADY STATE POWER CALCULATIONS

AVERAGE AND EFFECTIVE VALUES

11.1: $V_{\text{avg}} = 0.850\,\text{V}, V_{\text{eff}} = 2.345\,\text{V}$

11.2: $I_{\text{avg}} = 0.800\,\text{V}, V_{\text{eff}} = 2.450\,\text{A}$

11.3: $V_{\text{avg}} = 0.611\,\text{V}, V_{\text{eff}} = 2.271\,\text{V}$

11.4: $I_{\text{avg}} = -0.600\,\text{A}, I_{\text{eff}} = 1.915\,\text{V}$

11.5: $V_{\text{avg}} = 1.637\,\text{V}, V_{\text{eff}} = 2.828\,\text{V}$

11.6: $V_{\text{avg}} = 2/3\,\text{V}, V_{\text{eff}} = 1.856\,\text{V}$

AVERAGE POWER IN THE SINUSOIDAL STEADY STATE

11.7: $P_{\text{avg}} = 103.68\,\text{W}, PF = 0.800$

11.8: $P_{\text{avg}} = 3075\,\text{W}, PF = 0.6932$

11.9: $P_{\text{avg}} = 230.6\,\text{W}, PF = 0.9027$

11.10: $P_{\text{avg}} = 9192\,\text{W}, PF = 0.7365$

11.11: $C = 20\,\mu\text{F}$

11.12: $R = 600\,\Omega, C = 10\,\mu\text{F}$

11.13: $P_{10\,\Omega} = 1141.7\,\text{W}, P_{10\,\Omega} = = 933.9\,\text{W}, P_{30\,\Omega} = 306.8\,\text{W}$

11.14: $\mathbf{S} = 1556.9\underline{/-1.61°}\,\text{VA}$

11.15: $\mathbf{S} = 5971.4\underline{/35.84°}\,\text{VA}$

11.16: $\mathbf{S} = 12.21\underline{/8.42°}\,\text{VA}$

11.17: $\mathbf{S} = 2688\sqrt{2}\underline{/45°}\,\text{VA}$

11.18: $P = 360\,\text{W}, \mathbf{S} = 402.49\underline{/26.57°}\,\text{VA}$

$|S| = 402.49\,\text{VA}, PF = 0.8944$

POWER FACTOR CORRECTION

11.19 $P = 1479.85\,\text{W}, Q = 741.41\,\text{VAR}, |S| = 1655.19\,\text{VA}$

$PF = 0.8941, Q_{\text{corr}} = -255\,\text{VAR}, C = 10.82\,\mu\text{F}$

11.20 $P = 536\,\text{W}, Q = 292\,\text{VAR}, |S| = 610.38\,\text{VA}$

$PF = 0.8781, Q_{\text{corr}} = -88.7\,\text{VAR}, C = 23.53\,\mu\text{F}$

11.21 $P = 2520\,\text{W}, Q = 2200\,\text{VAR}, |S| = 3345.2\,\text{VA}$

$PF = 0.7533, Q_{\text{corr}} = -1126.5\,\text{VAR}, C = 44.20\,\mu\text{F}$

11.22 $P = 3120\,\text{W}, Q = 1760\,\text{VAR}, |S| = 3582.2\,\text{VA}$

$PF = 0.8710, Q_{\text{corr}} = -680.1\,\text{VAR}, C = 6.77\,\mu\text{F}$

MAXIMUM POWER TRANSFER

11.23: $\qquad Z_o = 50 + j108.22\,\Omega$

11.24: $\qquad Z_o = 10 + j24\,\Omega, P_o = 125\,\text{W}$

11.25: $\qquad Z_o = 20 + j15\,\Omega, P_o = 2000\,\text{W}$

11.26: $\qquad Z_o = 150 - j200\,\Omega, P_o = 341.6\,\text{W}$

CHAPTER TWELVE

THREE PHASE POWER SYSTEMS

BALANCED SYSTEMS

12.1: $P = 2160\,\text{W}$

12.2: $P = 2764.8\,\text{W}$

12.3: $\mathbf{i}_a = 30\underline{/0^\circ}\,\text{A},\ \mathbf{i}_b = 30\underline{/-120^\circ}\,\text{A},\ \mathbf{i}_c = 30\underline{/120^\circ}\,\text{A}$

12.4: $v_{an}(t) = 120.09\cos(377t + 0^\circ)\,\text{V},\ v_{bn}(t) = 120.09\cos(377t - 120^\circ)\,\text{V}$
$v_{cn}(t) = 120.09\cos(377t + 120^\circ)\,\text{V},\ v_{ab}(t) = 208\cos(377t + 30^\circ)\,\text{V}$
$v_{bc}(t) = 208\cos(377t - 90^\circ)\,\text{V},\ \mathbf{V}_{ca} = 208\cos(377t + 150^\circ)\,\text{V}$

12.5: $v_{an}(t) = 120.09\cos(377t + 0^\circ)\,\text{V},\ v_{bn}(t) = 120.09\cos(377t + 120^\circ)\,\text{V}$
$v_{cn}(t) = 120.09\cos(377t - 120^\circ)\,\text{V},\ v_{ab}(t) = 208\cos(377t - 30^\circ)\,\text{V}$
$v_{bc}(t) = 208\cos(377t + 90^\circ)\,\text{V},\ v_{ca}(t) = 208\cos(377t - 150^\circ)\,\text{V}$

12.6: $v_{ab}(t) = 208\cos(377t + 0^\circ)\,\text{V},\ v_{bc}(t) = 208\cos(377t + 120^\circ)\,\text{V}$
$\mathbf{V}_{ca} = 208\cos(377t - 120^\circ)\,\text{V}$

12.7: $v_{ab}(t) = 208\cos(377t + 0^\circ)\,\text{V},$
$v_{bc}(t) = 208\cos(377t + 120^\circ)\,\text{V}$
$v_{ca}(t) = 208\cos(377t - 120^\circ)\,\text{V}$

12.8: $v_{an}(t) = 254\cos(377t + 0^\circ)\,\text{V},\ v_{bn}(t) = 254\cos(377t - 120^\circ)\,\text{V}$
$v_{cn}(t) = 254\cos(377t + 120^\circ)\,\text{V},\ v_{ab}(t) = 440\cos(377t + 30^\circ)\,\text{V}$
$v_{bc}(t) = 440\cos(377t - 90^\circ)\,\text{V},\ v_{ca}(t) = 440\cos(377t + 150^\circ)\,\text{V}$
$i_{AN}(t) = 10.16\cos(377t - 36.87^\circ)\,\text{A},\ i_{BN}(t) = 10.16\cos(377t - 156.87^\circ)\,\text{A},$
$i_{CN}(t) = 10.16\cos(377t + 83.13^\circ)\,\text{A}$

12.9: $P = 6193.5\,\text{W}$

12.10: $\mathbf{V}_{AN} = \mathbf{V}_{an} = 138.6\underline{/0^\circ}\,\text{V},\ \mathbf{V}_{BN} = \mathbf{V}_{bn} = 138.6\underline{/-120^\circ}\,\text{V}$
$\mathbf{V}_{CN} = \mathbf{V}_{cn} = 138.6\underline{/120^\circ}\,\text{V},\ \mathbf{i}_{AN} = 4.62\underline{/-36.87^\circ}\,\text{A}$
$\mathbf{i}_{BN} = 4.62\underline{/-156.87^\circ}\,\text{A},\ \mathbf{i}_{CN} = 4.62\underline{/-83.13^\circ}\,\text{A}$
$P = 1536.8\,\text{W}$

12.11: $\mathbf{V}_{AN} = \mathbf{V}_{an} = 138.6\underline{/0^\circ}\,\text{V},\ \mathbf{V}_{BN} = \mathbf{V}_{bn} = 138.6\underline{/-120^\circ}\,\text{V}$
$\mathbf{V}_{CN} = \mathbf{V}_{cn} = 138.6\underline{/120^\circ}\,\text{V},\ \mathbf{i}_{AN} = 3.47\underline{/-53.13^\circ}\,\text{A}$
$\mathbf{i}_{BN} = 3.47\underline{/-172.13^\circ}\,\text{A},\ \mathbf{i}_{CN} = 4.47\underline{/66.87^\circ}\,\text{A}$
$P = 865.7\,\text{W}$

12.12: $\mathbf{V}_{AN} = 138.6\underline{/0^\circ}\,\text{V},\ \mathbf{V}_{BN} = 138.6\underline{/-120^\circ}\,\text{V}$
$\mathbf{V}_{CN} = 138.6\underline{/120^\circ}\,\text{V},\ \mathbf{V}_{AB} = 240\underline{/30^\circ}\,\text{V}$
$\mathbf{V}_{bc} = 240\underline{/-90^\circ}\,\text{V},\ \mathbf{V}_{AB} = 240\underline{/30^\circ}\,\text{V},\ \mathbf{V}_{CA} = 240\underline{/160^\circ}\,\text{V}$
$\mathbf{i}_{AB} = 8\underline{/-6.87^\circ}\,\text{A},\ \mathbf{i}_{BC} = 8\underline{/-126.87^\circ}\,\text{A},$
$\mathbf{i}_{CA} = 8\underline{/113.13^\circ}\,\text{A},\ P = 4608\,\text{W}$

12.13: $\mathbf{i}_{AB} = 4.80\underline{/13.74°}$ A, $\mathbf{i}_{BC} = 4.80\underline{/-106.26°}$ A
$\mathbf{i}_{CA} = 4.80\underline{/133.74°}$ A, $P = 3317.8$ W

12.14: $\mathbf{V}_{an} = 138.6\underline{/0°}$ V, $\mathbf{V}_{bn} = 138.6\underline{/-120°}$ V
$\mathbf{V}_{an} = 138.6\underline{/120°}$ V, $\mathbf{V}_{AB} = 240\underline{/30°}$ V,
$\mathbf{V}_{BC} = 240\underline{/-90°}$ V, $\mathbf{V}_{CA} = 240\underline{/150°}$ V
P 2660.4 W

12.15: $P = 1496.5$ W

12.16: $\mathbf{V}_{AB} = 240\underline{/0°}$ V, $\mathbf{V}_{BC} = 240\underline{/-120°}$ V
$\mathbf{V}_{CA} = 240\underline{/120°}$ V, $\mathbf{i}_{AB} = 8\underline{/-36.87°}$ A,
$\mathbf{V}_{BC} = 8\underline{/-156.87°}$ A, $\mathbf{V}_{CA} = 8\underline{/153.13°}$ A
$P = 4608$ W

12.17: $P = 3317.8$ W

UNBALANCED SYSTEMS

12.18: $P = 2857.2$ W

12.19: $P = 1346.9$ W

12.20: $\mathbf{i}_b = 24.25\underline{/-150°}$ A

12.21: $\mathbf{i}_1 = 15\underline{/-36.87°}$ A, $\mathbf{i}_2 = 10\underline{/-173.13°}$ A,
$\mathbf{i}_3 = 7.5\underline{/165°}$ A, $\mathbf{i}_n = 9.74\underline{/57.96°}$ A

12.22: $\mathbf{i}_1 = 15\underline{/-36.87°}$ A. $\mathbf{i}_2 = 10\underline{/66.87°}$ A,
$\mathbf{i}_3 = 7.5\underline{/-75°}$ A, $\mathbf{i}_n = 19.21\underline{/158.47°}$ A,

12.23: $P = 2694.3$ W

12.24: $PF = 0.8861$

12.25: $P = 3540$ W

12.26: $PF = 0.9909$

12.27: $P = 1055.9$ W

PROBLEM SOLUTIONS

CHAPTER ONE
INTRODUCTION AND
BASIC CONCEPTS

CHARGE AND CURRENT

1.1: How many electrons pass a given point in a conductor in 20 s if the conductor is carrying 10 A?

1 A is 1 C/s. Thus in 20 s, the charge flow is

$$Q = (20\,\text{s})(10\,\text{C/s}) = 200\,\text{C}$$

Because 1 electron possesses -1.602×10^{-19} C, 1 Coulomb consists of 6.242×10^{18} electrons. Then 200 Coulombs is equivalent to

$$(200\,\text{C})(6.242 \times 10^{18}\,\text{electrons}) = 1.248 \times 10^{21}\,\text{electrons} \Leftarrow$$

1.2: If 240 C pass through a wire in 12 s what is the current?

Because 1 A = 1 C/s

$$I = \frac{Q}{\Delta t} = \frac{240\,\text{C}}{12\,\text{s}} = 20\,\text{A} \Leftarrow$$

1.3: What current passes through a conductor if 3.2×10^{21} electrons flow through the conductor in 16 s?

1 electron possesses -1.602×10^{-19} C so that 3.2×10^{21} electrons is equivalent to 512.6 C. Hence in 16 s, the charge flow (the current) is

$$I = \frac{Q}{\Delta t} = \frac{512.6\,\text{C}}{16\,\text{s}} = 32.04\,\text{A} \Leftarrow$$

1.4: If the current in a conductor varies in accordance with the relationship

$$i = 4t + 24(1 - e^{-4t})\,\text{A}$$

how much charge flows from time, $t = 0$ to time, $t = 5\,s$?

Because, $i = dq/dt$,

$$q = \int i\,dt$$

Here

$$i = 4t + 24(1 - e^{-4t})\,\text{A}$$

133

so that

$$Q = \int_{t=0}^{t=5\,\text{s}} [4t + 24(1 - e^{-4t})\,\text{A}]\,dt$$

$$= \int_{t=0}^{t=5\,\text{s}} 4t\,dt + \int_{t=0}^{t=5\,\text{s}} 24\,dt - \int_{t=0}^{t=5\,\text{s}} 24e^{-4t}\,dt$$

$$= 2t^2 \Big|_{t=0}^{t=5\,\text{s}} + 24t \Big|_{t=0}^{t=5\,\text{s}} + 6e^{-4t} \Big|_{t=0}^{t=5\,\text{s}}$$

$$= 2(25 - 0) + 24(5 - 0) + 6(e^{-20} - 1)$$

$$= 50\,\text{C} + 120\,\text{C} - 6\,\text{C}$$

$$= 164\,\text{C} \Leftarrow$$

ENERGY, POWER, VOLTAGE, CURRENT AND CHARGE

1.5: How much work is done in moving $10\,\text{nC}$ of charge a distance of $68\,\text{cm}$ in the direction of a uniform electric field of having a field strength of $E = 80\,\text{kV/m}$?

$$F = EQ$$

$$= (80 \times 10^3\,\text{V/m})(100 \times 10^{-9}\,\text{C})$$

$$= 8 \times 10^{-3}\,\text{N}$$

and

$$W = Fr$$

$$= (8 \times 10^{-3}\,\text{N})(0.68\,\text{m})$$

$$= 5.44 \times 10^{-3}\,\text{J} \Leftarrow$$

1.6: A charge of $0.5\,\text{C}$ is brought from infinity to a point. Assume that infinity is at $0\,\text{V}$ and determine the voltage at the terminal point if $14.5\,\text{J}$ is required to move the charge.

$$V = \frac{W}{Q}$$

$$= \frac{14.5\,\text{J}}{0.5\,\text{C}}$$

$$= 29\,\text{V} \Leftarrow$$

1.7: If the potential difference between two points is $125\,\text{V}$, how much work is required to move a $3.2\,\text{C}$ charge?

$$W = QV$$

$$= (3.2\,\text{C})(125\,\text{V})$$
$$= 400\,\text{J} \Leftarrow$$

1.8: How many coulombs can be moved from point-A to point-B if $\Delta V_{AB} = 440\,\text{V}$ and a maximum of 842 J can be expended?

$$Q = \frac{W}{V}$$
$$= \frac{842\,\text{J}}{440\,\text{V}}$$
$$= 1.914\,\text{C} \Leftarrow$$

1.9: If 1 horsepower (hp) is equal to 0.746 kW, how much energy does a 20 hp motor deliver in 20 min?

$$P = (20\,\text{hp})(746\,\text{W/hp})$$
$$= 14,920\,\text{W}$$

and

$$W = (14,920\,\text{W})(20\,\text{min})(60\,\text{s/min})$$
$$= 1.7904 \times 10^7\,\text{J}$$

or

$$= 17.904\,\text{MJ} \Leftarrow$$

1.10: If a 150 W incandescent bulb operates at 120 V, how many coulombs and electrons flow through the bulb in 1 h?

With $V = 120\,\text{V}$ and $P = 150\,\text{W}$,

$$I = \frac{P}{V}$$
$$= \frac{150\,\text{W}}{120\,\text{V}}$$
$$= \frac{5}{4}\,\text{A}$$

Then

$$Q = I\Delta t$$
$$= \left(\frac{5}{4}\,\text{A}\right)(3600\,\text{s})$$
$$= 4500\,\text{C} \Leftarrow$$

and

$$\text{electrons} = \frac{Q}{1.602 \times 10^{-19}\,\text{C/electron}}$$
$$= \frac{4500\,\text{C}}{1.602 \times 10^{-19}\,\text{C/electron}}$$
$$= 28.09 \times 10^{21}\,\text{electrons} \Leftarrow$$

1.11: If a light bulb takes $1.2\,\text{A}$ at $120\,\text{V}$ and operates for $8\,\text{h/day}$, what is the cost of its operation for 30 days if power costs $\$0.21/\text{kWh}$?

$$P = VI$$
$$= (120\,\text{V})(1.2\,\text{A})$$
$$= 144\,\text{W}$$

Then

$$W = P\Delta t$$
$$= (144\,\text{W})(8\,\text{h/day})(30\,\text{days})$$
$$= 34{,}560\,\text{W-h} \longrightarrow 34.56\,\text{kWh}$$

At $\$0.21/\text{kWh}$, the cost will be

$$\text{cost} = (34.56\,\text{kWh})(\$0.21/\,\text{kWh}) = \$7.26 \Leftarrow$$

1.12: If 1 calorie (1 cal) is equal to $4.184\,\text{J}$ and it takes $1000\,\text{cal}$ to raise $1\,\text{kg}$ of water $1°\text{C}$. how much current is carried by a $120\,\text{V}$ heater if it is used to heat $4.82\,\text{kg}$ of water from $25°\text{C}$ to $45°\text{C}$ in $4\,\text{min}$?

With the recognition that heat is equivalent to energy

$$W = (4.82\,\text{kg})(1000\,\text{cal/kg-C})(45°\text{C} - 25°\text{C})$$

$$= 96,400 \, \text{cal}$$

$$= (96,400 \, \text{cal})(4.184 \, \text{J/cal})$$

$$= 403,338 \, \text{J}$$

Then

$$
\begin{aligned}
P &= \frac{W}{\Delta t} \\
&= \frac{403,778 \, \text{J}}{(4 \, \text{min})(60 \, \text{s/min})} \\
&= 1680.6 \, \text{W}
\end{aligned}
$$

and

$$
\begin{aligned}
I &= \frac{P}{V} \\
&= \frac{1680.6 \, \text{W}}{120 \, \text{V}} \\
&= 14 \, \text{A} \Leftarrow
\end{aligned}
$$

1.13: The waveforms for the current through and the voltage across a certain resistor are as shown in Fig 1.1. Determine the energy dissipated.

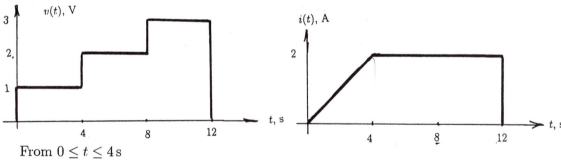

From $0 \leq t \leq 4 \, \text{s}$

$$v(t) = 1 \, \text{V}, \qquad i(t) = \frac{1}{2}t \, \text{A} \qquad \text{and} \qquad p(t) = \frac{1}{2}t \, \text{W}$$

From $4 \leq t \leq 8 \, \text{s}$

$$v(t) = 2 \, \text{V}, \qquad i(t) = 2 \, \text{A} \qquad \text{and} \qquad p(t) = 4 \, \text{W}$$

From $8 \leq t \leq 12 \, \text{s}$

$$v(t) = 3 \, \text{V}, \qquad i(t) = 2 \, \text{A} \qquad \text{and} \qquad p(t) = 6 \, \text{W}$$

The energy dissipated between 0 and 12 s is equal to

$$
\begin{aligned}
W(0,12) &= \int_{t=0}^{t=4\,\text{s}} p(\tau)\,d\tau + \int_{t=4\,\text{s}}^{t=8\,\text{s}} p(\tau)\,d\tau + \int_{t=8\,\text{s}}^{t=12\,\text{s}} p(\tau)\,d\tau \\
&= \int_{t=0}^{t=4\,\text{s}} \frac{1}{2}\tau\,d\tau + \int_{t=4\,\text{s}}^{t=8\,\text{s}} 4\,d\tau + \int_{t=8\,\text{s}}^{t=12\,\text{s}} 6\,d\tau \\
&= \left.\frac{1}{4}\tau^2\right|_{t=0}^{t=4\,\text{s}} + \left.4\tau\right|_{t=4}^{t=8\,\text{s}} + \left.6\tau\right|_{t=8}^{t=12\,\text{s}} \\
&= \frac{1}{4}(16-0) + 4(8-4) + 6(12-8) \\
&= 4\,\text{J} + 16\,\text{J} + 24\,\text{J} \\
&= 44\,\text{J} \Leftarrow
\end{aligned}
$$

1.14: The waveforms for the current through and the voltage across a certain resistor are as shown in Fig 1.2. Determine the energy dissipated.

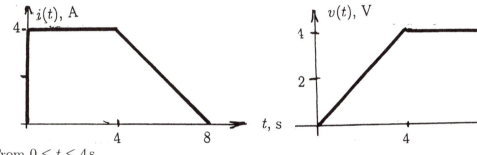

From $0 \le t \le 4\,\text{s}$

$$
v(t) = t\,\text{V}, \qquad i(t) = 4\,\text{A} \qquad \text{and} \qquad p(t) = 4t\,\text{W}
$$

From $4 \le t \le 8\,\text{s}$

$$
v(t) = 4\,\text{V}, \qquad i(t) = (-t+4)\,\text{A} \qquad \text{and} \qquad p(t) = (-4t+16)\,\text{W}
$$

The energy dissipated between 0 and 8 s is equal to

$$
\begin{aligned}
W(0,8) &= \int_{t=0}^{t=4\,\text{s}} p(\tau)\,d\tau + \int_{t=4\,\text{s}}^{t=8\,\text{s}} p(\tau)\,d\tau \\
&= \int_{t=0}^{t=4\,\text{s}} 4\tau\,d\tau - \int_{t=4\,\text{s}}^{t=8\,\text{s}} 4\tau\,d\tau + \int_{t=4\,\text{s}}^{t=8\,\text{s}} 16\,d\tau
\end{aligned}
$$

$$L = \sigma AR$$
$$= (5.805 \times 10^7 \, \text{U/m})(0.1016 \, \text{m})(1.08 \times 10^{-4} \, \text{m})(3.190 \times 10^{-4} \, \Omega)$$
$$= 0.2031 \, \text{m} \Leftarrow$$

1.19: What is the length of a rectangular aluminum bus bar having dimensions 16 cm by 1.4 cm, a conductivity of $3.61 \times 10^4 \, \text{U/m}$ and a resistance of $1.527 \, \Omega$?

With

$$A = (0.16 \, \text{m})(0.014 \, \text{m})$$
$$= 2.24 \times 10^{-3} \, \text{m}^2$$

then

$$L = \sigma AR$$
$$= (3.61 \times 10^4 \, \text{U/m})(2.24 \times 10^{-3} \, \text{m}^2)(1.527 \, \Omega)$$
$$= 123.48 \, \text{m} \Leftarrow$$

1.20: A metal wire having a resistance of $0.1567 \, \Omega$ at a temperature of 20°C is to be used in an application where its resistance can lie between $0.1314 \, \Omega$ and $0.1872 \, \Omega$. If its temperature coefficient of resistance is $0.00314 \, °\text{C}^{-1}$, find the permitted temperature extremes?

With the resistance at a particular temperature given by

$$R(T) = R_o[1 + \alpha(T - 20°\text{C})]$$

where R_o is the resistance at 20°C. Then

$$T = \frac{(R(T)/R_o) - 1}{\alpha} + 20°\text{C}$$

For $R(T) = 0.1872 \, \Omega$

$$T = \frac{(0.1872 \, \Omega/0.1567 \, \Omega) - 1}{0.00314 \, \Omega\text{-}°\text{C/m}} + 20°\text{C}$$
$$= 61.99°\text{C} + 20°\text{C}$$
$$= 81.99°\text{C} \Leftarrow$$

and for $R(T) = 0.1314 \, \Omega$

$$T = \frac{(0.1314 \, \Omega/0.1567 \, \Omega) - 1}{0.00314 \, \Omega\text{-}°\text{C/m}} + 20°\text{C}$$
$$= -51.42°\text{C} + 20°\text{C}$$
$$= -31.42°\text{C} \Leftarrow$$

1.21: Measurements taken on a conductor show a resistance of $8.24\,\Omega$ at $0°C$ and a resistance of $8.88\,\Omega$ at $20°C$. Determine the temperature coefficient of resistance at $20°C$.

Here, with the reference value of resistance taken at $T = 20°C$ and with $R(T)$ the value at $0°C$

$$
\begin{aligned}
R(T) &= R_o[1 + \alpha(T - 20°C)] \\
8.24\,\Omega &= (8.88\,\Omega)[1 + \alpha(0°C - 20°C)] \\
0.928 &= 1 - (20°C)\alpha \\
20\alpha &= 1 - 0.928 \\
&= 0.072 \\
\alpha &= 0.00360°C^{-1} \Leftarrow
\end{aligned}
$$

1.22: A light bulb has a filament with a "cold" resistance of $25\,\Omega$ and a temperature coefficient of resistance of $0.0048°C^{-1}$. If the bulb, while operating, draws $0.48\,A$ at $120\,V$, determine the "hot" resistance of the bulb and the operating temperature of its filament.

The "hot" resistance is the resistance that occurs at the operating temperature

$$
R(T) = \frac{V}{I} = \frac{120\,V}{0.48\,A} = 250\,\Omega \Leftarrow
$$

Then, with R_o taken as the "cold" resistance

$$
\begin{aligned}
R(T) &= R_o[1 + \alpha(T - 20°C)] \\
250\,\Omega &= (25\,\Omega)[1 + (0.0048°C^{-1})(T - 20°C)] \\
225\,\Omega &= (25\,\Omega)(0.0048°C^{-1})(T - 20°C) \\
&= (0.120\,\Omega\text{-}°C^{-1})(T - 20°C) \\
T - 20°C &= \frac{225\,\Omega}{0.120\,\Omega\text{-}°C^{-1}} \\
&= 1875°C \\
T &= 1895°C \Leftarrow
\end{aligned}
$$

1.23: An electric heater takes $1.8\,kW$ at $120\,V$ for $20\,min$ to boil a quantity of water. Find the current through the heater and its resistance.

$$
\begin{aligned}
I &= \frac{P}{V} \\
&= \frac{1800\,W}{120\,V}
\end{aligned}
$$

and
$$= 15 \, \text{A} \Leftarrow$$

$$
\begin{aligned}
R &= \frac{P}{I^2} \\
&= \frac{1800 \, \text{W}}{(15 \, \text{A})^2} \\
&= 8 \, \Omega \Leftarrow
\end{aligned}
$$

1.24: What is the resistance of the element in Problem 1.15?

$$R = \frac{P}{I^2} = \frac{135.3 \, \text{W}}{(5.208 \, \text{A})^2} = 5 \, \Omega \Leftarrow$$

APPLICATIONS OF OHM'S LAW

1.25: Two resistors are connected across the same voltage source. The length and diameter of resistor-1 are L_1 and d_1 respectively. Resistor-2 has a length of $0.40L_1$ and a diameter of $0.75d_1$. Determine the ratio of the currents and powers, I_1/I_2 and P_1/P_2.

Here
$$R_1 = \frac{4\rho L_1}{\pi d_1^2} \quad \text{and} \quad R_2 = \frac{4\rho L_2}{\pi d_2^2}$$

But $L_2 = 0.4L_1$ and $d_2 = 0.75L_1$ so that

$$R_2 = \frac{4\rho(0.4L_1)}{\pi(0.75d_1)^2} = \frac{2.844\rho L_1}{\pi d_1^2}$$

With $V_1 = V_2 = V$

$$I_1 = \frac{V}{R_1} = \frac{\pi V d_1^2}{4\rho L_1}$$

and
$$I_2 = \frac{V}{R_2} = \frac{0.352\pi V d_1^2}{\rho L_1}$$

Thus
$$\frac{I_1}{I_2} = \left(\frac{\pi V d_1^2}{4\rho L_1}\right)\left(\frac{2.844\rho L_1}{\pi V d_1^2}\right) = 0.711 \Leftarrow$$

and
$$\frac{P_1}{P_2} = \left(\frac{I_1}{I_2}\right)^2 = (0.711)^2 = 0.506 \Leftarrow$$

1.26: If the same current flows through the two resistors of Problem 1.25, determine the ratios, V_1/V_2 and P_1/P_2?

In Problem 1.25 it was shown that

$$R_2 = 0.711 R_1$$

With $I_1 = I_2 = I$

$$V_1 = R_1 I \qquad \text{and} \qquad V_2 = R_2 I$$

so that

$$\frac{V_1}{V_2} = \frac{R_1 I}{R_2 I} = \frac{R_2}{R_1} = 0.711 \Leftarrow$$

and because, in general, $P = V^2/R$

$$\frac{P_1}{P_2} = \left[\frac{(V_1)^2}{R_1} \right] \left[\frac{R_2}{(V_2)^2} \right] = \left(\frac{V_1}{V_2} \right)^2 \left(\frac{R_2}{R_1} \right)$$

Hence

$$\frac{P_1}{P_2} = (0.711)^2 (0.711) = 0.360 \Leftarrow$$

1.27: A $0.5\,\Omega$ resistor is connected across a $6\,\text{V}$ battery. Determine the current flowing through the resistor, the power absorbed by the resistor and the power delivered by the source.

Here

$$I = \frac{V}{R} = \frac{6\,\text{V}}{0.5\,\Omega} = 12\,\text{A} \Leftarrow$$

$$P_R = I^2 R = (12\,\text{A})^2 (0.5\,\Omega) = 72\,\text{W} \Leftarrow$$

and the power delivered by the source is

$$P_{\text{source}} = VI = (6\,\text{V})(12\,\text{A}) = 72\,\text{W} \Leftarrow$$

1.28: What is the length of the resistor in Problem 1.27 if it is an aluminum wire with a diameter of $16\,\text{BWG}$ $(1.651\,\text{mm})$?

$$R = \frac{\rho L}{A}$$

so that

$$L = \frac{RA}{\rho}$$

With the resistivity for aluminum taken at $2.72 \times 10^{-8}\Omega$-m

$$
\begin{aligned}
A &= \frac{\pi}{4}d^2 \\
&= \frac{\pi}{4}(0.001651\,\text{m}^2) \\
&= 2.141 \times 10^{-6}\,\text{m}^2 \\
L &= \frac{(0.5\,\Omega)(2.141 \times 10^{-6}\,\text{m}^2)}{2.72 \times 10^{-8}\Omega\text{-m}} \\
&= 39.35\,\text{m} \Leftarrow
\end{aligned}
$$

1.29: An unspecified resistance, R, and a $12\,\Omega$ resistor are both connected across a $24\,\text{V}$ source. The power dissipated by R is $12\,\text{W}$. Determine the value of R and the total power delivered by the source.

The power dissipated by R is $P = V^2/R$. Hence

$$
R = \frac{V^2}{P} = \frac{(24\,\text{V})^2}{12\,\text{W}} = 48\,\Omega \Leftarrow
$$

The power dissipated by the $12\,\Omega$ resistor is

$$
P = \frac{V^2}{R} = \frac{(24\,\text{V})^2}{12\,\Omega} = 48\,\text{W}
$$

Hence, because the total power delivered by the source must be equal to the power dissipated by the two resistors,

$$
P_{\text{source}} = 12\,\text{W} + 48\,\text{W} = 60\,\text{W} \Leftarrow
$$

1.30: Three resistors having identical magnitudes are fed, in turn, by a $12\,\text{A}$ current source. The power delivered by the current source is $1728\,\text{W}$. What is the magnitude of the resistances?

$$
P = 3I^2R = 1728\,\text{W}
$$

Thus

$$
3R = \frac{P}{I^2} = \frac{1728\,\text{W}}{(12\,\text{A})^2} = 12\,\Omega
$$

Therefore

$$
R = \frac{12\,\Omega}{3} = 4\,\Omega \Leftarrow
$$

1.31: In a certain network containing four resistors, the resistors dissipate powers (with the subscript corresponding to the value of the resistor) of

$$
\begin{aligned}
P_{30} &= 1750\,\text{W} \\
P_3 &= 1875\,\text{W} \\
P_4 &= 1406.25\,\text{W}
\end{aligned}
$$

and

$$
P_{12} = 468.75\,\text{W}
$$

Determine V_{30}, V_3, I_4 and I_{12}.

Here, V_{30} and V_3 can be found by using

$$
P = \frac{V^2}{R} \quad \text{so that} \longrightarrow V = \sqrt{PR}
$$

and I_4 and I_{12} can be found in a similar manner by using

$$
P = I^2 R \quad \text{so that} \longrightarrow I = \sqrt{\frac{P}{R}}
$$

Hence,

$$
V_{30} = \sqrt{(750\,\text{W})(30\,\Omega)} = \sqrt{22,500\,\text{V}^2} = 150\,\text{V} \Longleftarrow
$$

and

$$
V_3 = \sqrt{(1875\,\text{W})(3\,\Omega)} = \sqrt{5625\,\text{V}^2} = 75\,\text{V} \Longleftarrow \cdot
$$

Then

$$
I_4 = \sqrt{\frac{1406.25\,\text{W}}{4\,\Omega}} = \sqrt{351.56\,\text{A}^2} = 18.75\,\text{A} \Longleftarrow
$$

and

$$
I_{12} = \sqrt{\frac{468.75\,\text{W}}{12\,\Omega}} = \sqrt{39.06\,\text{A}^2} = 6.25\,\text{A} \Longleftarrow
$$

1.32: In a certain network containing four resistors, the resistors dissipate powers (with the subscript corresponding to the value of the resistor) of

$$
\begin{aligned}
P_5 &= 45\,\text{W} \\
P_1 &= 9\,\text{W} \\
P_3 &= 12\,\text{W}
\end{aligned}
$$

and

$$P_6 = 6\,\text{W}$$

Determine V_5, V_1, I_3 and I_6.

Here, V_5 and V_1 can be found by using

$$P = \frac{V^2}{R} \quad \text{so that} \longrightarrow V = \sqrt{PR}$$

and I_3 and I_6 can be found in a similar manner by using

$$P = I^2 R \quad \text{so that} \longrightarrow I = \sqrt{\frac{P}{R}}$$

Hence,

$$V_5 = \sqrt{(45\,\text{W})(5\,\Omega)} = \sqrt{225\,\text{V}^2} = 15\,\text{V} \Longleftarrow$$

and

$$V_1 = \sqrt{(9\,\text{W})(1\,\Omega)} = \sqrt{9\,\text{V}^2} = 3\,\text{V} \Longleftarrow$$

Then

$$I_3 = \sqrt{\frac{12\,\text{W}}{3\,\Omega}} = \sqrt{4\,\text{A}^2} = 2\,\text{A} \Longleftarrow$$

and

$$I_6 = \sqrt{\frac{6\,\text{W}}{6\,\Omega}} = \sqrt{1\,\text{A}^2} = 1\,\text{A} \Longleftarrow$$

CONTROLLED SOURCES

1.33: In the circuit of Fig 1.3, if the resistor dissipates 2304 W, determine the value of β,

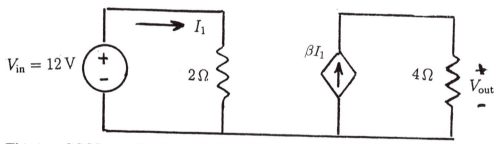

This is a CCCS and β is dimensionless. Here

$$I_1 = \frac{12\,\text{V}}{2\,\Omega} = 6\,\text{A}$$

$$\beta I_1 = \sqrt{\frac{2304\,\text{W}}{4\,\Omega}} = \sqrt{576\,\text{A}^2} = 24\,\text{A}$$

Then

$$\beta = \frac{24\,\text{A}}{I_1} = \frac{24\,\text{A}}{6\,\text{A}} = 4 \Leftarrow$$

1.34: In the circuit of Example 1.33, determine the voltage gain, $|V_{\text{out}}/V_{\text{in}}|$ and the power gain, $|P_{\text{out}}/P_{\text{in}}|$

Here, using the value of $\beta I_1 = 24\,\text{A}$ from Problem 1.33

$$V_{\text{out}} = (\beta I_1)(4\,\Omega) = (24\,\text{A})(4\,\Omega) = 96\,\text{V}$$

With $V_{\text{in}} = 12\,\text{V}$, the voltage gain will be

$$\left|\frac{V_{\text{out}}}{V_{\text{in}}}\right| = \frac{96\,\text{V}}{12\,\text{V}} = 8 \Leftarrow$$

The output power is specified in Problem 1.33 as 2304 W. The input power is

$$P_{\text{in}} = V_{\text{in}} I_1 = (12\,\text{V})(6\,\text{A}) = 72\,\text{W}$$

and the power gain is

$$\left|\frac{P_{\text{out}}}{P_{\text{in}}}\right| = \frac{2304\,\text{W}}{72\,\text{W}} = 32 \Leftarrow$$

1.35: If $\mu = 10$ in Fig 1.4, determine the value of R_1 that is required to deliver a power of 25,600 W to the 16 Ω resistor.

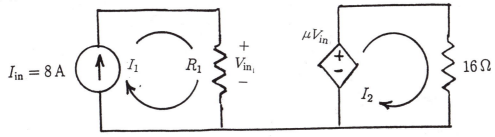

This is a VCVS and μ is dimensionless. Here

$$\frac{(10 V_{\text{in}})^2}{16\,\Omega} = 25,600\,\text{W}$$

$$(10 V_{\text{in}})^2 = (16\,\Omega)(25,600\,\text{W})$$

$$= 409,600\,\text{V}^2$$

$$10 V_{\text{in}} = \sqrt{409,600\,\text{V}^2}$$

and

$$V_{\text{in}} = 64\text{ V}$$

Then

$$R_1 = \frac{64\text{ V}}{8\text{ A}}$$
$$= 8\,\Omega \Leftarrow$$

1.36: Determine the current gain $|I_2/I_1|$ in the circuit of Problem 1.35.

Here, $I_1 = 8$ A and $V_{\text{in}} = 64$ V

$$I_2 = \frac{10V_{\text{in}}}{16\,\Omega} = \frac{640\text{ V}}{16\,\Omega} = 40\text{ A}$$

Hence, the current gain is

$$\left|\frac{I_2}{I_1}\right| = \frac{40\text{ A}}{8\text{ A}} = 5 \Leftarrow$$

1.37: In the circuit of Fig 1.5, $\beta = 8$ and $r_m = 8\,\Omega$. Determine the voltage gain, $|V_{\text{out}}/V_{\text{in}}|$ and the power gain, $|P_{\text{out}}/P_{\text{in}}|$.

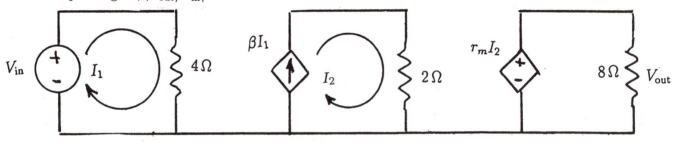

This problem involves a CCCS and a CCVS where β is dimensionless and where r_m is in Ω. Let $V_{\text{in}} = 10$ V. Then, by Ohm's law

$$I_1 = \frac{10\text{ V}}{4\,\Omega} = 2.5\text{ A}$$

$$I_2 = 8I_1 = 8(2.5\text{ A}) = 20\text{ A}$$

and

$$V_{\text{out}} = (8\,\Omega)I_2 = (8\,\Omega)(20\text{ A}) = 160\text{ V}$$

Thus, the voltage gain is

$$\left|\frac{V_{\text{out}}}{V_{\text{in}}}\right| = \frac{160\text{ V}}{10\text{ V}} = 16 \Leftarrow$$

With $V_{in} = 10 \text{V}$, the power delivered by the source is

$$P_{in} = \frac{(V_{in})^2}{4\,\Omega} = \frac{(10\,V)^2}{4\,\Omega} = 25 \text{ W}$$

and the power delivered to the $8\,\Omega$ resistor is

$$P_{out} = \frac{(V_{out})^2}{8\,\Omega} = \frac{(160\,V)^2}{8\,\Omega} = 3200 \text{ W}$$

This makes the power gain

$$\left| \frac{P_{out}}{P_{in}} \right| = \frac{3200 \text{ W}}{25 \text{ W}} = 128 \Leftarrow$$

1.38: In the circuit of Fig 1.6, $\mu = 8$ and $g_m = 0.5\,\text{U}$. Determine the current gain, $|I_3/I_{in}|$.

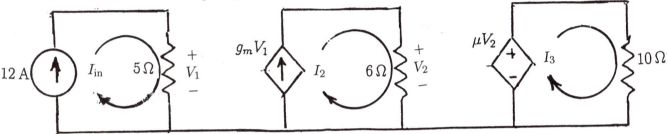

This problem involves a VCCS and a VCVS where μ is dimensionless and where g_m is in U. Here

$$V_1 = (5\,\Omega)(I_{in}) = (5\,\Omega)(12 \text{ A}) = 60 \text{ V}$$

$$I_2 = g_m V_1 = (0.5\,\text{U})(60 \text{ V}) = 30 \text{ A}$$

$$V_2 = (6\,\Omega)(I_2) = (6\,\Omega)(30 \text{ A}) = 180 \text{ V}$$

$$\mu V_2 = 8 V_2 = 8(180 \text{ V}) = 1440 \text{ V}$$

and

$$I_3 = \frac{\mu V_2}{10\,\Omega} = \frac{1440 \text{ V}}{10\,\Omega} = 144 \text{ A}$$

This makes the current gain

$$\left| \frac{I_3}{I_{in}} \right| = \frac{144 \text{ A}}{12 \text{ A}} = 12 \Leftarrow$$

$$v_2 = v_3$$
$$v_4 = v_5$$
$$v_7 = 12\,\text{V} \Leftarrow$$

Then, identify three closed paths, A, B and C, as shown and use KVL to write three equations and solve for the three additional unknown voltages.

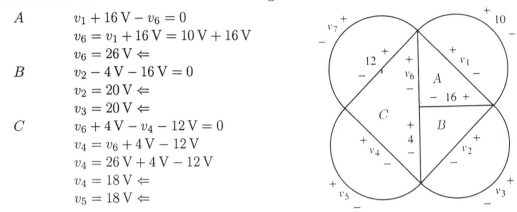

A $\quad v_1 + 16\,\text{V} - v_6 = 0$
$\quad\quad v_6 = v_1 + 16\,\text{V} = 10\,\text{V} + 16\,\text{V}$
$\quad\quad v_6 = 26\,\text{V} \Leftarrow$

B $\quad v_2 - 4\,\text{V} - 16\,\text{V} = 0$
$\quad\quad v_2 = 20\,\text{V} \Leftarrow$
$\quad\quad v_3 = 20\,\text{V} \Leftarrow$

C $\quad v_6 + 4\,\text{V} - v_4 - 12\,\text{V} = 0$
$\quad\quad v_4 = v_6 + 4\,\text{V} - 12\,\text{V}$
$\quad\quad v_4 = 26\,\text{V} + 4\,\text{V} - 12\,\text{V}$
$\quad\quad v_4 = 18\,\text{V} \Leftarrow$
$\quad\quad v_5 = 18\,\text{V} \Leftarrow$

2.6: Determine v_1 through v_6 in the network of Fig 2.6.

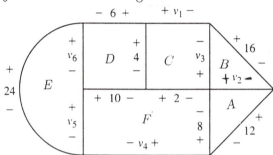

Identify six closed paths, A through F, as shown. Then use KVL to write six equations and solve for the six unknown voltages. Because a direct current application is not specified, use $v = v(t)$.

A $\quad v_2 + 12\,\text{V} + 8\,\text{V} = 0$
$\quad\quad v_2 = -20\,\text{V} \Leftarrow$

B $\quad v_3 + 16\,\text{V} - v_2 = 0$
$\quad\quad v_3 = v_2 - 16\,\text{V} = -20\,\text{V} - 16\,\text{V}$
$\quad\quad v_3 = -36\,\text{V} \Leftarrow$

C $\quad v_1 - v_3 - 2\,\text{V} - 4\,\text{V} = 0$
$\quad\quad v_1 = v_3 + 2\,\text{V} + 4\,\text{V}$
$\quad\quad v_1 = -36\,\text{V} + 6\,\text{V} = -30\,\text{V} \Leftarrow$

$$D \qquad -v_6 - 6\,\text{V} + 4\,\text{V} - 10\,\text{V} = 0$$
$$v_6 = -10\,\text{V} - 6\,\text{V} + 4\,\text{V}$$
$$v_6 = -12\,\text{V} \Leftarrow$$
$$E \qquad v_5 - 24\,\text{V} + v_6 = 0$$
$$v_5 = 24\,\text{V} - v_6 = 24\,\text{V} - (-12\,\text{V})$$
$$v_5 = 36\,\text{V} \Leftarrow$$
$$F \qquad v_4 - v_5 + 10\,\text{V} + 2\,\text{V} - 8\,\text{V} = 0$$
$$v_4 = v_5 - 10\,\text{V} - 2\,\text{V} + 8\,\text{V}$$
$$v_4 = 36\,\text{V} - 4\,\text{V} = 32\,\text{V} \Leftarrow$$

EQUIVALENT RESISTANCE PROBLEMS

2.7: Two resistors, R_1 and R_2 are connected in parallel. What is the value of the equivalent resistance if

$$
\begin{array}{ll}
\text{(a)} & R_1 = 20\,\Omega \text{ and } R_2 = 20\,\Omega \\
\text{(b)} & R_1 = 20\,\Omega \text{ and } R_2 = 80\,\Omega \\
\text{(c)} & R_1 = 20\,\Omega \text{ and } R_2 = \infty\,\Omega \\
\text{(d)} & R_1 = 20\,\Omega \text{ and } R_2 = 0\,\Omega \\
\text{(e)} & R_1 = 20\,\Omega \text{ and } R_2 = 50\,\Omega
\end{array}
$$

(a)

$$R_{eq} = \frac{R_1 R_2}{R_1 + R_2} = \frac{(20\,\Omega)(20\,\Omega)}{20\,\Omega + 20\,\Omega} = \frac{400\,\Omega^2}{40\,\Omega} = 10\,\Omega \Leftarrow$$

(b)

$$R_{eq} = \frac{R_1 R_2}{R_1 + R_2} = \frac{(20\,\Omega)(80\,\Omega)}{20\,\Omega + 80\,\Omega} = \frac{1600\,\Omega^2}{100\,\Omega} = 16\,\Omega \Leftarrow$$

(c)

$$R_{eq} = \frac{1}{\dfrac{1}{R_1} + \dfrac{1}{R_2}} = \frac{1}{\dfrac{1}{20\,\Omega} + \dfrac{1}{\infty}} = 20\,\Omega \Leftarrow$$

(d)

$$R_{eq} = \frac{R_1 R_2}{R_1 + R_2} = \frac{(20\,\Omega)(0\,\Omega)}{20\,\Omega + 0\,\Omega} = \frac{0\,\Omega^2}{20\,\Omega} = 0\,\Omega \Leftarrow$$

(e)

$$R_{eq} = \frac{R_1 R_2}{R_1 + R_2} = \frac{(20\,\Omega)(50\,\Omega)}{20\,\Omega + 50\,\Omega} = \frac{1000\,\Omega^2}{70\,\Omega} = 14.29\,\Omega \Leftarrow$$

2.8: One of the resistances in the network shown in Fig 2.7 is blurred because of coffee spillage. If the equivalent resistance of the network is $R_{eq} = 20\,\Omega$, what is the value of the blurred resistor?

Designate the blurred resistor as R. Then

$$R_{eq} = 20\,\Omega = 4\,\Omega + \frac{(12\,\Omega)(3\,\Omega + R)}{12\,\Omega + 3\,\Omega + R} + 8\,\Omega$$

$$\frac{(36\,\Omega^2 + 12R\,\Omega)}{15\,\Omega + R} = 8\,\Omega$$

$$36\,\Omega^2 + 12R\,\Omega = 120\,\Omega^2 + 8R\,\Omega$$

$$4R\,\Omega = 84\,\Omega^2$$

or

$$R = 21\,\Omega \Leftarrow$$

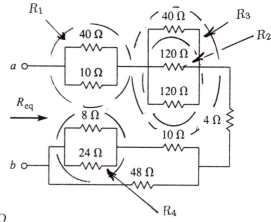

2.9: Determine the equivalent resistance looking into terminals a-b of the network shown in Fig 2.8.

Redraw the circuit showing combinations of resistances. Here

$$R_1 = \frac{(40\,\Omega)(10\,\Omega)}{40\,\Omega + 10\,\Omega} = \frac{400\,\Omega^2}{50\,\Omega} = 8\,\Omega$$

$$R_2 = \frac{(120\,\Omega)(120\,\Omega)}{120\,\Omega + 120\,\Omega} = \frac{14,400\,\Omega^2}{240\,\Omega} = 60\,\Omega$$

$$R_3 = \frac{(40\,\Omega)(60\,\Omega)}{40\,\Omega + 60\,\Omega} = \frac{2400\,\Omega^2}{100\,\Omega} = 24\,\Omega$$

and

$$R_4 = \frac{(8\,\Omega)(24\,\Omega)}{8\,\Omega + 24\,\Omega} = \frac{192\,\Omega^2}{32\,\Omega} = 6\,\Omega$$

Now, the picture and a further modification is

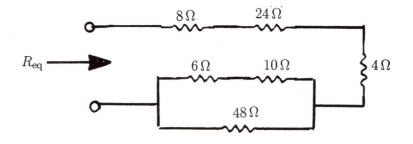

The resistance of the $16\,\Omega$ and $48\,\Omega$ combination is

$$R = \frac{(16\,\Omega)(48\,\Omega)}{16\,\Omega + 48\,\Omega} = \frac{728\,\Omega^2}{64\,\Omega} = 12\,\Omega$$

Therefore, the equivalent resistance looking into terminals $a\text{-}b$ is

$$R_{\text{eq}} = 8\,\Omega + 24\,\Omega + 12\,\Omega + 4\,\Omega = 48\,\Omega \Leftarrow$$

2.10: Determine the equivalent resistance looking into terminals $a\text{-}b$ of the network shown in Fig 2.9.

Redraw the circuit showing combinations of resistances.

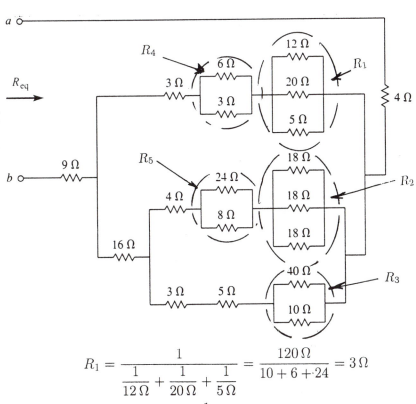

Here

$$R_1 = \frac{1}{\dfrac{1}{12\,\Omega} + \dfrac{1}{20\,\Omega} + \dfrac{1}{5\,\Omega}} = \frac{120\,\Omega}{10 + 6 + 24} = 3\,\Omega$$

$$R_2 = \frac{1}{3}(18\,\Omega) = 6\,\Omega$$

$$R_3 = \frac{(40\,\Omega)(10\,\Omega)}{40\,\Omega + 10\,\Omega} = \frac{400\,\Omega^2}{50\,\Omega} = 8\,\Omega$$

$$R_4 = \frac{(6\,\Omega)(3\,\Omega)}{6\,\Omega + 3\,\Omega} = \frac{18\,\Omega^2}{9\,\Omega} = 2\,\Omega$$

and

$$R_5 = \frac{(24\,\Omega)(8\,\Omega)}{24\,\Omega + 8\,\Omega} = \frac{192\,\Omega^2}{32\,\Omega} = 6\,\Omega$$

Now, the picture with two further combinations is

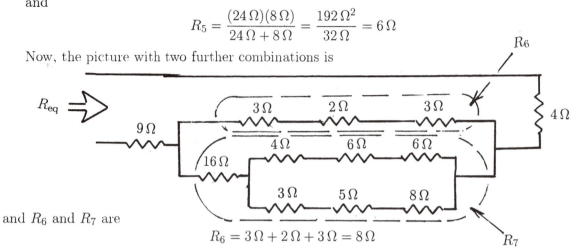

and R_6 and R_7 are

$$R_6 = 3\,\Omega + 2\,\Omega + 3\,\Omega = 8\,\Omega$$

and

$$R_7 = 16\,\Omega + \frac{1}{2}(16\,\Omega) = 16\,\Omega + 8\,\Omega = 24\,\Omega$$

Then

$$
\begin{aligned}
R_{eq} &= 9\,\Omega + \frac{(8\,\Omega)(24\,\Omega)}{8\,\Omega + 24\,\Omega} + 4\,\Omega \\
&= 9\,\Omega + \frac{192\,\Omega^2}{32\,\Omega} + 4\,\Omega \\
&= 9\,\Omega + 6\,\Omega + 4\,\Omega \\
&= 19\,\Omega \Leftarrow
\end{aligned}
$$

2.11: Determine the equivalent resistance looking into terminals *a-b* of the network shown in Fig 2.10.

Redraw the circuit showing combinations of resistances.

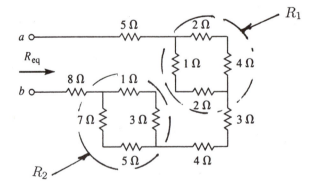

Here

$$R_1 = \frac{(2\,\Omega + 1\,\Omega)(2\,\Omega + 4\,\Omega)}{2\,\Omega + 1\,\Omega + 2\,\Omega + 4\,\Omega} = \frac{18\,\Omega^2}{9\,\Omega} = 2\,\Omega$$

and

$$R_2 = \frac{(1\,\Omega + 3\,\Omega)(7\,\Omega + 5\,\Omega)}{1\,\Omega + 3\,\Omega + 7\,\Omega + 5\,\Omega} = \frac{48\,\Omega^2}{16\,\Omega} = 3\,\Omega$$

Thus

$$R_{\text{eq}} = 5\,\Omega + 2\,\Omega + 3\,\Omega + 4\,\Omega + 3\,\Omega + 8\,\Omega = 25\,\Omega \Leftarrow$$

2.12: Find the equivalent resistance looking into terminals a-b in Fig 2.11.

Redraw the circuit showing combinations of resistances.

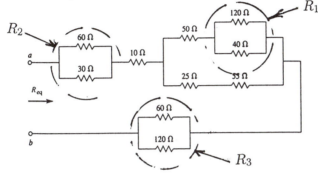

Here

$$R_1 = \frac{(120\,\Omega)(40\,\Omega)}{120\,\Omega + 40\,\Omega} = \frac{4800\,\Omega^2}{160\,\Omega} = 30\,\Omega$$

$$R_2 = \frac{(60\,\Omega)(30\,\Omega)}{60\,\Omega + 30\,\Omega} = \frac{1800\,\Omega^2}{90\,\Omega} = 20\,\Omega$$

and

$$R_3 = \frac{(120\,\Omega)(60\,\Omega)}{120\,\Omega + 60\,\Omega} = \frac{7200\,\Omega^2}{180\,\Omega} = 40\,\Omega$$

The picture is now

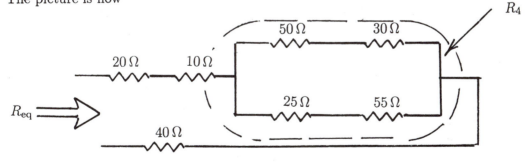

and the parallel combination containing the four resistors is evaluated as

$$R_4 = \frac{(50\,\Omega + 30\,\Omega)(25\,\Omega + 55\,\Omega)}{50\,\Omega + 30\,\Omega + 55\,\Omega + 25\,\Omega} = \frac{6400\,\Omega^2}{160\,\Omega} = 40\,\Omega$$

The equivalent resistance is

$$\begin{aligned} R_{\text{eq}} &= R_2 + 10\,\Omega + R_4 + R_3 \\ &= 20\,\Omega + 10\,\Omega + 40\,\Omega + 40\,\Omega \\ &= 110\,\Omega \end{aligned}$$

OHM'S LAW PROBLEMS

These problems include applications of KCL, KVL, current dividers and voltage dividers.

2.13: Two resistors of $4\,\Omega$ and $12\,\Omega$ are connected in series to a 24 V source. Determine (a) the current flow and (b) the voltage across each resistor.

(a) The equivalent resistance is

$$R_{\text{eq}} = 4\,\Omega + 12\,\Omega = 16\,\Omega$$

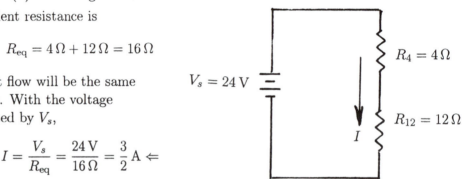

and the current flow will be the same in each resistor. With the voltage source designated by V_s,

$$I = \frac{V_s}{R_{\text{eq}}} = \frac{24\,\text{V}}{16\,\Omega} = \frac{3}{2}\,\text{A} \Leftarrow$$

(b) Designate each resistor with a subscript that represents its resistance value. Then, the voltage across each resistor will be

$$V_4 = IR_4 = \left(\frac{3}{2}\,\text{A}\right)(4\,\Omega) = 6\,\text{V} \Leftarrow$$

and

$$V_{12} = IR_{12} = \left(\frac{3}{2}\,\text{A}\right)(12\,\Omega) = 18\,\text{V} \Leftarrow$$

Note that by KVL, the sum of the voltage drops across each resistor must equal the source voltage

$$V_s = V_4 + V_{12} = 6\,\text{V} + 18\,\text{V} = 24\,\text{V} \;\checkmark$$

2.14: Determine the power drawn by each resistor in Problem 2.13.

Again, with each resistor designated by its resistance value

$$P_4 = \frac{V_4^2}{R_4} = \frac{(6\,\text{V})^2}{4\,\Omega} = \frac{36\,\text{V}^2}{4\,\Omega} = 9\,\text{W} \Leftarrow$$

and

$$P_{12} = \frac{V_{12}^2}{R_{12}} = \frac{(18\,\text{V})^2}{12\,\Omega} = \frac{324\,\text{V}^2}{12\,\Omega} = 27\,\text{W} \Leftarrow$$

Note that the power drawn from the voltage source is the sum of the powers dissipated by each resistor.

$$P_s = IV_s = \left(\frac{3}{2}\,\text{A}\right)(24\,\text{V}) = 36\,\text{W} = P_4 + P_{12}\;\checkmark$$

Moreover, note that

$$P_4 = I^2 R_4 = \left(\frac{3}{2}\,\text{A}\right)^2 (4\,\Omega) = 9\,\text{W}$$

and

$$P_{12} = I^2 R_{12} = \left(\frac{3}{2}\,\text{A}\right)^2 (12\,\Omega) = 27\,\text{W}$$

2.15: Use the voltage divider principle to determine the voltage across each resistor in the series connection of the $4\,\Omega$ and $12\,\Omega$ resistors of Problem 2.13

With the resistance designations corresponding to the resistance values

$$R_{eq} = 4\,\Omega + 12\,\Omega = 16\,\Omega$$

and with the source voltage at $V_s = 24\,\text{V}$

$$V_4 = \left(\frac{R_4}{R_{eq}}\right) V_s = \left(\frac{4\,\Omega}{16\,\Omega}\right)(24\,\text{V}) = 6\,\text{V} \Leftarrow$$

and

$$V_{12} = \left(\frac{R_{12}}{R_{eq}}\right) V_s = \left(\frac{12\,\Omega}{16\,\Omega}\right)(24\,\text{V}) = 18\,\text{V} \Leftarrow$$

2.16: A $4\,\Omega$ resistor is connected in parallel to a $12\,\Omega$ resistor and the combination is connected to a $12\,\text{A}$ source. Determine (a) the voltage across the combination and (b) the current through each resistor.

(a) Here, the equivalent resistance is

$$R_{eq} = \frac{(4\,\Omega)(12\,\Omega)}{4\,\Omega + 12\,\Omega} = \frac{48\,\Omega^2}{16\,\Omega} = 3\,\Omega$$

and, the current source designated by I_s, the voltage across the parallel combination (and the voltage across each resistor) will be

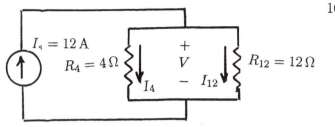

$$V = I_s R_{eq} = (12\,\text{A})(3\,\Omega) = 36\,\text{V} \Leftarrow$$

(b) Designate each resistor with a subscript that represents its resistance value. Then, the current through each resistor will be

$$I_4 = \frac{V}{R_4} = \frac{36\,\text{V}}{4\,\Omega} = 9\,\text{A} \Leftarrow$$

and

$$I_{12} = \frac{V}{R_{12}} = \left(\frac{36\,\text{V}}{12\,\Omega}\right) = 3\,\text{A} \Leftarrow$$

Note that by KCL, the sum of the individual currents in each resistor must equal the source current

$$I_s = I_4 + I_{12} = 9\,\text{A} + 3\,\text{A} = 12\,\text{A} \checkmark$$

2.17: Determine the power drawn by each resistor in Problem 2.16.

Again, with each resistor designated by its resistance value

$$P_4 = \frac{V^2}{R_4} = \frac{(36\,\text{V})^2}{4\,\Omega} = \frac{1296\,\text{V}^2}{4\,\Omega} = 324\,\text{W} \Leftarrow$$

and

$$P_{12} = \frac{V^2}{R_{12}} = \frac{(36\,\text{V})^2}{12\,\Omega} = \frac{1296\,\text{V}^2}{12\,\Omega} = 108\,\text{W} \Leftarrow$$

Note that the power drawn from the current source is the sum of the powers dissipated by each resistor.

$$P_s = I_s V = (12\,\text{A})(36\,\text{V}) = 432\,\text{W} = P_4 + P_{12} \checkmark$$

Also, note that

$$P_4 = I_4^2 R_4 = (9\,\text{A})^2(4\,\Omega) = 324\,\text{W}$$

and

$$P_{12} = I_{12}^2 R_{12} = (3\,\text{A})^2(12\,\Omega) = 108\,\text{W}$$

2.18: Use the current divider principle to determine the current through each resistor in the parallel connection of the $4\,\Omega$ and $12\,\Omega$ resistors of Problem 2.16.

162

With the resistance (and conductance) designations corresponding to the resistance values

$$G_4 = \frac{1}{R_4} = \frac{1}{4\,\Omega} = 0.2500\,\text{℧}$$

and

$$G_{12} = \frac{1}{R_{12}} = \frac{1}{12\,\Omega} = 0.0833\,\text{℧}$$

Thus

$$G_{eq} = 0.2500\,\text{℧} + 0.0833\,\text{℧} = 0.3333\,\text{℧}$$

which is the reciprocal of R_{eq}

$$R_{eq} = \frac{1}{G_{eq}} = \frac{1}{0.3333\,\text{℧}} = 3\,\Omega$$

With the source current at $I_s = 12\,\text{A}$

$$I_4 = \left(\frac{G_4}{G_{eq}}\right) I_s = \left(\frac{0.2500\,\text{℧}}{0.3333\,\text{℧}}\right)(12\,\text{A}) = 9\,\text{A} \Leftarrow$$

and

$$I_{12} = \left(\frac{G_{12}}{G_{eq}}\right) I_s = \left(\frac{0.0833\,\text{℧}}{0.3333\,\text{℧}}\right)(12\,\text{A}) = 3\,\text{A} \Leftarrow$$

Note that the same result can be obtained using resistance values

$$I_4 = \left(\frac{R_{12}}{R_{eq}}\right) I_s = \left(\frac{12\,\Omega}{16\,\Omega}\right)(12\,\text{A}) = 9\,\text{A}$$

and

$$I_{12} = \left(\frac{R_4`}{R_{eq}}\right) I_s = \left(\frac{4\,\Omega}{16\,\Omega}\right)(12\,\text{A}) = 3\,\text{A}$$

2.19: In the network of Fig 2.12, designate each resistor by its resistance value and find the current in all four of the resistors.

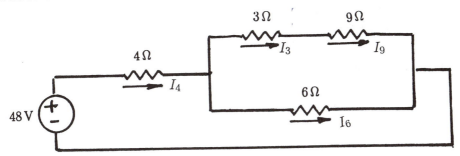

Redraw Fig 2.12 to show the four currents. The equivalent resistance seen by the voltage source is

$$R_{eq} = 4\,\Omega + \frac{(3\,\Omega + 9\,\Omega)(6\,\Omega)}{3\,\Omega + 6\,\Omega + 9\,\Omega} = 4\,\Omega + \frac{72\,\Omega^2}{18\,\Omega} = 4\,\Omega + 4\,\Omega = 8\,\Omega$$

The current through R_4 is

$$I_4 = \frac{V}{R_{eq}} = \frac{48\,\text{V}}{8\,\Omega} = 6\,\text{A} \Leftarrow$$

By the current divider principle

$$I_6 = \left(\frac{R_3 + R_9}{R_3 + R_6 + R_9}\right) I_4 = \left(\frac{3\,\Omega + 9\,\Omega}{3\,\Omega + 6\,\Omega + 9\,\Omega}\right)(6\,\text{A}) = \left(\frac{12\,\Omega}{18\,\Omega}\right)(6\,\text{A}) = 4\,\text{A} \Leftarrow$$

and then by KCL

$$I_3 = I_9 = I_4 - I_6 = 6\,\text{A} - 4\,\text{A} = 2\,\text{A} \Leftarrow$$

2.20: Determine the power dissipated by each resistor in the network of Problem 2.19.

Because the current flowing through each resistor is known, the simplest approach appears to be $P = I^2R$. Hence

$$P_3 = I_3^2 R_3 = (2\,\text{A})^2(3\,\Omega) = 12\,\text{W} \Leftarrow$$
$$P_4 = I_4^2 R_4 = (6\,\text{A})^2(4\,\Omega) = 144\,\text{W} \Leftarrow$$
$$P_6 = I_6^2 R_6 = (4\,\text{A})^2(6\,\Omega) = 96\,\text{W} \Leftarrow$$

and

$$P_9 = I_9^2 R_9 = (2\,\text{A})^2(9\,\Omega) = 36\,\text{W} \Leftarrow$$

The total power dissipated by all of the resistors is

$$P = 12\,\text{W} + 144\,\text{W} + 96\,\text{W} + 36\,\text{W} = 288\,\text{W}$$

and it is observed that this power is equal to the power delivered by the source

$$P = V_s I_4 = (48\,\text{V})(6\,\text{A}) = 288\,\text{W} \ \sqrt{}$$

2.21: In the series network shown in Fig 2.13, $R_{eq} = 10\,\Omega$, $V_2 = 24\,\text{V}$ and $P_3 = 16\,\text{W}$. Determine the values of R_1, R_2 and R_3?

In this series circuit, the current that flows through all of the resistances is

$$I = \frac{40\,\text{V}}{R_{eq}}$$
$$= \frac{40\,\text{V}}{10\,\Omega}$$
$$= 4\,\text{A}$$

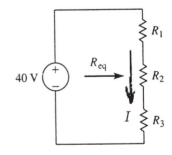

Then
$$P_3 = I^2 R_3$$

and with $P_3 = 16\,\mathrm{W}$

$$R_3 = \frac{P_3}{I^2} = \frac{16\,\mathrm{W}}{(4\,\mathrm{A})^2} = 1\,\Omega \Leftarrow$$

Because $V_2 = R_2 I$ and $V_2 = 24\,\mathrm{V}$

$$R_2 = \frac{V_2}{I} = \frac{24\,\mathrm{V}}{4\,\mathrm{A}} = 6\,\Omega \Leftarrow$$

and because $R_{\mathrm{eq}} = 10\,\Omega$

$$R_1 = R_{\mathrm{eq}} - R_2 - R_3 = 10\,\Omega - 6\,\Omega - 1\,\Omega = 3\,\Omega \Leftarrow$$

2.22: Figure 2.14 shows four resistors with subscripts corresponding to their resistance values. The right hand leg containing the $2\,\Omega$ and $4\,\Omega$ resistors is designated as R_s (s for series combination) and the current entering the parallel combination of the $3\,\Omega, 2\,\Omega$ and $4\,\Omega$ resistors is designated as I_p. If the current entering the network is $I = 6\,\mathrm{A}$, determine the current through and the voltage across each resistor.

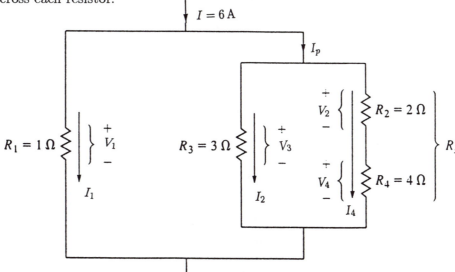

Here
$$R_s = 2\,\Omega + 4\,\Omega = 6\,\Omega$$

and the resistance of the parallel combination of R_s and the $3\,\Omega$ resistor is

$$R_p = \frac{(3\,\Omega)R_s}{3\,\Omega + R_s} = \frac{(3\,\Omega)(6\,\Omega)}{3\,\Omega + 6\,\Omega} = \frac{18\,\Omega^2}{9\,\Omega} = 2\,\Omega$$

By current division

$$I_1 = \left(\frac{R_p}{1\,\Omega + R_p}\right) I = \left(\frac{2\,\Omega}{1\,\Omega + 2\,\Omega}\right) (6\,\text{A}) = 4\,\text{A} \Leftarrow$$

and by KCL

$$I_p = 6\,\text{A} - I_1 = 6\,\text{A} - 4\,\text{A} = 2\,\text{A}$$

Once again, employ current division to find

$$I_3 = \left(\frac{R_s}{3\,\Omega + R_s}\right) I_p = \left(\frac{6\,\Omega}{3\,\Omega + 6\,\Omega}\right) (2\,\text{A}) = \frac{4}{3}\,\text{A} \Leftarrow$$

and again by KCL

$$I_s = I_p - I_3 = 2\,\text{A} - \frac{4}{3}\,\text{A} = \frac{2}{3}\,\text{A}$$

Thus

$$I_2 = I_4 = \frac{2}{3}\,\text{A} \Leftarrow$$

The four voltage drops are

$$V_1 = R_1 I_1 = (1\,\Omega)(4\,\text{A}) = 4\,\text{V} \Leftarrow$$

$$V_2 = R_2 I_2 = (2\,\Omega)\left(\frac{2}{3}\,\text{A}\right) = \frac{4}{3}\,\text{V} \Leftarrow$$

$$V_3 = R_3 I_3 = (3\,\Omega)\left(\frac{4}{3}\,\text{A}\right) = 4\,\text{V} \Leftarrow$$

and

$$V_4 = R_4 I_4 = (4\,\Omega)\left(\frac{2}{3}\,\text{A}\right) = \frac{8}{3}\,\text{V} \Leftarrow$$

Observe that by KVL

$$\begin{aligned} -V_3 + V_2 + V_4 &= 0 \\ -4\,\text{V} + \frac{4}{3}\,\text{V} + \frac{8}{3}\,\text{V} &= 0 \\ 0 &= 0 \checkmark \end{aligned}$$

and

$$\begin{aligned} -V_1 + V_3 &= 0 \\ -4\,\text{V} + 4\,\text{V} &= 0 \\ 0 &= 0 \checkmark \end{aligned}$$

2.23: Find the current through and the voltage across each resistor in Fig 2.15.

The resistances may be designated in accordance with their resistance values and the network may be redrawn to show R_a, R_b and R_p

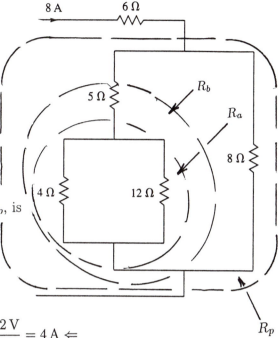

$$R_a = \frac{(12\,\Omega)(4\,\Omega)}{12\,\Omega + 4\,\Omega} = \frac{48\,\Omega^2}{16\,\Omega} = 3\,\Omega$$

$$R_b = 3\,\Omega + 5\,\Omega = 8\,\Omega$$

and

$$R_p = \frac{1}{2}(8\,\Omega) = 4\,\Omega$$

The voltage across the parallel combination, R_p, is

$$V_p = R_p I = (4\,\Omega)(8\,\text{A}) = 32\,\text{V}$$

so that

$$V_8 = 32\,\text{V} \Leftarrow$$

and

$$I_8 = \frac{V_8}{R_8} = \frac{32\,\text{V}}{8\,\Omega} = 4\,\text{A} \Leftarrow$$

Then by KCL

$$
\begin{aligned}
I_5 &= I - I_8 \\
&= 8\,\text{A} - 4\,\text{A} \\
&= 4\,\text{A} \Leftarrow
\end{aligned}
$$

and

$$V_5 = R_5 I_5 = (5\,\Omega)(4\,\text{A}) = 20\,\text{V} \Leftarrow$$

By current division

$$I_4 = \left(\frac{R_{12}}{R_4 + R_{12}}\right) I_5 = \left(\frac{12\,\Omega}{4\,\Omega + 12\,\Omega}\right)(4\,\text{A}) = 3\,\text{A} \Leftarrow$$

and

$$I_{12} = \left(\frac{R_4}{R_4 + R_{12}}\right) I_5 = \left(\frac{4\,\Omega}{4\,\Omega + 12\,\Omega}\right)(4\,\text{A}) = 1\,\text{A} \Leftarrow$$

Then

$$V_4 = R_4 I_4 = (4\,\Omega)(3\,\text{A}) = 12\,\text{V} \Leftarrow$$

and

$$V_{12} = R_{12}I_{12} = (12\,\Omega)(1\,\text{A}) = 12\,\text{V} \Leftarrow$$

Finally

$$V_6 = R_6I_6 = (6\,\Omega)(8\,\text{A}) = 48\,\text{V} \Leftarrow$$

and

$$I_6 = I = 8\,\text{A}$$

2.24: In the network of Fig 2.16, the resistance designators correspond to the resistance values. Determine the currents, I_1 and I_2 and the voltage across R_6 without using current and voltage division.

Five resistances have been placed on the circuit in order to assist the development of R_{eq}.

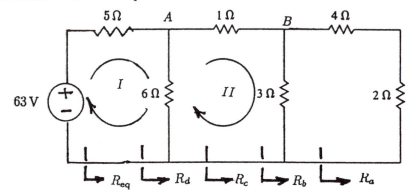

Here

$$R_a = 4\,\Omega + 2\,\Omega = 6\,\Omega$$

$$R_b = \frac{(3\,\Omega)(6\,\Omega)}{3\,\Omega + 6\,\Omega} = \frac{18\,\Omega^2}{9\,\Omega} = 2\,\Omega$$

$$R_c = 1\,\Omega + 2\,\Omega = 3\,\Omega$$

$$R_d = \frac{(6\,\Omega)(3\,\Omega)}{6\,\Omega + 3\,\Omega} = \frac{18\,\Omega^2}{9\,\Omega} = 2\,\Omega$$

and

$$R_{eq} = 5\,\Omega + R_d = 5\,\Omega + 2\,\Omega = 7\,\Omega$$

With $V_s = 63\,\text{V}$, the current through the $5\,\Omega$ resistor is obtained first

$$I_5 = \frac{V_s}{R_{eq}} = \frac{63\,\text{V}}{7\,\Omega} = 9\,\text{A}$$

and then V_5 will be

$$V_5 = R_5I_5 = (5\,\Omega)(9\,\text{A}) = 45\,\text{V}$$

Application of KVL around the loop designated by -I in the network diagram gives

$$V_5 + V_6 - 63\,\text{V} = 0$$

so that

$$V_6 = 63\,\text{V} - V_5 = 63\,\text{V} - 45\,\text{V} = 18\,\text{V} \Leftarrow$$

Then

$$I_6 = \frac{V_6}{R_6} = \frac{18\,\text{V}}{6\,\Omega} = 3\,\text{A}$$

Next apply KCL to point-A which has been inserted on the circuit diagram

$$I_1 - I_5 + I_6 = 0$$

so that

$$I_1 = I_5 - I_6 = 9\,\text{A} - 3\,\text{A} = 6\,\text{A} \Leftarrow$$

and

$$V_1 = R_1 I_1 = (6\,\text{A})(1\,\Omega) = 6\,\text{V}$$

Once again, apply KVL, but this time around the loop designated by -II in the network diagram

$$V_1 + V_3 - V_6 = 0$$

so that

$$V_3 = V_6 - V_1 = 18\,\text{V} - 6\,\text{V} = 12\,\text{V}$$

and then

$$I_3 = \frac{V_3}{R_3} = \frac{12\,\text{V}}{3\,\Omega} = 4\,\text{A}$$

Finally, an application of KCL at point-B gives

$$I_4 + I_3 - I_1 = 0$$

or

$$I_4 = I_1 - I_3 = 6\,\text{A} - 4\,\text{A} = 2\,\text{A}$$

and because $I_2 = I_4$

$$I_2 = 2\,\text{A} \Leftarrow$$

2.25: In the network of Problem 2.24, determine the currents, I_1 and I_2 and the voltage across R_6 by employing current and voltage division in place of KVL and KCL.

Third, at point-C

$$I_3 = \left(\frac{R_6}{R_6 + R_c}\right) I_9 = \left(\frac{6\,\Omega}{6\,\Omega + 6\,\Omega}\right)(12\,\text{A}) = \left(\frac{1}{2}\right)(12\,\text{A}) = 6\,\text{A}$$

and finally, at point-D

$$I_4 = \left(\frac{R_a}{R_4 + R_a}\right) I_3 = \left(\frac{12\,\Omega}{4\,\Omega + 12\,\Omega}\right)(6\,\text{A}) = \left(\frac{3}{4}\right)(6\,\text{A}) = \frac{9}{2}\,\text{A} \Leftarrow$$

By KCL

$$I_7 - I_3 + I_4 = 0$$

or

$$I_7 = I_3 - I_4 = 6\,\text{A} - \frac{9}{2}\,\text{A} = \frac{3}{2}\,\text{A}$$

and

$$V_7 = R_7 I_7 = (7\,\Omega)\left(\frac{3}{2}\right)\,\text{A} = \frac{21}{2}\,\text{V} \Leftarrow$$

2.27: Find I_9, I_4 and V_7 in the network shown in Fig 2.17 without using current division.

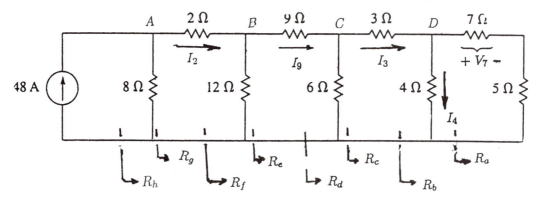

It is noted that the resistance designators correspond to the resistance values. Eight resistances have been placed on the circuit in order to assist the development of R_{eq} and it was established in Problem 2.26 that the eight resistances, R_a through R_h are equal to

$$
\begin{array}{ll}
R_a = 12\,\Omega & R_e = 12\,\Omega \\
R_b = 3\,\Omega & R_f = 6\,\Omega \\
R_c = 6\,\Omega & R_g = 8\,\Omega \\
R_d = 3\,\Omega & R_h = 4\,\Omega
\end{array}
$$

The voltage across the $8\,\Omega$ resistor is equal to the voltage across $R_h = R_{\text{eq}}$. Hence with $I_s = 48\,\text{A}$, at point-A

$$V_8 = R_{\text{eq}} I_s = (4\,\Omega)(48\,\text{A}) = 192\,\text{V}$$

Three voltage divisions at points-B, -C and -D with $V_8 = 192\,\text{V}$ can be used. First, at point-B

$$V_{12} = \left(\frac{R_f}{R_2 + R_f}\right) V_8 = \left(\frac{6\,\Omega}{2\,\Omega + 6\,\Omega}\right)(192\,\text{V}) = \left(\frac{3}{4}\right)(192\,\text{V}) = 144\,\text{V}$$

Second, at point-C

$$V_6 = \left(\frac{R_d}{R_9 + R_d}\right) V_{12} = \left(\frac{3\,\Omega}{9\,\Omega + 3\,\Omega}\right)(144\,\text{V}) = \left(\frac{1}{4}\right)(144\,\text{A}) = 36\,\text{V}$$

and finally, at point-D

$$V_4 = \left(\frac{R_b}{R_3 + R_b}\right) V_6 = \left(\frac{3\,\Omega}{3\,\Omega + 3\,\Omega}\right)(36\,\text{V}) = \left(\frac{1}{2}\right)(36\,\text{V}) = 18\,\text{V}$$

One more voltage divider at point-D gives

$$V_7 = \left(\frac{R_7}{R_7 + R_5}\right) V_4 = \left(\frac{7\,\Omega}{5\,\Omega + 7\,\Omega}\right)(18\,\text{V}) = \left(\frac{7}{12}\right)(18\,\text{V}) = \frac{21}{2}\,\text{V} \Leftarrow$$

The current through the $4\,\Omega$ resistor is

$$I_4 = \frac{V_4}{R_4} = \frac{18\,\text{V}}{4\,\Omega} = \frac{9}{2}\,\text{A} \Leftarrow$$

The voltage drop across R_9 will be

$$V_9 = V_{12} - V_6 = 144\,\text{V} - 36\,\text{V} = 108\,\text{V}$$

so that

$$I_9 = \frac{V_9}{R_9} = \frac{108\,\text{V}}{9\,\Omega} = 12\,\text{A} \Leftarrow$$

2.28: Find I_3, I_4, V_{12} and V_{14} in the network shown in Fig 2.18.

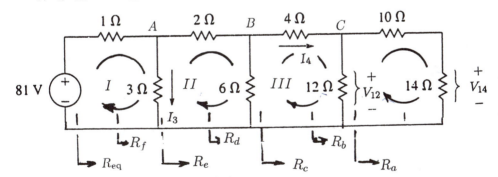

It is noted that the resistance designators correspond to the resistance values. Six resistances have been placed on the circuit in order to assist the development of R_{eq}. Note also the presence of points-A, B and C and loops I, II, III and IV.

Here

$$R_a = R_{10} + R_{14} = 10\,\Omega + 14\,\Omega = 24\,\Omega$$

$$R_b = \frac{R_{12}R_a}{R_{12} + R_a} = \frac{(12\,\Omega)(24\,\Omega)}{12\,\Omega + 24\,\Omega} = \frac{288\,\Omega^2}{36\,\Omega} = 8\,\Omega$$

$$R_c = R_4 + R_b = 4\,\Omega + 8\,\Omega = 12\,\Omega$$

$$R_d = \frac{R_6 R_c}{R_6 + R_c} = \frac{(6\,\Omega)(12\,\Omega)}{6\,\Omega + 12\,\Omega} = \frac{72\,\Omega^2}{18\,\Omega} = 4\,\Omega$$

$$R_e = R_2 + R_d = 2\,\Omega + 4\,\Omega = 6\,\Omega$$

$$R_f = \frac{R_3 R_e}{R_3 + R_e} = \frac{(3\,\Omega)(6\,\Omega)}{3\,\Omega + 6\,\Omega} = \frac{18\,\Omega^2}{9\,\Omega} = 2\,\Omega$$

and

$$R_{eq} = R_1 + R_f = 1\,\Omega + 2\,\Omega = 3\,\Omega$$

Then

$$I_1 = \frac{V_1}{R_{eq}} = \frac{81\,\text{V}}{3\,\Omega} = 27\,\text{A}$$

and

$$V_1 = R_1 I_1 = (1\,\Omega)(27\,\text{A}) = 27\,\text{V}$$

KVL around loop-I then gives

$$V_1 + V_3 - 81\,\text{V} = 0$$

or

$$V_3 = 81\,\text{V} - V_1 = 81\,\text{V} - 27\,\text{V} = 54\,\text{V}$$

and

$$I_3 = \frac{V_3}{R_3} = \frac{54\,\text{V}}{3\,\Omega} = 18\,\text{A} \Leftarrow$$

KCL at point-A then gives

$$I_1 - I_3 - I_2 = 0$$

so that

$$I_2 = I_1 - I_3 = 27\,\text{A} - 18\,\text{A} = 9\,\text{A}$$

and this gives

$$V_2 = R_2 I_2 = (2\,\Omega)(9\,\text{A}) = 18\,\text{V}$$

Then KVL around loop-II provides

$$V_2 + V_6 - V_3 = 0$$

or

$$V_6 = V_3 - V_2 = 54\,\text{V} - 18\,\text{V} = 36\,\text{V}$$

and

$$I_6 = \frac{V_6}{R_6} = \frac{36\,\text{V}}{6\,\Omega} = 6\,\text{A}$$

Now apply KCL at point-B

$$I_4 + I_6 - I_2 = 0$$

so that

$$I_4 = I_2 - I_6 = 9\,\text{A} - 6\,\text{A} = 3\,\text{A} \Leftarrow$$

and this gives

$$V_4 = R_4 I_4 = (4\,\Omega)(3\,\text{A}) = 12\,\text{V}$$

Then using KVL around loop-III

$$V_4 + V_{12} - V_6 = 0$$

or

$$V_{12} = V_6 - V_4 = 36\,\text{V} - 12\,\text{V} = 24\,\text{V} \Leftarrow$$

and

$$I_{12} = \frac{V_{12}}{R_{12}} = \frac{24\,\text{V}}{12\,\Omega} = 2\,\text{A}$$

A final application of KCL, this time at point-C, gives

$$I_{10} + I_{12} - I_4 = 0$$

so that

$$I_{10} = I_4 - I_{12} = 3\,\text{A} - 2\,\text{A} = 1\,\text{A}$$

and this gives

$$V_{10} = R_{10} I_{10} = (10\,\Omega)(1\,\text{A}) = 10\,\text{V}$$

and a final application of KVL, this time around loop-IV, gives V_{14}

$$V_{10} + V_{14} - V_{12} = 0$$

or

$$V_{14} = V_{12} - V_{10} = 24\,\text{V} - 10\,\text{V} = 14\,\text{V} \Leftarrow$$

2.29: Find I_6, I_{10} and V_2 in the network shown in Fig 2.19.

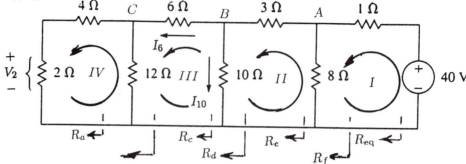

It is noted that the resistance designators correspond to the resistance values. Six resistances have been placed on the circuit in order to assist the development of R_{eq}. Note also the presence of points-A, B and C and loops I, II, III and IV. The fact that the voltage source is on the right is noted. Here

$$R_a = R_4 + R_2 = 4\,\Omega + 2\,\Omega = 6\,\Omega$$

$$R_b = \frac{R_{12}R_a}{R_{12} + R_a} = \frac{(12\,\Omega)(6\,\Omega)}{12\,\Omega + 6\,\Omega} = \frac{72\,\Omega^2}{18\,\Omega} = 4\,\Omega$$

$$R_c = R_6 + R_b = 6\,\Omega + 4\,\Omega = 10\,\Omega$$

$$R_d = \frac{R_{10}R_c}{R_{10} + R_c} = \frac{(10\,\Omega)(10\,\Omega)}{10\,\Omega + 10\,\Omega} = \frac{100\,\Omega^2}{20\,\Omega} = 5\,\Omega$$

$$R_e = R_3 + R_d = 3\,\Omega + 5\,\Omega = 8\,\Omega$$

$$R_f = \frac{R_8R_e}{R_5 + R_e} = \frac{(8\,\Omega)(8\,\Omega)}{8\,\Omega + 8\,\Omega} = \frac{64\,\Omega^2}{16\,\Omega} = 4\,\Omega$$

and

$$R_{eq} = R_1 + R_f = 1\,\Omega + 4\,\Omega = 5\,\Omega$$

Then with $V_s = 40\,V$

$$I_1 = \frac{V_s}{R_{eq}} = \frac{40\,V}{5\,\Omega} = 8\,A$$

and

$$V_1 = R_1 I_1 = (1\,\Omega)(8\,A) = 8\,V$$

KVL around loop-I then gives

$$V_1 + V_8 - 40\,V = 0$$

or

$$V_8 = 40\,V - V_1 = 40\,V - 8\,V = 32\,V$$

and

$$I_8 = \frac{V_8}{R_8} = \frac{32\,\text{V}}{8\,\Omega} = 4\,\text{A}$$

KCL at point-A then gives

$$I_1 - I_3 - I_8 = 0$$

so that

$$I_3 = I_1 - I_8 = 8\,\text{A} - 4\,\text{A} = 4\,\text{A}$$

and this gives

$$V_3 = R_3 I_3 = (3\,\Omega)(4\,\text{A}) = 12\,\text{V}$$

Then KVL around loop-II provides

$$V_3 + V_{10} - V_8 = 0$$

or

$$V_{10} = V_8 - V_3 = 32\,\text{V} - 12\,\text{V} = 20\,\text{V}$$

and

$$I_{10} = \frac{V_{10}}{R_{10}} = \frac{20\,\text{V}}{10\,\Omega} = 2\,\text{A} \Leftarrow$$

Now apply KCL at point-B

$$I_3 - I_{10} - I_6 = 0$$

so that

$$I_6 = I_3 - I_{10} = 4\,\text{A} - 2\,\text{A} = 2\,\text{A} \Leftarrow$$

and this gives

$$V_6 = R_6 I_6 = (6\,\Omega)(2\,\text{A}) = 12\,\text{V}$$

Then using KVL around loop-III

$$V_6 + V_{12} - V_{10} = 0$$

or

$$V_{12} = V_{10} - V_6 = 20\,\text{V} - 12\,\text{V} = 8\,\text{V}$$

The voltage, V_2 can be obtained by voltage division

$$V_2 = \left(\frac{R_2}{R_2 + R_4}\right) V_{12} = \left(\frac{2\,\Omega}{2\,\Omega + 4\,\Omega}\right)(8\,\text{V})$$

or

$$V_2 = \left(\frac{1}{3}\right)(8\,\text{V}) = \frac{8}{3}\,\text{V} \Leftarrow$$

2.30: Find I_9, I_{40} and V_7 in the network shown in Fig 2.20.

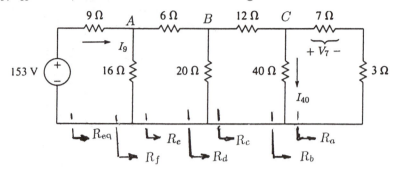

It is noted that the resistance designators correspond to the resistance values. Six resistances have been placed on the circuit in order to assist the development of R_{eq}. Note also the presence of points-A, B and C. Here

$$R_a = R_7 + R_3 = 7\,\Omega + 3\,\Omega = 10\,\Omega$$

$$R_b = \frac{R_{40}R_a}{R_{40} + R_a} = \frac{(40\,\Omega)(10\,\Omega)}{40\,\Omega + 10\,\Omega} = \frac{400\,\Omega^2}{50\,\Omega} = 8\,\Omega$$

$$R_c = R_{12} + R_b = 12\,\Omega + 8\,\Omega = 20\,\Omega$$

$$R_d = \frac{R_{20}R_c}{R_{20} + R_c} = \frac{(20,\Omega)(20\,\Omega)}{20\,\Omega + 20\,\Omega} = \frac{400\,\Omega^2}{40\,\Omega} = 10\,\Omega$$

$$R_e = R_6 + R_d = 6\,\Omega + 10\,\Omega = 16\,\Omega$$

$$R_f = \frac{R_{16}R_e}{R_{16} + R_e} = \frac{(16\,\Omega)(16\,\Omega)}{16\,\Omega + 16\,\Omega} = \frac{256\,\Omega^2}{32\,\Omega} = 8\,\Omega$$

and

$$R_{eq} = R_9 + R_f = 9\,\Omega + 8\,\Omega = 17\,\Omega$$

Then

$$I_9 = \frac{V_s}{R_{eq}} = \frac{153\,\text{V}}{17\,\Omega} = 9\,\text{A} \Leftarrow$$

and three current divisions can be employed to determine I_6, I_{12} and I_{40}. At point-A

$$I_6 = \left(\frac{R_{16}}{R_{16} + R_e}\right) I_9 = \left(\frac{16\,\Omega}{16\,\Omega + 16\,\Omega}\right)(9\,\text{A}) = \left(\frac{1}{2}\right)(9\,\text{A}) = \frac{9}{2}\,\text{A}$$

At point-B

$$I_{12} = \left(\frac{R_{20}}{R_{20} + R_c}\right) I_6 = \left(\frac{20\,\Omega}{20\,\Omega + 20\,\Omega}\right)\left(\frac{9}{2}\,\text{A}\right) = \left(\frac{1}{2}\right)\left(\frac{9}{2}\,\text{A}\right) = \frac{9}{4}\,\text{A}$$

and at point-C

$$I_{40} = \left(\frac{R_a}{R_{40} + R_a}\right) I_{12} = \left(\frac{10\,\Omega}{40\,\Omega + 10\,\Omega}\right) \left(\frac{9}{4}\,A\right) = \left(\frac{1}{5}\right) \left(\frac{9}{4}\,A\right) = \frac{9}{20}\,A \Leftarrow$$

$$I_7 = \left(\frac{R_{40}}{R_{40} + R_a}\right) I_{12} = \left(\frac{40\,\Omega}{40\,\Omega + 10\,\Omega}\right) \left(\frac{9}{4}\,A\right) = \left(\frac{4}{5}\right) \left(\frac{9}{4}\,A\right) = \frac{9}{5}\,A$$

and

$$V_7 = R_7 I_7 = (7\,\Omega) \left(\frac{9}{5}\,A\right) = \frac{63}{5}\,V \Leftarrow$$

2.31: Find I_{13}, I_{14}, V_6 and V_7 in the network shown in Fig 2.21.

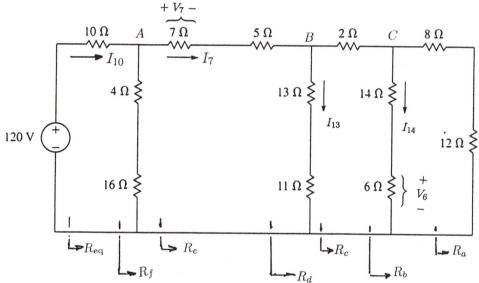

It is noted that the resistance designators correspond to the resistance values. Six resistances have been placed on the circuit in order to assist the development of R_{eq}. Note also the presence of points-A, B and C. Here

$$R_a = R_8 + R_{12} = 8\,\Omega + 12\,\Omega = 20\,\Omega$$

$$R_b = \left(\frac{1}{2}\right) R_a = \left(\frac{1}{2}\right) (20\,\Omega) = 10\,\Omega$$

$$R_c = R_2 + R_b = 2\,\Omega + 10\,\Omega = 12\,\Omega$$

$$R_d = \frac{(R_{13} + R_{11})R_c}{R_{13} + R_{11} + R_c} = \frac{(13\,\Omega + 11\,\Omega)(12\,\Omega)}{13\,\Omega + 11\,\Omega + 12\,\Omega} = \frac{288\,\Omega^2}{36\,\Omega} = 8\,\Omega$$

$$R_e = R_7 + R_5 + R_d = 7\,\Omega + 5\,\Omega + 8\,\Omega = 20\,\Omega$$

$$R_f = \frac{(R_4 + R_{16})R_e}{R_4 + R_{16} + R_e} = \frac{(4\,\Omega + 16\,\Omega)(20\,\Omega)}{4\,\Omega + 16\,\Omega + 20\,\Omega} = \frac{400\,\Omega^2}{40\,\Omega} = 10\,\Omega$$

and

$$R_{eq} = R_{10} + R_f = 10\,\Omega + 10\,\Omega = 20\,\Omega$$

Then

$$I_{10} = \frac{V_s}{R_{eq}} = \frac{120\,\text{V}}{20\,\Omega} = 6\,\text{A}$$

and two current divisions can be employed to determine I_{13} and I_{14}. At point-A

$$I_7 = \left(\frac{R_4 + R_{16}}{R_4 + R_{16} + R_e}\right) I_{10} = \left(\frac{4\,\Omega + 16\,\Omega}{4\,\Omega + 16\,\Omega + 20\,\Omega}\right)(6\,\text{A}) = \left(\frac{1}{2}\right)(6\,\text{A}) = 3\,\text{A}$$

and

$$V_7 = R_7 I_7 = (7\,\Omega)(3\,\text{A}) = 21\,\text{V} \Leftarrow$$

At point-B

$$I_{13} = \left(\frac{R_c}{R_{11} + R_{13} + R_c}\right) I_7 = \left(\frac{12\,\Omega}{11\,\Omega + 13\,\Omega + 12\,\Omega}\right)(3\,\text{A}) = \left(\frac{1}{3}\right)(3\,\text{A}) = 1\,\text{A} \Leftarrow$$

and by KCL with $I_7 = I_5$

$$I_5 - I_{13} - I_2 = 0$$

or

$$I_2 = I_5 - I_{13} = 3\,\text{A} - 1\,\text{A} = 2\,\text{A}$$

One more current divider at point-C gives

$$I_{14} = \left(\frac{R_a}{R_{14} + R_6 + R_a}\right) I_2 = \left(\frac{20\,\Omega}{14\,\Omega + 6\,\Omega + 20\,\Omega}\right)(2\,\text{A}) = \left(\frac{1}{2}\right)(2\,\text{A}) = 1\,\text{A} \Leftarrow$$

and with $I_6 = I_{14}$

$$V_6 = R_6 I_6 = (6\,\Omega)(1\,\text{A}) = 6\,\text{V} \Leftarrow$$

2.32: Find I_5, V_{20} and the power dissipated in the $24\,\Omega$ resistor in Fig 2.22.

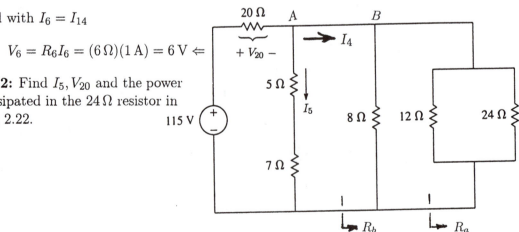

It is noted that the resistance designators correspond to the resistance values. Three resistances have been placed on the circuit in order to assist the development of R_{eq}. Note also the presence of points-A and B. Here

$$R_a = \frac{R_{12}R_{24}}{R_{12} + R_{24}} = \frac{(12\,\Omega)(24\,\Omega)}{12\,\Omega + 24\,\Omega} = \frac{288\,\Omega^2}{36\,\Omega} = 8\,\Omega$$

$$R_b = \left(\frac{1}{2}\right)R_a = \left(\frac{1}{2}\right)(8\,\Omega) = 4\,\Omega$$

$$R_c = \frac{(R_5 + R_7)R_b}{R_5 + R_7 + R_b} = \frac{(5\,\Omega + 7\,\Omega)(4\,\Omega)}{5\,\Omega + 7\,\Omega + 4\,\Omega} = \frac{48\,\Omega^2}{16\,\Omega} = 3\,\Omega$$

and

$$R_{\text{eq}} = R_{20} + R_c = 20\,\Omega + 3\,\Omega = 23\,\Omega$$

Thus with $V_s = 115\,\text{V}$

$$I_{20} = \frac{V_s}{R_{20}} = \frac{115\,\text{V}}{23\,\Omega} = 5\,\text{A}$$

and

$$V_{20} = R_{20}I_{20} = (20\,\Omega)(5\,\text{A}) = 100\,\text{V} \Leftarrow$$

Current division at point-A yields

$$I_5 = \left(\frac{R_b}{R_5 + R_7 + R_b}\right)I_{20} = \left(\frac{4\,\Omega}{5\,\Omega + 7\,\Omega + 4\,\Omega}\right)(5\,\text{A}) = \left(\frac{1}{4}\right)(5\,\text{A}) = \frac{5}{4}\,\text{A} \Leftarrow$$

and by KCL

$$I_4 + I_5 - I_{20} = 0$$

or

$$I_4 = I_{20} - I_5 = 5\,\text{A} - \frac{5}{4}\,\text{A} = \frac{15}{4}\,\text{A}$$

With

$$G_8 = \frac{1}{R_8} = \frac{1}{8\,\Omega} = 0.1250\,\mho$$

$$G_{12} = \frac{1}{R_{12}} = \frac{1}{12\,\Omega} = 0.0833\,\mho$$

and

$$G_{24} = \frac{1}{R_{24}} = \frac{1}{24\,\Omega} = 0.0417\,\mho$$

then

$$\sum G = G_8 + G_{12} + G_{24}$$
$$= 0.1250\,\text{℧} + 0.0833\,\text{℧} + 0.0427\,\text{℧}$$
$$= 0.2500\,\text{℧}$$

By a current divider at point-C

$$I_{24} = \left(\frac{G_{24}}{\sum G}\right) I_4 = \left(\frac{0.0417\,\text{℧}}{0.2500\,\text{℧}}\right)\left(\frac{15}{4}\,\text{A}\right) = \frac{5}{8}\,\text{A}$$

The power in R_{24} is

$$P_{24} = I_{24}^2 R_{24} = \left(\frac{5}{8}\,\text{A}\right)^2 (24\,\Omega) = \frac{75}{8}\,\text{W} \quad \text{or} \quad 9.375\,\text{W} \Leftarrow$$

CHAPTER THREE
NODAL AND LOOP ANALYSES

SINGLE NODE PROBLEMS

3.1: Use nodal analysis to determine the voltage at point-1 (Node-1) in the network of Fig 3.1.

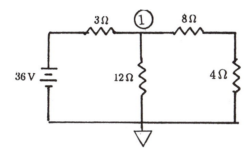

KCL at node-1 gives

$$\frac{V_1 - 36\,\text{V}}{3} + \frac{V_1}{12} + \frac{V_1}{8+4} = 0$$

$$\left(\frac{1}{3} + \frac{1}{12} + \frac{1}{12}\right) V_1 = \frac{36\,\text{V}}{3} = 12\,\text{A}$$

$$\frac{V_1}{2} = 12\,\text{A}$$

$$V_1 = 24\,\text{V} \Leftarrow$$

Note that

$$R_a = 8\,\Omega + 4\,\Omega = 12\,\Omega$$

$$R_b = \left(\frac{1}{2}\right)(12\,\Omega) = 6\,\Omega$$

and

$$R_{\text{eq}} = 3\,\Omega + 6\,\Omega = 9\,\Omega$$

Then by voltage division

$$V_1 = \frac{R_b}{R_{\text{eq}}}(36\,\text{V}) = \left(\frac{6\,\Omega}{9\,\Omega}\right)(36\,\text{V}) = 24\,\text{V}\,\checkmark$$

183

3.2: Use nodal analysis to determine the voltage at point-1 (Node-1) in the network of Fig 3.2.

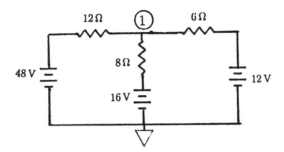

KCL at node-1 gives

$$\frac{V_1 - 48\,\text{V}}{12} + \frac{V_1 - 16\,\text{V}}{8} + \frac{V_1 - 12\,\text{V}}{6} = 0$$

$$\left(\frac{1}{12} + \frac{1}{8} + \frac{1}{6}\right) V_1 = \frac{48\,\text{V}}{12} + \frac{16\,\text{V}}{8} + \frac{12\,\text{V}}{6}$$

$$\left(\frac{2 + 3 + 4}{24}\right) V_1 = 4\,\text{A} + 2\,\text{A} + 2\,\text{A} = 8\,\text{A}$$

$$V_1 = \left(\frac{24}{9}\right)(8\,\text{A}) = \frac{64}{3}\,\text{V} \Leftarrow$$

3.3: Use nodal analysis to determine the voltage at point-1 (Node-1) in the network of Fig 3.3.

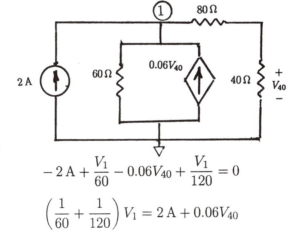

KCL at node-1 gives

$$-2\,\text{A} + \frac{V_1}{60} - 0.06V_{40} + \frac{V_1}{120} = 0$$

$$\left(\frac{1}{60} + \frac{1}{120}\right) V_1 = 2\,\text{A} + 0.06V_{40}$$

But V_{40} is related to V_1 by a simple voltage divider

$$V_{40} = \left(\frac{40\,\Omega}{40\,\Omega + 80\,\Omega}\right) V_1 = \frac{V_1}{3}$$

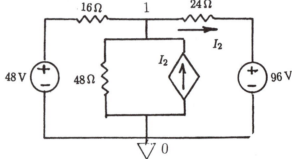

so that

$$\frac{V_1}{40} = 2\,\text{A} + 0.06\left(\frac{V_1}{3}\right)$$

or

$$0.025V_1 = 2\,\text{A} + 0.020V_1$$

Hence

$$0.005V_1 = 2\,\text{A}$$

$$V_1 = \frac{2\,\text{A}}{0.005} = 400\,\text{V} \Leftarrow$$

3.4: Using nodal analysis, determine I_2 in the network of Fig 3.4.

KCL at node-1 gives

$$\frac{V_1 - 48\,\text{V}}{16} + \frac{V_1}{48} - I_2 + \frac{V_1 - 96\,\text{V}}{24} = 0$$

$$\left(\frac{1}{16} + \frac{1}{48} + \frac{1}{24}\right)V_1 = \frac{48\,\text{V}}{16} + \frac{96\,\text{V}}{24} + I_2$$

$$\left(\frac{3+1+2}{48}\right)V_1 = 3\,\text{A} + 4\,\text{A} + I_2$$

$$\frac{V_1}{8} = 7\,\text{A} + I_2$$

But

$$I_2 = \frac{V_1 - 96\,\text{V}}{24} = \frac{V_1}{24} - 4\,\text{A}$$

so that

$$\frac{V_1}{8} = 7\,\text{A} + \frac{V_1}{24} - 4\,\text{A}$$

$$\left(\frac{1}{8} - \frac{1}{24}\right)V_1 = \frac{1}{12}V_1 = 7\,\text{A} - 4\,\text{A} = 3\,\text{A}$$

and

$$V_1 = 12(3\,\text{A}) = 36\,\text{V}$$

Therefore,

$$I_2 = \frac{V_1 - 96\,\text{V}}{24} = \frac{36\,\text{V} - 96\,\text{V}}{24} = -\frac{60\,\text{V}}{24} = -\frac{5}{2}\,\text{A} \Leftarrow$$

The current is flowing from right to left in the network diagram.

MULTIPLE NODE PROBLEMS

3.5: Use nodal analysis to determine the current flowing downward in the $8\,\Omega$ resistor in the network of Fig 3.5.

Because all sources are current sources and there are no controlled sources, the node equations may be written by inspection. With node-0 as the reference node, KCL gives for node-1

$$\left(\frac{1}{4\,\Omega} + \frac{1}{16\,\Omega}\right) V_1 - \frac{1}{16\,\Omega} V_2 = 2\,\text{A} + 8\,\text{A}$$

and for node-2

$$-\frac{1}{16\,\Omega} V_1 + \left(\frac{1}{16\,\Omega} + \frac{1}{8\,\Omega}\right) V_2 = 2\,\text{A} - 8\,\text{A}$$

In the matrix form, $\mathbf{GV} = \mathbf{I}_s$

$$\begin{bmatrix} 0.3125 & -0.0625 \\ -0.0625 & 0.1875 \end{bmatrix} \begin{bmatrix} V_1 \\ V_2 \end{bmatrix} = \begin{bmatrix} 10\,\text{A} \\ -6\,\text{A} \end{bmatrix}$$

and a matrix inversion gives

$$\begin{bmatrix} V_1 \\ V_2 \end{bmatrix} = \begin{bmatrix} 0.3125 & -0.0625 \\ -0.0625 & 0.1875 \end{bmatrix}^{-1} \begin{bmatrix} 10\,\text{A} \\ -6\,\text{A} \end{bmatrix}$$

$$= \begin{bmatrix} 3.4286 & 1.1429 \\ 1.1429 & 5.7143 \end{bmatrix} \begin{bmatrix} 10\,\text{A} \\ -6\,\text{A} \end{bmatrix}$$

or

$$\begin{bmatrix} V_1 \\ V_2 \end{bmatrix} = \begin{bmatrix} 27.4286\,\text{V} \\ -22.8571\,\text{V} \end{bmatrix}$$

The current through the 8 Ω resistor will be

$$I_8 = \frac{V_2}{8\,\Omega} = \frac{-22.8571\,\text{V}}{8\,\Omega} = -2.8571\,\text{A} \quad \text{or} \quad -20/7\,\text{A} \Leftarrow$$

This current is flowing upward through the 8 Ω resistor.

3.6: Determine the node voltages, V_1 and V_2, in the network shown in Fig 3.6.

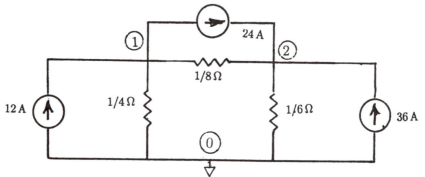

Because all sources are current sources and there are no controlled sources, the node equations may be written by inspection. With node-0 as the reference node,

$$(4+8)V_1 - 8V_2 = 12\,\text{A} - 24\,\text{A}$$
$$-8V_1 + (8+6)V_2 = 24\,\text{A} + 36\,\text{A}$$

or in matrix form

$$\begin{bmatrix} 12 & -8 \\ -8 & 14 \end{bmatrix} \begin{bmatrix} V_1 \\ V_2 \end{bmatrix} = \begin{bmatrix} -12\,\text{A} \\ 60\,\text{A} \end{bmatrix}$$

By a matrix inversion

$$\begin{bmatrix} V_1 \\ V_2 \end{bmatrix} = \begin{bmatrix} 12 & -8 \\ -8 & 14 \end{bmatrix}^{-1} \begin{bmatrix} -12\,\text{A} \\ 60\,\text{A} \end{bmatrix}$$

$$= \frac{1}{104} \begin{bmatrix} 14 & 8 \\ 8 & 12 \end{bmatrix} \begin{bmatrix} -12\,\text{A} \\ 60\,\text{A} \end{bmatrix} = \begin{bmatrix} 3\,\text{V} \\ 6\,\text{V} \end{bmatrix} \Leftarrow$$

3.7: Use nodal analysis to determine the node voltages, V_1 and V_2 and then find the current through the 4 Ω resistor in the network of Fig 3.7.

First, note that

$$12I_a = 12\left(\frac{V_1 - V_2}{4}\right) = 3(V_1 - V_2)$$

Then, two node equations can be written using KCL

$$\frac{V_1 - 16\,\text{V}}{2} - 4\,\text{A} + \frac{V_1}{6} + \frac{V_1 - V_2}{4} = 0$$

and

$$\frac{V_2 - V_1}{4} + \frac{V_2 - 12I_a}{1} + \frac{V_2 - 36\,\text{V}}{3} = 0$$

The second node equation can be modified to read

$$\frac{V_2 - V_1}{4} + \frac{V_2 - 3(V_1 - V_2)}{1} + \frac{V_2 - 36\,\text{V}}{3} = 0$$

or

$$\frac{V_2 - V_1}{4} + \frac{4V_2 - 3V_1}{1} + \frac{V_2 - 36\,\text{V}}{3} = 0$$

The node equations are

$$\left(\frac{1}{2} + \frac{1}{6} + \frac{1}{4}\right)V_1 - \frac{1}{4}V_2 = 12\,\text{A}$$

and

$$-\left(\frac{1}{4} + 3\right)V_1 + \left(\frac{1}{4} + 4 + \frac{1}{3}\right)V_2 = 12\,\text{A}$$

or

$$\frac{11}{12}V_1 - \frac{1}{4}V_2 = 12\,\text{A}$$

and

$$-\frac{13}{4}V_1 + \frac{55}{12}V_2 = 12\,\text{A}$$

These can be written in matrix form (after a multiplication by 12) as

$$\begin{bmatrix} 11 & -3 \\ -39 & 55 \end{bmatrix} \begin{bmatrix} V_1 \\ V_2 \end{bmatrix} = \begin{bmatrix} 144\,\text{A} \\ 144\,\text{A} \end{bmatrix}$$

Because all sources are current sources and there are no controlled sources, the node equations may be written by inspection. With node-0 as the reference node,

$$(2+4+16)V_1 - 4V_2 - 16V_3 = 8\,\text{A}$$
$$-4V_1 + (4+2+8)V_2 - 4V_3 = 12\,\text{A}$$
$$-16V_1 - 4V_2 + (4+2+16)V_3 = -I_s$$

or in matrix form

$$\begin{bmatrix} 22 & -4 & -16 \\ -4 & 16 & -4 \\ -16 & -4 & 22 \end{bmatrix} \begin{bmatrix} V_1 \\ V_2 \\ V_3 \end{bmatrix} = \begin{bmatrix} 8\,\text{A} \\ 12\,\text{A} \\ -I_s \end{bmatrix}$$

A Cramer's rule solution yields

$$V_2 = \frac{\begin{vmatrix} 22 & 8 & -16 \\ -4 & 12 & -4 \\ -16 & -I_s & 22 \end{vmatrix}}{\begin{vmatrix} 22 & -4 & -16 \\ -4 & 16 & -4 \\ -16 & -4 & 22 \end{vmatrix}} = 36\,\text{V}$$

Thus

$$\frac{5808 + 512 - 64I_s - 3072 - 88I_s + 704}{2432} = 36$$

$$3952 - 152I_s = 87,552$$
$$-152I_s = 83,600$$
$$I_s = -576\,\text{A} \Leftarrow$$

The direction of I_s is upward in the network diagram.

3.10: In the network of Fig 3.10, determine the current I_{40} using nodal analysis.

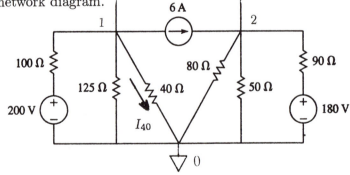

The two node equations can be written using KCL with node-0 as the reference node

$$\frac{V_1 - 200\,\text{V}}{100} + 6\,\text{A} + \frac{V_1}{125} + \frac{V_1 - V_2}{60} + \frac{V_1}{40} = 0$$

and

$$\frac{V_2 - V_1}{60} + \frac{V_2}{80} + \frac{V_2}{50} + \frac{V_2 - 180\,\text{V}}{90} - 6\,\text{A} = 0$$

or

$$\left(\frac{1}{100} + \frac{1}{125} + \frac{1}{60} + \frac{1}{40}\right) V_1 - \frac{1}{60} V_2 = 2\,\text{A} - 6\,\text{A} = -4\,\text{A}$$

and

$$-\frac{1}{60} V_1 + \left(\frac{1}{60} + \frac{1}{80} + \frac{1}{50} + \frac{1}{90}\right) V_2 = 6\,\text{A} + 2\,\text{A} = 8\,\text{A}$$

The node equations can be written in matrix form

$$\begin{bmatrix} 0.0597 & -0.0167 \\ -0.0167 & 0.0603 \end{bmatrix} \begin{bmatrix} V_1 \\ V_2 \end{bmatrix} = \begin{bmatrix} -4\,\text{A} \\ 8\,\text{A} \end{bmatrix}$$

and a matrix inversion provides

$$\begin{bmatrix} V_1 \\ V_2 \end{bmatrix} = \begin{bmatrix} 0.0597 & -0.0167 \\ -0.0167 & 0.0603 \end{bmatrix}^{-1} \begin{bmatrix} -4\,\text{A} \\ 8\,\text{A} \end{bmatrix}$$

or

$$\begin{bmatrix} V_1 \\ V_2 \end{bmatrix} = \begin{bmatrix} 18.157 & 5.029 \\ 5.029 & 17.976 \end{bmatrix} \begin{bmatrix} -4\,\text{A} \\ 8\,\text{A} \end{bmatrix} = \begin{bmatrix} -32.40\,\text{V} \\ 123.70\,\text{V} \end{bmatrix} \Leftarrow$$

The current, I_{40}, will be

$$I_{40} = \frac{V_1}{40\,\Omega} = \frac{-32.40\,\text{V}}{40\,\Omega} = -0.81\,\text{A} \Leftarrow$$

The branch current, I_{40} with magnitude of 0.81 A is directed upward in the network diagram.

3.11: Determine the node voltages, V_1 and V_2 in the network shown in Fig 3.11.

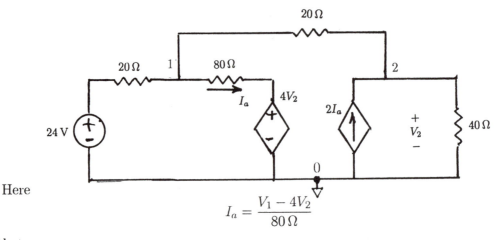

Here

$$I_a = \frac{V_1 - 4V_2}{80\,\Omega}$$

so that

$$2I_a = 2\frac{V_1 - 4V_2}{80\,\Omega} = \frac{V_1 - 4V_2}{40\,\Omega}$$

The node equations are found from an application of KCL. With node-0 selected as the reference node, for node-1

$$\frac{V_1 - 24\,\text{V}}{20} + \frac{V_1 - 4V_2}{80} + \frac{V_1 - V_2}{20} = 0$$

and for node-2

$$\frac{V_2}{40} + \frac{V_2 - V_1}{20} - 2I_a = 0$$

or

$$\frac{V_2}{40} + \frac{V_2 - V_1}{20} - \frac{V_1 - 4V_2}{40} = 0$$

These can be rearranged to

$$\left(\frac{1}{20} + \frac{1}{80} + \frac{1}{20}\right) V_1 - \left(\frac{1}{20} + \frac{1}{20}\right) V_2 = \frac{24\,\text{V}}{20\,\Omega} = \frac{6}{5}\,\text{A}$$

and

$$-\left(\frac{1}{20} + \frac{1}{40}\right) V_1 + \left(\frac{1}{40} + \frac{1}{20} + \frac{1}{10}\right) V_2 = 0$$

The node equations can be written in matrix form after simplification

$$\begin{bmatrix} 9 & -8 \\ -3 & 7 \end{bmatrix} \begin{bmatrix} V_1 \\ V_2 \end{bmatrix} = \begin{bmatrix} 96\,\text{A} \\ 0 \end{bmatrix}$$

and a matrix inversion provides

$$\begin{bmatrix} V_1 \\ V_2 \end{bmatrix} = \begin{bmatrix} 9 & -8 \\ -3 & 7 \end{bmatrix}^{-1} \begin{bmatrix} 96\,\text{A} \\ 0 \end{bmatrix} = \frac{1}{39}\begin{bmatrix} 7 & 8 \\ 3 & 9 \end{bmatrix}\begin{bmatrix} 96\,\text{A} \\ 0 \end{bmatrix} = \begin{bmatrix} 224/13\,\text{V} \\ 96/13\,\text{V} \end{bmatrix} \Leftarrow$$

3.12: Use nodal analysis to determine the current in the $1/8\,\Omega$ resistor in the network of Fig 3.12.

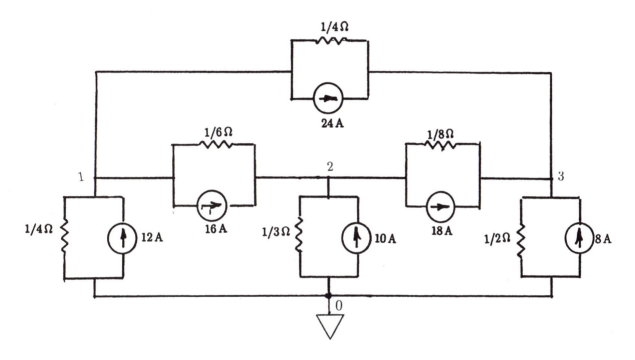

Because all sources are current sources and there are no controlled sources, the node equations may be written by inspection. With node-0 as the reference node,

$$(4+6+4)V_1 - 6V_2 - 4V_3 = 12\,\text{A} - 16\,\text{A} - 24\,\text{A}$$
$$-6V_1 + (6+8+3)V_2 - 8V_3 = 16\,\text{A} + 10\,\text{A} - 18\,\text{A}$$
$$-4V_1 - 8V_2 + (4+8+2)V_3 = 8\,\text{A} + 18\,\text{A} + 24\,\text{A}$$

or in matrix form

$$\begin{bmatrix} 14 & -6 & -4 \\ -6 & 17 & -8 \\ -4 & -8 & 14 \end{bmatrix}\begin{bmatrix} V_1 \\ V_2 \\ V_3 \end{bmatrix} = \begin{bmatrix} -28\,\text{A} \\ 8\,\text{A} \\ 50\,\text{A} \end{bmatrix}$$

and a matrix inversion yields

$$\begin{bmatrix} V_1 \\ V_2 \\ V_3 \end{bmatrix} = \begin{bmatrix} 14 & -6 & -4 \\ -6 & 17 & -8 \\ -4 & -8 & 14 \end{bmatrix}^{-1} \begin{bmatrix} -28\,\text{A} \\ 8\,\text{A} \\ 50\,\text{A} \end{bmatrix}$$

or

$$\begin{bmatrix} V_1 \\ V_2 \\ V_3 \end{bmatrix} = \begin{bmatrix} 0.1364 & 0.0909 & 0.0909 \\ 0.0909 & 0.1411 & 0.1066 \\ 0.0909 & 0.1066 & 0.1583 \end{bmatrix} \begin{bmatrix} -28\,\text{A} \\ 8\,\text{A} \\ 50\,\text{A} \end{bmatrix} = \begin{bmatrix} 1.455\,\text{V} \\ 3.912\,\text{V} \\ 6.223\,\text{V} \end{bmatrix}$$

The current in the $1/8\,\Omega$ resistor, designated as I_8, will be

$$\begin{aligned} I_8 &= G_8(V_3 - V_2) \\ &= (8\,\mho)(6.223\,\text{V} - 3.912\,\text{V}) \\ &= 18.49\,\text{A} \end{aligned}$$

and this current is directed as shown in the network diagram.

3.13: Determine the node voltages, V_1, V_2 and V_3 in the network of Fig 3.13.

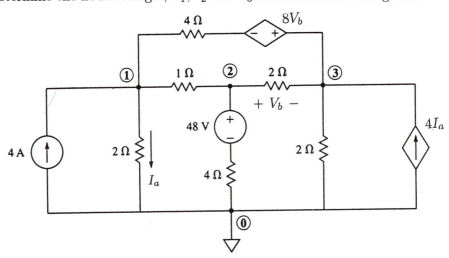

Note first that $V_b = V_2 - V_3$ and $I_a = V_1/2$ so that

$$8V_b = 8(V_2 - V_3) \qquad \text{and} \qquad 4I_a = 2V_1$$

Then, the three node equations, obtained by a systematic application of KVL are

$$-4\,\text{A} + \frac{1}{2}V_1 + \frac{V_1 - V_2}{1} - \frac{V_1 - (V_3 - 8V_b)}{4} = 0$$

$$\frac{V_2 - V_1}{1} + \frac{V_2 - V_3}{2} + \frac{V_2 - 48\,\text{V}}{4} = 0$$

and

$$\frac{V_3 - V_2}{2} + \frac{V_3}{2} - 4I_a + \frac{V_3 - (V_1 - 8V_b)}{4} = 0$$

With $8V_b = 8(V_2 - V_3)$ and $4I_a = 2V_1$, these simplify to

$$\left(\frac{1}{2} + 1 + \frac{1}{4}\right) V_1 - V_2 - \frac{1}{4}V_3 = 4\,\text{A} - 2(V_2 - V_3)$$

$$-V_1 + \left(1 + \frac{1}{2} + \frac{1}{4}\right) V_2 - \frac{1}{2}V_3 = 12\,\text{A}$$

and

$$-\frac{1}{4}V_1 - \frac{1}{2}V_2 + \left(\frac{1}{2} + \frac{1}{2} + \frac{1}{4}\right) V_3 = 2V_1 + 2(V_2 - V_3)$$

The final simplification is

$$\frac{7}{4}V_1 + V_2 - \frac{9}{4}V_3 = 4\,\text{A}$$

$$-V_1 + \frac{7}{4}V_2 - \frac{1}{2}V_3 = 12\,\text{A}$$

and

$$-\frac{9}{4}V_1 - \frac{5}{2}V_2 + \frac{13}{4}V_3 = 0$$

and in matrix form

$$\begin{bmatrix} 7 & 4 & -9 \\ -4 & 7 & -2 \\ -9 & -10 & 13 \end{bmatrix} \begin{bmatrix} V_1 \\ V_2 \\ V_3 \end{bmatrix} = \begin{bmatrix} 16\,\text{A} \\ 48\,\text{A} \\ 0 \end{bmatrix}$$

A matrix inversion yields

$$\begin{bmatrix} V_1 \\ V_2 \\ V_3 \end{bmatrix} = \begin{bmatrix} 7 & 4 & -9 \\ -4 & 7 & -2 \\ -9 & -10 & 13 \end{bmatrix}^{-1} \begin{bmatrix} 16\,\text{A} \\ 48\,\text{A} \\ 0 \end{bmatrix}$$

$$\begin{bmatrix} V_1 \\ V_2 \\ V_3 \end{bmatrix} = - \begin{bmatrix} 0.4733 & 0.2533 & 0.3667 \\ 0.4667 & 0.0667 & 0.3333 \\ 0.6867 & 0.2267 & 0.4333 \end{bmatrix} \begin{bmatrix} 16\,\text{A} \\ 48\,\text{A} \\ 0 \end{bmatrix}$$

or

$$\begin{bmatrix} V_1 \\ V_2 \\ V_3 \end{bmatrix} = \begin{bmatrix} -19.733\,\text{V} \\ -10.667\,\text{V} \\ -21.867\,\text{V} \end{bmatrix} \Leftarrow$$

SUPERNODE PROBLEMS

3.14: Determine the node voltages, V_1, V_2 and V_3 in the network of Fig 3.14.

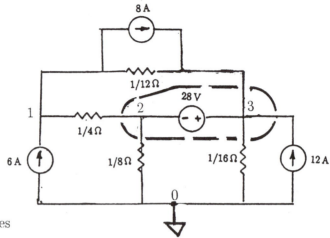

For node-1, KCL gives

$$-6\,\text{A} + 4(V_1 - V_2) + 12(V_1 - V_3) + 8\,\text{A} = 0$$

and for the indicated supernode

$$4(V_2 - V_1) + 8V_2 + 16V_3 + 12(V_3 - V_1) - 12\,\text{A} - 8\,\text{A} = 0$$

and within the supernode

$$V_3 - V_2 = 28\,\text{V}$$

This permits writing the set of node equations as

$$(4 + 12)V_1 - 4V_2 - 12V_3 = -2\,\text{A}$$

$$-(4 + 12)V_1 + (4 + 8)V_2 + (16 + 12)V_3 = 20\,\text{A}$$

and

$$-V_2 + V_3 = 28\,\text{V}$$

In matrix form

$$\begin{bmatrix} 16 & -4 & -12 \\ -16 & 12 & 28 \\ 0 & -1 & 1 \end{bmatrix} \begin{bmatrix} V_1 \\ V_2 \\ V_3 \end{bmatrix} = \begin{bmatrix} -2\,\text{A} \\ 20\,\text{A} \\ 28\,\text{V} \end{bmatrix}$$

and a matrix inversion yields

$$\begin{bmatrix} V_1 \\ V_2 \\ V_3 \end{bmatrix} = \begin{bmatrix} 16 & -4 & -12 \\ -16 & 12 & 28 \\ 0 & -1 & 1 \end{bmatrix}^{-1} \begin{bmatrix} -2\,\text{A} \\ 20\,\text{A} \\ 28\,\text{V} \end{bmatrix}$$

or

$$\begin{bmatrix} V_1 \\ V_2 \\ V_3 \end{bmatrix} = \begin{bmatrix} 0.1042 & 0.0417 & 0.0833 \\ 0.0417 & 0.0417 & -0.6667 \\ 0.0417 & 0.0417 & 0.3333 \end{bmatrix} \begin{bmatrix} -2\,\mathrm{A} \\ 20\,\mathrm{A} \\ 28\,\mathrm{V} \end{bmatrix} = \begin{bmatrix} 2.958\,\mathrm{V} \\ -17.917\,\mathrm{V} \\ 10.083\,\mathrm{V} \end{bmatrix} \Leftarrow$$

3.15: Determine the node voltages, V_1, V_2, V_3 and V_4 in the network of Fig 3.15.

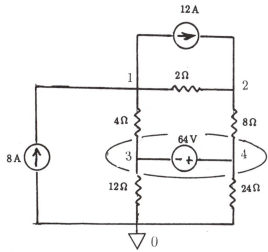

For nodes-1 and -2, KCL gives

$$-8\,\mathrm{A} + \frac{V_1 - V_3}{4} + 12\,\mathrm{A} + \frac{V_1 - V_2}{2} = 0$$

and

$$-12\,\mathrm{A} + \frac{V_2 - V_1}{2} + \frac{V_2 - V_4}{8} = 0$$

For the indicated supernode

$$\frac{V_3 - V_1}{4} + \frac{V_3}{12} + \frac{V_4 - V_2}{8} + \frac{V_4}{24} = 0$$

and within the supernode

$$V_4 - V_3 = 64\,\mathrm{V}$$

The set of node equation is

$$\left(\frac{1}{4} + \frac{1}{2}\right) V_1 - \frac{1}{2}V_2 - \frac{1}{4}V_3 = -4\,\mathrm{A}$$

$$-\frac{1}{2}V_1 + \left(\frac{1}{2} + \frac{1}{8}\right) V_2 - \frac{1}{8}V_4 = 12\,\mathrm{A}$$

$$-\frac{1}{4}V_1 - \frac{1}{8}V_2 + \left(\frac{1}{4} + \frac{1}{12}\right)V_3 + \left(\frac{1}{8} + \frac{1}{24}\right)V_4 = 0$$

and

$$-V_3 + V_4 = 64\,\text{V}$$

In matrix form after simplification

$$\begin{bmatrix} 3 & -2 & -1 & 0 \\ -4 & 5 & 0 & -1 \\ -6 & -3 & 8 & 4 \\ 0 & 0 & -1 & 1 \end{bmatrix} \begin{bmatrix} V_1 \\ V_2 \\ V_3 \\ V_4 \end{bmatrix} = \begin{bmatrix} -16\,\text{A} \\ 96\,\text{A} \\ 0 \\ 64\,\text{V} \end{bmatrix}$$

and a matrix inversion yields

$$\begin{bmatrix} V_1 \\ V_2 \\ V_3 \\ V_4 \end{bmatrix} = \begin{bmatrix} 3 & -2 & -1 & 0 \\ -4 & 5 & 0 & -1 \\ -6 & -3 & 8 & 4 \\ 0 & 0 & -1 & 1 \end{bmatrix}^{-1} \begin{bmatrix} -16\,\text{A} \\ 96\,\text{A} \\ 0 \\ 64\,\text{V} \end{bmatrix}$$

$$\begin{bmatrix} V_1 \\ V_2 \\ V_3 \\ V_4 \end{bmatrix} = \begin{bmatrix} 2.7143 & 1.2857 & 0.3333 & -0.0476 \\ 2.5714 & 1.4286 & 0.3333 & 0.0952 \\ 2.0000 & 1.0000 & 0.3333 & -0.3333 \\ 2.0000 & 1.0000 & 0.3333 & 0.6667 \end{bmatrix} \begin{bmatrix} -16\,\text{A} \\ 96\,\text{A} \\ 0 \\ 64\,\text{V} \end{bmatrix}$$

or

$$\begin{bmatrix} V_1 \\ V_2 \\ V_3 \\ V_4 \end{bmatrix} = \begin{bmatrix} 76.952\,\text{V} \\ 102.095\,\text{V} \\ 42.667\,\text{V} \\ 106.667\,\text{V} \end{bmatrix} \Leftarrow$$

200

3.16: Determine the node voltages, V_1, V_2 and V_3 in the network of Fig 3.16.

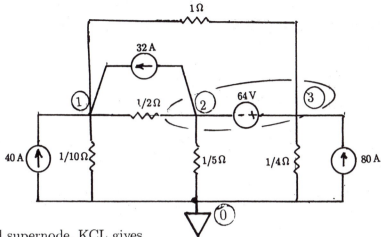

For the indicated supernode, KCL gives

$$-80\,\text{A} + 2(V_2 - V_1) + 5V_2 + 1(V_3 - V_1) + 4V_3 + 32\,\text{A} = 0$$

or

$$-3V_1 + 7V_2 + 5V_3 = 48\,\text{A}$$

However, inside of the supernode

$$V_3 - V_2 = 64\,\text{V} \qquad \text{or} \qquad V_3 = V_2 + 64\,\text{V}$$

so that

$$-3V_1 + 7V_2 + 5(V_2 + 64\,\text{V}) = 48\,\text{A}$$

or

$$-3V_1 + 12V_2 = -272\,\text{A}$$

Next, consider node-1, and again employ KCL

$$-40\,\text{A} + 10V_1 - 32\,\text{A} + 1(V_1 - V_3) + 2(V_1 - V_2) = 0$$

or

$$13V_1 - 2V_2 - V_3 = 72\,\text{A}$$

Then with $V_3 = V_2 + 64\,\text{V}$

$$13V_1 - 2V_2 - (V_2 + 64\,\text{V}) = 72\,\text{A}$$

or

$$13V_1 - 3V_2 = 136 \, \text{A}$$

Thus, there are two node equations, one for the supernode and one for node-1

$$
\begin{aligned}
-3V_1 + 12V_2 &= -272 \, \text{A} \\
13V_1 - 3V_2 &= 136 \, \text{A}
\end{aligned}
$$

and these can be put into matrix form

$$
\begin{bmatrix} -3 & 12 \\ 13 & -3 \end{bmatrix}
\begin{bmatrix} V_1 \\ V_2 \end{bmatrix}
=
\begin{bmatrix} -272 \, \text{A} \\ 136 \, \text{A} \end{bmatrix}
$$

and a matrix inversion yields

$$
\begin{bmatrix} V_1 \\ V_2 \end{bmatrix}
=
\begin{bmatrix} -3 & 12 \\ 13 & -3 \end{bmatrix}^{-1}
\begin{bmatrix} -272 \, \text{A} \\ 136 \, \text{A} \end{bmatrix}
$$

or

$$
\begin{bmatrix} V_1 \\ V_2 \end{bmatrix}
=
\begin{bmatrix} 0.0204 & 0.0816 \\ 0.0884 & 0.0204 \end{bmatrix}
\begin{bmatrix} -272 \, \text{A} \\ 136 \, \text{A} \end{bmatrix}
=
\begin{bmatrix} 5.551 \, \text{V} \\ -21.279 \, \text{V} \end{bmatrix} \Leftarrow
$$

3.17: Determine the node voltages, V_1, V_2 and V_3 in the network of Fig 3.17.

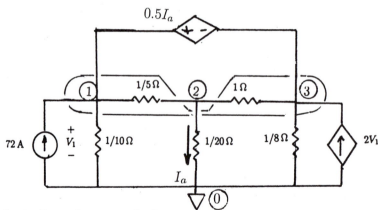

Use KCL for the indicated supernode first

$$-72 \, \text{A} + 10V_1 + 5(V_1 - V_2) + 1(V_3 - V_2) + 8V_3 - 2V_1 = 0$$

and after adjustment

$$13V_1 - 6V_2 + 9V_3 = 72 \, \text{A}$$

With

$$I_a = 20V_2 \quad \text{so that} \quad \frac{1}{2}I_a = 10V_2$$

and

$$V_1 - V_3 = 10V_2 \quad \text{so that} \quad V_1 = V_3 + 10V_2$$

the KVL equation for the supernode becomes

$$13(V_3 + 10V_2) - 6V_2 + 9V_3 = 72\,\text{A}$$

or

$$124V_2 + 22V_3 = 72\,\text{A}$$

At node-2, KCL gives

$$5(V_2 - V_1) + 1(V_2 - V_3) + 20V_2 = 0$$

or

$$-5V_1 + 26V_2 - V_3 = 0$$

and with $V_1 = V_3 + 10V_2$

$$-5(V_3 + 10V_2) + 26V_2 - V_3 = 0$$

or

$$-24V_2 - 6V_3 = 0$$

This shows that

$$V_3 = -4V_2$$

and from the supernode equation

$$124V_2 + 22(-4V_2) = 124V_2 - 88V_2 \ = \ 72\,\text{A}$$
$$36V_2 \ = \ 72\,\text{A}$$

or

$$V_2 = 2\,\text{V} \Leftarrow$$

Then

$$V_3 = -4V_2 = -4(2\,\text{V}) = -8\,\text{V} \Leftarrow$$

and

$$\begin{aligned} V_1 &= V_3 + 10V_2 \\ &= -8\,\text{V} + 10(2\,\text{V}) \\ &= 12\,\text{V} \Leftarrow \end{aligned}$$

found from an application of KCL. With node-0 selected as the reference node, the three node equations can be written as

$$\frac{V_a - 63\,V}{5} + \frac{1}{6}V_a + \frac{V_a - V_b}{1} = 0$$

$$\frac{V_b - V_a}{1} + \frac{1}{3}V_b + \frac{V_b - V_c}{4} = 0$$

and

$$\frac{V_c - V_b}{4} + \frac{1}{2}V_c = 0$$

or

$$\left(\frac{1}{5} + \frac{1}{6} + 1\right)V_a \quad - V_b \qquad\qquad\qquad = \frac{63}{5}\,A$$

$$- V_a \qquad + \left(1 + \frac{1}{3} + \frac{1}{4}\right)V_b \quad - \frac{1}{4}V_c \quad = \quad 0$$

$$- \frac{1}{4}V_b \qquad + \left(\frac{1}{4} + \frac{1}{2}\right)V_c \quad = \quad 0$$

These may be written in matrix form. After simplification

$$\begin{bmatrix} 41 & -30 & 0 \\ -12 & 19 & -3 \\ 0 & -1 & 3 \end{bmatrix} \begin{bmatrix} V_a \\ V_b \\ V_c \end{bmatrix} = \begin{bmatrix} 378\,A \\ 0 \\ 0 \end{bmatrix}$$

A matrix inversion provides the three node voltages

$$\begin{bmatrix} V_a \\ V_b \\ V_c \end{bmatrix} = \begin{bmatrix} 41 & -30 & 0 \\ -12 & 19 & -3 \\ 0 & -1 & 3 \end{bmatrix}^{-1} \begin{bmatrix} 378\,A \\ 0 \\ 0 \end{bmatrix}$$

or

$$\begin{bmatrix} V_a \\ V_b \\ V_c \end{bmatrix} = \begin{bmatrix} 0.0476 & 0.0794 & 0.0794 \\ 0.0317 & 0.1085 & 0.1085 \\ 0.0106 & 0.0362 & 0.3695 \end{bmatrix} \begin{bmatrix} 378\,A \\ 0 \\ 0 \end{bmatrix} = \begin{bmatrix} 18\,V \\ 12\,V \\ 4\,V \end{bmatrix} \Leftarrow$$

The voltage across R_6 will be equal to V_a

$$V_6 = V_a = 18\,V \Leftarrow$$

and the currents are obtained from the node voltages

$$I_1 = \frac{V_a - V_b}{1\,\Omega} = \frac{18\,V - 12\,V}{1\,\Omega} = 6\,A \Leftarrow$$

and

$$I_2 = \frac{V_c}{2\,\Omega} = \frac{4\,\text{V}}{2\,\Omega} = 2\,\text{A} \Leftarrow$$

These values check the values in Problems 2.24 and 2.25.

3.20: Use nodal analysis to determine I_3, I_4, V_{12} and V_{14} in the network shown in Fig 2.18 (This is a repetition of Problem 2.28).

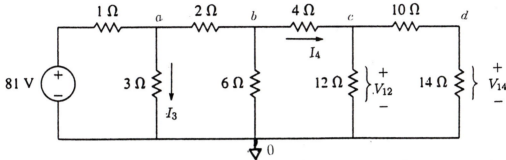

Because the resistances are designated with numerals representing their magnitude, the node voltages are represented with letter subscripts (V_a through V_d). The node equations are found from an application of KCL and with node-0 selected as the reference node, the four node equations can be written as

$$\frac{V_a - 81\,\text{V}}{1} + \frac{1}{3}V_a + \frac{V_a - V_b}{2} = 0$$

$$\frac{V_b - V_a}{2} + \frac{1}{6}V_b + \frac{V_b - V_c}{4} = 0$$

$$\frac{V_c - V_b}{4} + \frac{1}{12}V_c + \frac{V_c - V_d}{10} = 0$$

and

$$\frac{V_d - V_c}{10} + \frac{1}{14}V_d = 0$$

or

$$
\begin{aligned}
\left(1 + \frac{1}{3} + \frac{1}{2}\right) V_a & \quad - \frac{1}{2}V_b & & & = 81\,\text{A} \\
-\frac{1}{2}V_a & \quad + \left(\frac{1}{2} + \frac{1}{6} + \frac{1}{4}\right) V_b & \quad -\frac{1}{4}V_c & & = 0 \\
& \quad -\frac{1}{4}V_b & \quad + \left(\frac{1}{4} + \frac{1}{12} + \frac{1}{10}\right) V_c & \quad -\frac{1}{10}V_d & = 0 \\
& & \quad -\frac{1}{10}V_c & \quad + \left(\frac{1}{10} + \frac{1}{14}\right) V_d & = 0
\end{aligned}
$$

$$I_5 = \frac{V_a - V_b}{5\,\Omega} = \frac{15\,\text{V} - 8.75\,\text{V}}{5\,\Omega} = \frac{6.25\,\text{V}}{5\,\Omega} = \frac{5}{4}\,\text{A} \Leftarrow$$

and

$$P_{24} = \frac{(V_{24})^2}{24\,\Omega} = \frac{(15\,\text{V})^2}{24\,\Omega} = 9.375\,\text{W} \Leftarrow$$

These values check the values in Problem 2.32.

SINGLE LOOP ANALYSIS

3.22: Determine the current, I, in the network shown in Fig 3.18.

By KVL

$$
\begin{aligned}
-64\,\text{V} + 40I + 20I + 40\,\text{V} &= 0 \\
60I &= 24\,\text{V} \\
I &= \frac{24\,\text{V}}{60\,\Omega} \\
&= 400\,\text{mA} \Leftarrow
\end{aligned}
$$

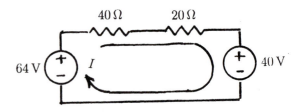

3.23: Determine the current, I, in the network shown in Fig 3.19.

By KVL

$$-120\,\text{V} + 400I + 100(I + I_a) = 0$$

$$
\begin{aligned}
500I &= 120\,\text{V} - 100(1\,\text{A}) \\
&= 20\,\text{V} \\
I &= 40\,\text{mA} \Leftarrow
\end{aligned}
$$

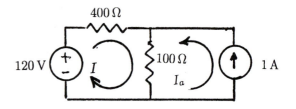

3.24: Determine the current, I, in the network shown in Fig 3.20.

Here

$$I_a = 0.80\,\text{A} \qquad \text{and} \qquad I_b = 0.10\,\text{A}$$

Then, by KVL

$$100I + 250(I + I_b) + 50(I - I_a) = 0$$

$$
\begin{aligned}
400I &= 50I_a - 250I_b \\
&= 50(0.80\,\text{A}) - 250(0.10\,\text{A}) \\
&= 40\,\text{V} - 25\,\text{V} \\
I &= \frac{15\,\text{V}}{400\,\Omega} \\
&= 37.5\,\text{mA} \Leftarrow
\end{aligned}
$$

3.25: Determine the current, I, and the equivalent resistance seen by the 102 V source in Fig 3.21.

By KVL

$$-102\,\text{V} + 4I + 32I + 8V_a = 0$$

But $V_a = 4I$ so that

$$
\begin{aligned}
36I &= 102\,\text{V} - 8(4I) \\
68I &= 102\,\text{V} \\
I &= \frac{102\,\text{V}}{68\,\Omega} \\
&= 1.50\,\text{A} \Leftarrow
\end{aligned}
$$

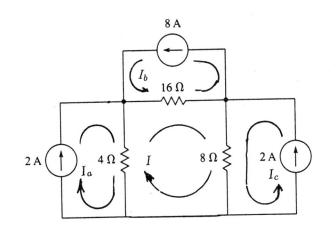

The equivalent resistance seen by the independent 102 V source will be

$$R_{\text{eq}} = \frac{V_s}{I} = \frac{102\,\text{V}}{1.50\,\text{A}} = 68\,\Omega \Leftarrow$$

3.26: Use loop analysis to determine the current flowing in the $8\,\Omega$ resistor in Fig 3.5 (This is a repeat of Problem 3.5).

The single loop (mesh) current is designated by the clockwise current, I, and the three current sources contribute I_a, I_b and I_c. By KVL

$$4(I - I_a) + 16(I + I_b) + 8(I + I_c) = 0$$

$$(4 + 16 + 8)I = 4I_a - 16I_b - 8I_c$$

$$28I = 4(2\,\text{A}) - 16(8\,\text{A}) - 8(2\,\text{A})$$

$$
\begin{aligned}
28I &= -136\,\text{V} \\
I &= \frac{-136\,\text{V}}{28\,\Omega} \\
&= -\frac{34}{7}\,\Omega
\end{aligned}
$$

The current through the $8\,\Omega$ resistor will be

$$I_8 = I - I_c = -\frac{34}{7}\,\text{A} + \frac{14}{7}\,\text{A} = -\frac{20}{7}\,\text{A} \Leftarrow$$

The network contains five meshes. Because all sources are voltage sources and there are no controlled sources, the five mesh equations may be written by inspection in the form $\mathbf{RI} = \mathbf{V}_s$. The elements on the principal diagonal of the coefficient matrix, $\mathbf{R}$, derive from the sum of the resistances in the individual meshes. With all elements in ohms, the principal diagonal elements are

$$
\begin{aligned}
r_{11} &= 4\,\Omega + 8\,\Omega + 10\,\Omega + 2\,\Omega + 8\,\Omega = 32\,\Omega \\
r_{22} &= 10\,\Omega + 16\,\Omega + 8\,\Omega + 2\,\Omega + 6\,\Omega = 42\,\Omega \\
r_{33} &= 12\,\Omega + 8\,\Omega + 4\,\Omega = 24\,\Omega \\
r_{44} &= 2\,\Omega + 6\,\Omega + 10\,\Omega + 4\,\Omega = 22\,\Omega
\end{aligned}
$$

and

$$
r_{55} = 2\,\Omega + 4\,\Omega + 10\,\Omega = 16\,\Omega
$$

Then, the off diagonal elements are the individual conductances between the nodes

$$
\begin{array}{ll}
r_{12} = r_{21} = 10\,\Omega & r_{13} = r_{31} = 8\,\Omega \\
r_{14} = r_{41} = 2\,\Omega & r_{15} = r_{51} = 0\,\Omega \\
r_{23} = r_{32} = 0\,\Omega & r_{24} = r_{42} = 6\,\Omega \\
r_{25} = r_{52} = 2\,\Omega & r_{34} = r_{43} = 4\,\Omega \\
r_{35} = r_{53} = 0\,\Omega & r_{45} = r_{54} = 10\,\Omega
\end{array}
$$

The source terms are equal to the sum of the voltage *rises* in each mesh

$$
\begin{aligned}
V_{s1} &= 48\,\text{V} - 40\,\text{V} = 8\,\text{V} \\
V_{s2} &= 40\,\text{V} - 80\,\text{V} = -40\,\text{V} \\
V_{s3} &= 64\,\text{V} \\
V_{s4} &= 0\,\text{V} \\
V_{s5} &= 36\,\text{V}
\end{aligned}
$$

The mesh equations, in matrix form, are

$$
\begin{bmatrix}
32 & -10 & -8 & -2 & 0 \\
-10 & 42 & 0 & -6 & -2 \\
-8 & 0 & 24 & -4 & 0 \\
-2 & -6 & -4 & 22 & -10 \\
0 & -2 & 0 & -10 & 16
\end{bmatrix}
\begin{bmatrix}
I_1 \\ I_2 \\ I_3 \\ I_4 \\ I_5
\end{bmatrix}
=
\begin{bmatrix}
8 \\ -40 \\ 64 \\ 0 \\ 36
\end{bmatrix} \Leftarrow
$$

3.30: Determine the value of V_s that will produce a current of $I_2 = 1.532\,\text{A}$ in the network of Fig 3.24.

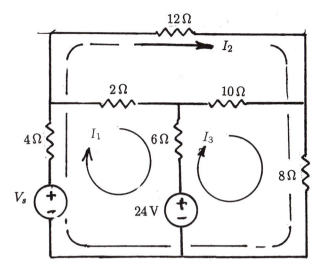

Because there are no currents or controlled sources present, the mesh and loop currents may be represented with numeral subscripts (I_1 through I_3) and may be written by inspection. There are two clockwise mesh currents, I_1 and I_3 and one clockwise loop current, I_2. Both the mesh and loop equations are written from a systematic application of KVL

$$-V_s + 4(I_1 + I_2) + 2I_1 + 6(I_1 - I_3) + 24\,\text{V} = 0$$
$$-V_s + 4(I_2 + I_1) + 12I_2 + 8(I_2 + I_3) = 0$$
$$-24\,\text{V} + 6(I_3 - I_1) + 10I_3 + 8(I_3 + I_2) = 0$$

After rearrangement, these become

$$
\begin{aligned}
12I_1 &+ 4I_2 &- 6I_3 &= V_s - 24\,\text{V} \\
4I_1 &+ 24I_2 &+ 8I_3 &= V_s \\
-6I_1 &+ 8I_2 &+ 24I_3 &= 24\,\text{V}
\end{aligned}
$$

In matrix form

$$
\begin{bmatrix} 12 & 4 & -6 \\ 4 & 24 & 8 \\ -6 & 8 & 24 \end{bmatrix}
\begin{bmatrix} I_1 \\ I_2 \\ I_3 \end{bmatrix} =
\begin{bmatrix} V_s - 24\,\text{V} \\ V_s \\ 24\,\text{V} \end{bmatrix}
$$

A Cramer's rule solution yields

$$
I_2 = \frac{\begin{vmatrix} 12 & V_s - 24 & -6 \\ 4 & V_s & 8 \\ -6 & 24 & 24 \end{vmatrix}}{\begin{vmatrix} 12 & 4 & -6 \\ 4 & 24 & 8 \\ -6 & 8 & 24 \end{vmatrix}} = 1.532\,\text{A}
$$

In matrix form

$$\begin{bmatrix} 38 & -24 & 0 & 0 \\ -24 & 39 & -4 & -3 \\ 0 & -4 & 22 & -6 \\ 0 & -3 & -6 & 9 \end{bmatrix} \begin{bmatrix} I_a \\ I_b \\ I_c \\ I_d \end{bmatrix} = \begin{bmatrix} 270\,\text{V} \\ 0 \\ 0 \\ 0 \end{bmatrix}$$

and a matrix inversion yields

$$\begin{bmatrix} I_a \\ I_b \\ I_c \\ I_d \end{bmatrix} \begin{bmatrix} 38 & -24 & 0 & 0 \\ -24 & 39 & -4 & -3 \\ 0 & -4 & 22 & -6 \\ 0 & -3 & -6 & 9 \end{bmatrix}^{-1} \begin{bmatrix} 270\,\text{V} \\ 0 \\ 0 \\ 0 \end{bmatrix}$$

or

$$\begin{bmatrix} I_a \\ I_b \\ I_c \\ I_d \end{bmatrix} = \begin{bmatrix} 0.0455 & 0.0303 & 0.0101 & 0.0168 \\ 0.0303 & 0.0480 & 0.0160 & 0.0267 \\ 0.0101 & 0.0160 & 0.0609 & 0.0459 \\ 0.0168 & 0.0267 & 0.0459 & 0.1506 \end{bmatrix} \begin{bmatrix} 270\,\text{V} \\ 0 \\ 0 \\ 0 \end{bmatrix} = \begin{bmatrix} 12.273\,\text{A} \\ 8.182\,\text{A} \\ 2.727\,\text{A} \\ 4.546\,\text{A} \end{bmatrix}$$

The currents sought are

$$I_6 = I_d - I_c = 4.546\,\text{A} - 2.727\,\text{A} = 1.819\,\text{A} \Leftarrow$$

and

$$I_4 = I_b - I_c = 8.182\,\text{A} - 2.727\,\text{A} = 5.455\,\text{A} \Leftarrow$$

The voltage drop across the $12\,\Omega$ resistor is

$$V_{12} = 12 I_c = 12(2.727\,\text{A}) = 32.724\,\text{V} \Leftarrow$$

The current through the $24\,\Omega$ resistor is

$$I_{24} = I_a - I_b = 12.273\,\text{A} - 8.182\,\text{A} = 4.091\,\text{A}$$

and the power dissipated by the $24\,\Omega$ resistor is

$$P_{24} = I_{24}^2 R_{24} = (4.091\,\text{A})^2 (24\,\Omega) = 401.7\,\text{W} \Leftarrow$$

3.33: Use mesh analysis to find the current through, the voltage across and the power dissipated by the $6\,\Omega$ resistor in Fig 3.27.

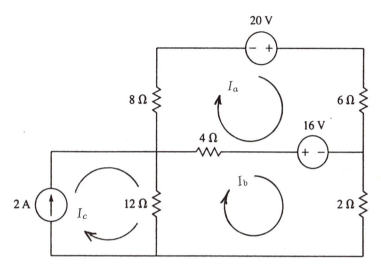

Because the resistances are designated with numerals representing their magnitude, the two unknown mesh currents are designated by letter subscripts (I_a and I_b) and the mesh with the known mesh current is $I_c = 2\,\text{A}$. The two equations may be written by a repeated application of KVL

$$-20\,\text{V} + 6I_a - 16\,\text{V} + 4(I_a - I_b) + 8I_a = 0$$

$$16\,\text{V} + 2I_b + 12(I_b - I_c) + 4(I_b - I_a) = 0$$

and with $I_c = 2\,\text{A}$, these simplify to

$$
\begin{aligned}
18I_a \quad - 4I_b &= \quad 36\,\text{V} \\
-4I_a \quad + 18I_b &= \quad 8\,\text{V}
\end{aligned}
$$

In matrix form

$$
\begin{bmatrix} 18 & -4 \\ -4 & 18 \end{bmatrix}
\begin{bmatrix} I_a \\ I_b \end{bmatrix}
=
\begin{bmatrix} 36\,\text{V} \\ 8\,\text{V} \end{bmatrix}
$$

and a matrix inversion yields

$$
\begin{bmatrix} I_a \\ I_b \end{bmatrix}
=
\begin{bmatrix} 18 & -4 \\ -4 & 18 \end{bmatrix}^{-1}
\begin{bmatrix} 36\,\text{V} \\ 8\,\text{V} \end{bmatrix}
=
\begin{bmatrix} 0.0584 & 0.0130 \\ 0.0130 & 0.0584 \end{bmatrix}
\begin{bmatrix} 36\,\text{V} \\ 8\,\text{V} \end{bmatrix}
=
\begin{bmatrix} 2.208\,\text{A} \\ 0.935\,\text{A} \end{bmatrix}
$$

For the $6\,\Omega$ resistor

$$I_6 = I_a = 2.208\,\text{A} \Leftarrow$$

$$V_6 = 6I_a = 6(2.208\,\text{A}) = 13.248\,\text{V} \Leftarrow$$

$$3(I_b - I_a) + 2I_b + 4I_b = 0$$

$$1I_c = 6\,\text{V}$$

In matrix form, the mesh equations are

$$\begin{bmatrix} 4 & -3 & -1 \\ -3 & 9 & 0 \\ 0 & 0 & 1 \end{bmatrix} \begin{bmatrix} I_a \\ I_b \\ I_c \end{bmatrix} = \begin{bmatrix} 0 \\ 0 \\ 6\,\text{V} \end{bmatrix}$$

and a matrix inversion yields the current vector

$$\begin{bmatrix} I_a \\ I_b \\ I_c \end{bmatrix} = \begin{bmatrix} 4 & -3 & -1 \\ -3 & 9 & 0 \\ 0 & 0 & 1 \end{bmatrix}^{-1} \begin{bmatrix} 0 \\ 0 \\ 6\,\text{V} \end{bmatrix}$$

or

$$\begin{bmatrix} I_a \\ I_b \\ I_c \end{bmatrix} = \begin{bmatrix} 0.3333 & 0.1111 & 0.3333 \\ 0.1111 & 0.1481 & 0.1111 \\ 0 & 0 & 1.0000 \end{bmatrix} \begin{bmatrix} 0 \\ 0 \\ 6\,\text{V} \end{bmatrix} = \begin{bmatrix} 2\,\text{A} \\ 2/3\,\text{A} \\ 6\,\text{A} \end{bmatrix}$$

The currents in each resistor are

$$I_1 = I_c - I_a = 6\,\text{A} - 2\,\text{A} = 4\,\text{A} \Leftarrow$$

$$I_2 = I_b = \frac{2}{3}\,\text{A} \Leftarrow$$

$$I_3 = I_a - I_b = 2\,\text{A} - \frac{2}{3}\,\text{A} = \frac{4}{3}\,\text{A} \Leftarrow$$

$$I_4 = I_b = \frac{2}{3}\,\text{A} \Leftarrow$$

and the voltages across each resistor are

$$V_1 = R_1 I_1 = (1\,\Omega)(4\,\text{A}) = 4\,\text{V} \Leftarrow$$

$$V_2 = R_2 I_2 = (2\,\Omega)\left(\frac{2}{3}\,\text{A}\right) = \frac{4}{3}\,\text{V} \Leftarrow$$

$$V_2 = R_3 I_3 = (3\,\Omega)\left(\frac{4}{3}\,\text{A}\right) = 4\,\text{V} \Leftarrow$$

$$V_4 = R_4 I_4 = (4\,\Omega)\left(\frac{2}{3}\,\text{A}\right) = \frac{8}{3}\,\text{V} \Leftarrow$$

3.36: Use mesh analysis to determine the currents, I_1 and I_2 and the voltage across R_6 in the network of Fig 2.16 (This is a repeat of Problems 2.24 and 3.19).

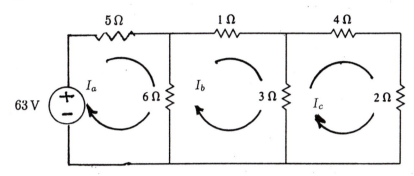

Because the resistances are designated with numerals representing their magnitude, the mesh currents are designated by letter subscripts (I_a, I_b and I_c). Because all sources are voltage sources and there are no controlled sources, the three mesh equations may be written by inspection via a repeated application of KVL

$$
\begin{aligned}
11I_a &- 6I_b & &= 63\,\text{V} \\
-6I_a &+ 10I_b &- 3I_c &= 0 \\
&- 3I_b &+ 9I_c &= 0
\end{aligned}
$$

In matrix form

$$
\begin{bmatrix} 11 & -6 & 0 \\ -6 & 10 & -3 \\ 0 & -3 & 9 \end{bmatrix}
\begin{bmatrix} I_a \\ I_b \\ I_c \end{bmatrix}
=
\begin{bmatrix} 63\,\text{V} \\ 0 \\ 0 \end{bmatrix}
$$

and a matrix inversion yields

$$
\begin{bmatrix} I_a \\ I_b \\ I_c \end{bmatrix}
\begin{bmatrix} 11 & -6 & 0 \\ -6 & 10 & -3 \\ 0 & -3 & 9 \end{bmatrix}^{-1}
\begin{bmatrix} 63\,\text{V} \\ 0 \\ 0 \end{bmatrix}
$$

or

$$
\begin{bmatrix} I_a \\ I_b \\ I_c \end{bmatrix}
=
\begin{bmatrix} 0.1429 & 0.0952 & 0.0317 \\ 0.0952 & 0.1746 & 0.0582 \\ 0.0317 & 0.0582 & 0.1305 \end{bmatrix}
\begin{bmatrix} 63\,\text{V} \\ 0 \\ 0 \end{bmatrix}
=
\begin{bmatrix} 9\,\text{A} \\ 6\,\text{A} \\ 2\,\text{A} \end{bmatrix}
$$

The currents sought are

$$I_1 = I_b = 6\,\text{A} \Leftarrow$$

and

$$I_2 = I_c = 2\,\text{A} \Leftarrow$$

and the voltage across R_6 is

$$V_6 = R_6 I_6 = R_6(I_a - I_b) = (6\,\Omega)(9\,\text{A} - 6\,\text{A}) = 18\,\text{V} \Leftarrow$$

3.37: Use mesh analysis to find I_3, I_4, V_{12} and V_{14} in the network shown in Fig 2.18 (This is a repeat of Problems 2.28 and 3.20).

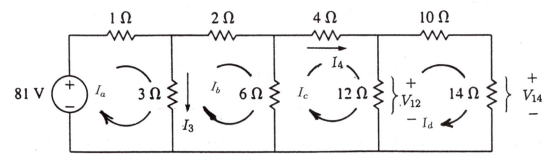

Because the resistances are designated with numerals representing their magnitude, the mesh currents are designated by letter subscripts (I_a through I_d). Because all sources are voltage sources and there are no controlled sources, the four mesh equations may be written by inspection via a repeated application of KVL

$$
\begin{aligned}
4I_a &- 3I_b & & & &= 81\,\text{V} \\
-3I_a &+ 11I_b &- 6I_c & & &= 0 \\
&- 6I_b &+ 22I_c &- 12I_d &= 0 \\
& &- 12I_c &+ 36I_d &= 0
\end{aligned}
$$

In matrix form

$$
\begin{bmatrix}
4 & -3 & 0 & 0 \\
-3 & 11 & -6 & 0 \\
0 & -6 & 22 & -12 \\
0 & 0 & -12 & 36
\end{bmatrix}
\begin{bmatrix}
I_a \\ I_b \\ I_c \\ I_d
\end{bmatrix}
=
\begin{bmatrix}
81\,\text{V} \\ 0 \\ 0 \\ 0
\end{bmatrix}
$$

and a matrix inversion yields

$$
\begin{bmatrix}
I_a \\ I_b \\ I_c \\ I_d
\end{bmatrix}
=
\begin{bmatrix}
4 & -3 & 0 & 0 \\
-3 & 11 & -6 & 0 \\
0 & -6 & 22 & -12 \\
0 & 0 & -12 & 36
\end{bmatrix}^{-1}
\begin{bmatrix}
81\,\text{V} \\ 0 \\ 0 \\ o
\end{bmatrix}
$$

or

$$
\begin{bmatrix}
I_a \\ I_b \\ I_c \\ I_d
\end{bmatrix}
=
\begin{bmatrix}
0.3333 & 0.1111 & 0.0370 & 0.0123 \\
0.1111 & 0.1481 & 0.0494 & 0.0165 \\
0.0370 & 0.0494 & 0.0720 & 0.0240 \\
0.0123 & 0.0165 & 0.0240 & 0.0358
\end{bmatrix}
\begin{bmatrix}
81\,\text{V} \\ 0 \\ 0 \\ 0
\end{bmatrix}
=
\begin{bmatrix}
27\,\text{A} \\ 9\,\text{A} \\ 3\,\text{A} \\ 1\,\text{A}
\end{bmatrix}
$$

The currents sought are

$$I_3 = I_a - I_b = 27\,\text{A} - 9\,\text{A} = 18\,\text{A} \Leftarrow$$

and

$$I_4 = I_c = 3\,\text{A} \Leftarrow$$

and the voltages required are

$$V_{12} = R_{12}I_{12} = R_{12}(I_c - I_d) = (12\,\Omega)(3\,\text{A} - 1\,\text{A}) = 24\,\text{V} \Leftarrow$$

and

$$V_{14} = R_{14}I_{14} = R_{14}I_d = (14\,\Omega)(1\,\text{A}) = 14\,\text{V} \Leftarrow$$

3.38: Use mesh analysis to determine I_5, V_{20} and the power dissipated in the $24\,\Omega$ resistor in Fig 2.22 (This is a repeat of Problems 2.32 and 3.21).

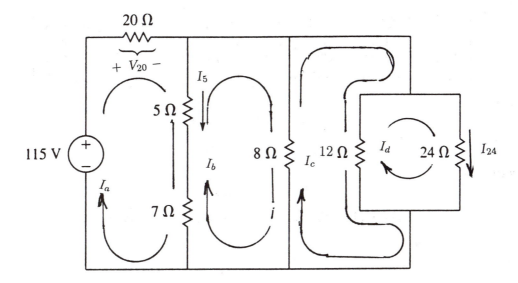

Because the resistances are designated with numerals representing their magnitude, the mesh currents are designated by letter subscripts (I_a through I_d). Because all sources are voltage sources and there are no controlled sources, the four mesh equations may be written by inspection via a repeated application of KVL

$$
\begin{array}{rrrrl}
32I_a & -12I_b & & & = & 115\,\text{V} \\
-12I_a & +20I_b & -8I_c & & = & 0 \\
& -8I_b & +20I_c & -12I_d & = & 0 \\
& & -12I_c & +36I_d & = & 0
\end{array}
$$

In matrix form

$$\begin{bmatrix} 32 & -12 & 0 & 0 \\ -12 & 20 & -8 & 0 \\ 0 & -8 & 20 & -12 \\ 0 & 0 & -12 & 36 \end{bmatrix} \begin{bmatrix} I_a \\ I_b \\ I_c \\ I_d \end{bmatrix} = \begin{bmatrix} 115\,\text{V} \\ 0 \\ 0 \\ 0 \end{bmatrix}$$

and a matrix inversion yields

$$\begin{bmatrix} I_a \\ I_b \\ I_c \\ I_d \end{bmatrix} = \begin{bmatrix} 32 & -12 & 0 & 0 \\ -12 & 20 & -8 & 0 \\ 0 & -8 & 20 & -12 \\ 0 & 0 & -12 & 36 \end{bmatrix}^{-1} \begin{bmatrix} 115\,\text{V} \\ 0 \\ 0 \\ 0 \end{bmatrix}$$

or

$$\begin{bmatrix} I_a \\ I_b \\ I_c \\ I_d \end{bmatrix} = \begin{bmatrix} 0.0435 & 0.0326 & 0.0163 & 0.0054 \\ 0.0326 & 0.0870 & 0.0435 & 0.0145 \\ 0.0163 & 0.0435 & 0.0842 & 0.0281 \\ 0.0054 & 0.0145 & 0.0281 & 0.0371 \end{bmatrix} \begin{bmatrix} 115\,\text{V} \\ 0 \\ 0 \\ 0 \end{bmatrix} = \begin{bmatrix} 5.000\,\text{A} \\ 3.750\,\text{A} \\ 1.875\,\text{A} \\ 0.625\,\text{A} \end{bmatrix}$$

Here

$$I_5 = I_a - I_b = 5.00\,\text{A} - 3.75\,\text{A} = 1.25\,\text{A} \Leftarrow$$

$$V_{20} = I_a R_{20} = (5.00\,\text{A})(20\,\Omega) = 100\,\text{V} \Leftarrow$$

and with

$$I_{24} = I_d = 0.625\,\text{A}$$

$$P_{24} = (I_{24})^2 R_{24} = (0.625\,\text{A})^2 (24\,\Omega) = 9.375\,\text{W} \Leftarrow$$

CHAPTER FOUR
THE OPERATIONAL AMPLIFIER

ANALYSIS USING THE IDEAL OP-AMP MODEL

4.1: Determine the voltage gain and the input resistance for each of the ideal operational amplifier circuits of Fig 4.1.

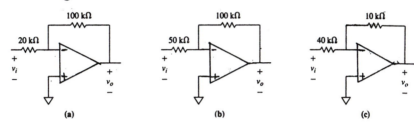

(a)   (b)   (c)

All of the op-amps shown in Fig 4.1 are in the inverting configuration and the gain in all three cases will be given by

$$G = -\frac{R_f}{R_1}$$

(a)

$$G = -\frac{R_f}{R_1} = -\frac{100\,\text{k}\Omega}{20\,\text{k}\Omega} = -5 \Leftarrow$$

and

$$R_{\text{in}} = R_1 = 20\,\text{k}\Omega \Leftarrow$$

(b)

$$G = -\frac{R_f}{R_1} = -\frac{100\,\text{k}\Omega}{50\,\text{k}\Omega} = -2 \Leftarrow$$

and

$$R_{\text{in}} = R_1 = 50\,\text{k}\Omega \Leftarrow$$

(c)

$$G = -\frac{R_f}{R_1} = -\frac{10\,\text{k}\Omega}{40\text{k}\Omega} = -0.25 \Leftarrow$$

and

$$R_{\text{in}} = R_1 = 40\,\text{k}\Omega \Leftarrow$$

4.2: Determine the voltage gain and the input resistance for each of the ideal operational amplifier circuits of Fig 4.2.

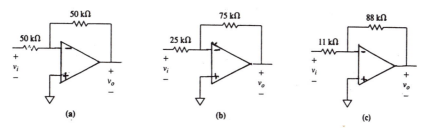

(a)   (b)   (c)

All of the op-amps shown in Fig 4.2 are in the inverting configuration and the gain in all three cases will be given by

$$G = -\frac{R_f}{R_1}$$

(a)

$$G = -\frac{R_f}{R_1} = -\frac{50\,\text{k}\Omega}{50\,\text{k}\Omega} = -1 \Leftarrow$$

and

$$R_{\text{in}} = R_1 = 50\,\text{k}\Omega \Leftarrow$$

(b)

$$G = -\frac{R_f}{R_1} = -\frac{75\,\text{k}\Omega}{25\,\text{k}\Omega} = -3 \Leftarrow$$

and

$$R_{\text{in}} = R_1 = 25\,\text{k}\Omega \Leftarrow$$

(c)

$$G = -\frac{R_f}{R_1} = -\frac{88\,\text{k}\Omega}{11\,\text{k}\Omega} = -8 \Leftarrow$$

and

$$R_{\text{in}} = R_1 = 11\,\text{k}\Omega \Leftarrow$$

4.3: Determine the voltage gain of the ideal operational amplifier circuit of Fig 4.3.

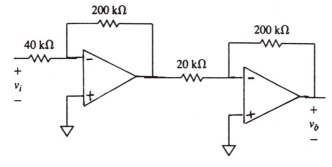

Both of the op-amps shown in Fig 4.3 are in the inverting configuration. With v_m as the output of the first op-amp

$$v_m = -\frac{R_{f,1}}{R_{1,1}}v_i = -\frac{200\,\text{k}\Omega}{40\,\text{k}\Omega} = -5v_i$$

Then, with v_m taken as the input to the second op-amp

$$v_o = -\frac{R_{f,2}}{R_{1,2}}v_m = -\frac{200\,\text{k}\Omega}{20\,\text{k}\Omega} = -10v_m$$

But $v_m = -5v_i$ so that

$$\frac{v_o}{v_i} = -10(-5) = 50 \Leftarrow$$

4.4: Determine the ratio, v_o/v_i, in the ideal operational amplifier circuit of Fig 4.4.

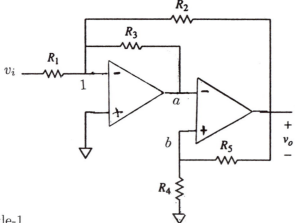

By KCL at node-1

$$\frac{v_1 - v_i}{R_1} + \frac{v_1 - v_o}{R_2} + \frac{v_1 - v_a}{R_3} = 0$$

and because $v_1 = v_- = v_+ = 0$ for the first op-amp

$$-\frac{v_i}{R_1} = \frac{v_o}{R_2} + \frac{v_a}{R_3}$$

Using a voltage divider,

$$v_b = \left(\frac{R_4}{R_4 + R_5}\right)v_o$$

and because $v_a = v_- = v_b = v_+$ for the second op-amp

$$-\frac{v_i}{R_1} = \frac{v_o}{R_2} + \frac{1}{R_3}\left(\frac{R_4}{R_4 + R_5}\right) v_o$$

Then

$$\frac{v_o}{v_i} = -\frac{1}{R_1\left[\dfrac{1}{R_2} + \dfrac{R_4}{R_3(R_4 + R_5)}\right]}$$

or

$$\frac{v_o}{v_i} = -\frac{R_2 R_3(R_4 + R_5)}{R_1(R_3 R_4 + R_3 R_5 + R_2 R_4)} \Longleftarrow$$

4.5: For the ideal operational amplifier shown in Fig 4.5, determine the gain $G = v_o/v_i$, for the conditions of (a) $R_1 = 100\,\text{k}\Omega$ and $R_f = 900\,\text{k}\Omega$, (b) $R_1 = 50\,\text{k}\Omega$ and $R_f = 750\,\text{k}\Omega$ and (c) $R_1 = 100\,\text{k}\Omega$ and $R_f = 1.1\,\text{M}\Omega$.

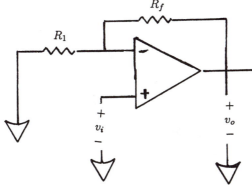

The op-amp shown in Fig 4.5 is in the non-inverting configuration and the gain in all three cases will be given by

$$G = 1 + \frac{R_f}{R_1}$$

(a)

$$G = 1 + \frac{R_f}{R_1} = 1 + \frac{900\,\text{k}\Omega}{100\,\text{k}\Omega} = 1 + 9 = 10 \Longleftarrow$$

(b)

$$G = 1 + \frac{R_f}{R_1} = 1 + \frac{750\,\text{k}\Omega}{50\,\text{k}\Omega} = 1 + 15 = 16 \Longleftarrow$$

(c)

$$G = 1 + \frac{R_f}{R_1} = 1 + \frac{1.1\,\text{M}\Omega}{100\,\text{k}\Omega} = 1 + 11 = 12 \Longleftarrow$$

4.6: For the ideal operational amplifier arrangement shown in Fig 4.6, determine the output voltage, v_o if the input voltage, v_i, is equal to 1.20 V.

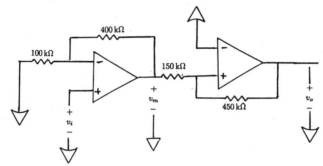

These op-amps are in the non-inverting configuration. With v_m designated as the output of the first op-amp

$$v_m = \left(1 + \frac{R_{f,1}}{R_{1,1}}\right) v_i = \left(1 + \frac{400\,\text{k}\Omega}{100\,\text{k}\Omega}\right) (1.2\,\text{V}) = (1+4)(1.2\,\text{V}) = 6\,\text{V}$$

Then, with v_m designated as the input to the second op-amp

$$v_o = \left(1 + \frac{R_{f,2}}{R_{1,2}}\right) v_i = \left(1 + \frac{450\,\text{k}\Omega}{150\,\text{k}\Omega}\right) (6\,\text{V}) = (1+3)(6\,\text{V}) = 24\,\text{V} \Leftarrow$$

4.7: Figure 4.7 shows an ideal operational amplifier connected as a *differential amplifier*. Both the inverting and non-inverting terminals are used. Determine an expression for the output voltage, v_o, as a function of v_{i1} and v_{i2} [$v_o = f(v_{i1}, v_{12})$]. Then determine $v_o = v_o = f(v_{i1}, v_{12})$ if $R_1 = R_2$.

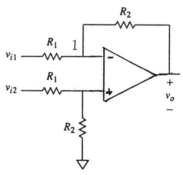

For the non-inverting terminal at node-2, KCL gives

$$\frac{v_+}{R_2} + \frac{v_+ - v_{i2}}{R_2} = 0$$

232

or

$$\left(\frac{1}{R_1} + \frac{1}{R_2}\right) v_+ = \frac{v_{i2}}{R_1}$$

so that

$$v_+ = \left(\frac{R_2}{R_1 + R_2}\right) v_{i2}$$

For the inverting terminal at node-1, KCL provides

$$\frac{v_- - v_{i1}}{R_1} + \frac{v_- - v_o}{R_2} = 0$$

or

$$\left(\frac{1}{R_1} + \frac{1}{R_2}\right) v_- = \frac{v_o}{R_2} + \frac{v_{i1}}{R_1}$$

so that v_o is given by

$$v_o = \left(\frac{R_1 + R_2}{R_1}\right) v_- - \frac{R_2}{R_1} v_{i1} = \left(1 + \frac{R_2}{R_1}\right) v_- - \frac{R_2}{R_1} v_{i1}$$

But $v_+ = v_-$ so that

$$v_o = \left(1 + \frac{R_2}{R_1}\right) \left(\frac{R_2}{R_1 + R_2}\right) v_{i2} - \frac{R_2}{R_1} v_{i1} \Leftarrow$$

If all R's are equal, the foregoing reduces to

$$v_o = v_{i2} - v_{i1} \Leftarrow$$

4.8: Figure 4.8 shows an ideal operational amplifier that can be used as a strain gage which is based on the fact that the the value resistance, ΔR, will change slightly when the resistor is bent or twisted. Determine the value of ΔR as a function of the input voltage, v_i, and the two resistances, R_a and R_b.

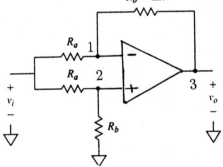

Here, node equations at points-1 and -2 give

$$\frac{v_- - v_i}{R_a} + \frac{v_- - v_o}{R_b + \Delta R} = 0$$

and

$$\frac{v_+}{R_b} + \frac{v_+ - v_i}{R_a} = 0$$

Then

$$v_- = \left(\frac{R_b + \Delta R}{R_a + R_b + \Delta R}\right) v_i + \left(\frac{R_a}{R_a + R_b + \Delta R}\right) v_o$$

or

$$v_- = \left[\frac{R_a}{R_a + R_b + \Delta R} - 1\right] v_i + \left(\frac{R_a}{R_a + R_b + \Delta R}\right) v_o$$

and

$$v_+ = \left(\frac{R_b}{R_a + R_b}\right) v_i$$

Because $v_+ = v_-$

$$\left(\frac{R_b}{R_a + R_b} + \frac{R_a}{R_a + R_b + \Delta R} - 1\right) v_i = \left(\frac{R_a}{R_a + R_b + \Delta R}\right) v_o$$

and after clearing of fractions

$$-R_a \Delta R v_i = R_a(R_a + R_b) v_o$$

or

$$v_o = -\left(\frac{\Delta R}{R_a + R_b}\right) v_i$$

and

$$\Delta R = -(R_a + R_b)\frac{v_o}{v_i} \Longleftarrow$$

4.9: Figure 4.9 shows how an ideal operational amplifier can be put together by using resistors that have relatively small resistance values. If $R_1 = 2000\,\text{Om}$, determine the value of a single feedback resistor to produce a gain of -1200 and then, with $R_1 = 2000\,\Omega$ and $R_b = 50\,\Omega$, determine the value of R_a to provide a gain of -1200.

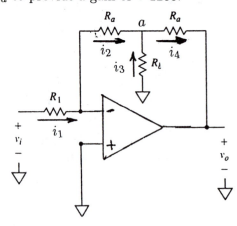

This ideal op-amp is in the inverting configuration. If $R_1 = 2000\,\Omega$ and $v_o/v_i = -1200$, then

$$R_f = -R_1 \frac{v_o}{v_i} = -(2000\,\Omega)(-1200) = 2.4 \times 10^6\,\Omega = 2.4\,\text{M}\Omega \Leftarrow$$

With $v_+ = v_- = 0$

$$i_1 = \frac{v_i - v_-}{R_1} = \frac{v_i}{R_1}$$

and because the ideal op-amp does not draw current

$$i_2 = i_1 = \frac{v_i}{R_1}$$

The voltage at node-a will be

$$v_a = v_1 - R_a i_2 = 0 - R_a i_2 = -\frac{R_a}{R_1} v_i$$

and then, noting that i_3 flows upward

$$i_3 = \frac{0 - v_a}{R_b} = \left(\frac{R_a}{R_1 R_b}\right) v_i$$

An application of KCL at node-a gives

$$i_4 = i_2 + i_3 = \frac{1}{R_1} v_i + \left(\frac{R_a}{R_1 R_b}\right) v_i = \left(\frac{R_a + R_b}{R_1 R_b}\right) v_i$$

so that

$$v_o = v_a - R_a i_4 = -\frac{R_a}{R_1} v_i - \left[\frac{R_a(R_a + R_b)}{R_1 R_b}\right] v_i$$

or

$$v_o = -\frac{R_a}{R_1} \left[1 + \frac{R_a + R_b}{R_b}\right] v_i \Leftarrow$$

If $v_o/v_i = -1200$, $R_b = 50\,\Omega$ and $R_1 = 2000\,\Omega$

$$-1200 = -\frac{R_a}{2000\,\Omega} \left[1 + \frac{R_a + 50\,\Omega}{50\,\Omega}\right]$$

and hence

$$R_a = 10{,}900\,\Omega \Leftarrow$$

Notice that four resistors with values less than $12{,}000\,\Omega$ can be used to take the place of a single feedback resistor of $2.4\,\text{M}\Omega$.

4.10: Figure 4.10 illustrates the use of an ideal operational amplifier as a *negative impedance converter*. Determine the input resistance.

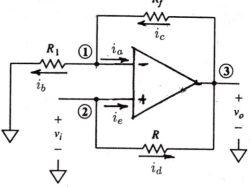

In this case, the input resistance is defined as

$$R_{\text{in}} \equiv \frac{v_i}{i_d}$$

and because the ideal op-amp does not draw current at either input terminal and because $v_+ = v_-$

$$i_b = \frac{v_i}{R_1}$$

Then, KCL at node-1 gives

$$i_c = i_b = \frac{v_i}{R_1}$$

By KVL

$$v_o = v_i + R_f i_c = v_i + \frac{R_f}{R_1} v_i = \left(1 + \frac{R_f}{R_1}\right) v_i$$

and by Ohm's law

$$i_d = \frac{v_i - v_o}{R} = \frac{v_i}{R} - \frac{1}{R}\left(1 + \frac{R_f}{R_1}\right) v_i = -\left(\frac{R_f}{RR_1}\right) v_i$$

Hence, the input resistance is

$$R_{\text{in}} = \frac{v_i}{i_d} = -\frac{R_1}{R_f} R \Leftarrow$$

4.11: An ideal operational amplifier in the inverting configuraton is to have a gain of -125 and an input resistance as high as possible. If no resistance in the op-amp circuit is to have a value higher that $5\,\text{M}\Omega$, how is this design achieved by using just two resistors.

Here R_{in} as high as possible has the implication that

$$R = R_1 = \text{high}$$

With the gain

$$G = -\frac{R_f}{R_1}$$

R_f should be set at $5\,\text{M}\Omega$. Thus

$$R_f = 5\,\text{M}\Omega$$

and

$$R_1 = -\frac{R_f}{G} = -\frac{5\,\text{M}\Omega}{-125} = 40,000\,\Omega \Leftarrow$$

4.12: In the cascade of ideal operational amplifiers shown in Fig 4.11, if $v_i = 2\,\text{V}$ and $v_o = 30\,\text{V}$, determine the value of R.

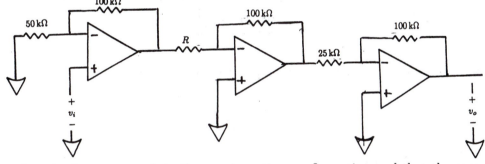

Observe that the first op-amp is in the non-inverting configuration and that the second and third op-amps are inverting op-amps. With v_{01} at the output of the first op-amp and v_{02} at the output of the second op-amp

$$\frac{v_o}{v_1} = \left(\frac{v_{o1}}{v_i}\right)\left(\frac{v_{o2}}{v_{o1}}\right)\left(\frac{v_o}{v_{o2}}\right)$$

or

$$\frac{v_o}{v_1} = \left(1 + \frac{R_{f,1}}{R_{1,1}}\right)\left(-\frac{R_{f,2}}{R_{1,2}}\right)\left(-\frac{R_{f,3}}{R_{1,3}}\right)$$

Thus

$$\frac{30\,\text{V}}{2\,\text{V}} = \left(1 + \frac{100\,\text{k}\Omega}{50\,\text{k}\Omega}\right)\left(-\frac{100\,\text{k}\Omega}{R}\right)\left(-\frac{100\,\text{k}\Omega}{25\,\text{k}\Omega}\right)$$

The two operational amplifiers are in the inverting configuration and they are summing amplifiers. With v_m as the output voltage of the first op-amp

$$
\begin{aligned}
v_m &= -\left(\frac{R_{f,1}}{R_{1,1}} + \frac{R_{f,1}}{R_{1,2}} + \frac{R_{f,1}}{R_{1,3}}\right) \\
&= -\left[\frac{1\,\text{M}\Omega}{200\,\text{k}\Omega}(2\,\text{V}) + \frac{1\,\text{M}\Omega}{500\,\text{k}\Omega}(4\,\text{V}) + \frac{1\,\text{M}\Omega}{250\,\text{k}\Omega}(3\,\text{V})\right] \\
&= -[5(2\,\text{V}) + 2(4\,\text{V}) + 4(3\,\text{V})] \\
&= -[10\,\text{V} + 8\,\text{V} + 12\,\text{V}] \\
&= -30\,\text{V}
\end{aligned}
$$

Then, for the second op-amp with v_m as one of the inputs

$$
\begin{aligned}
v_o &= -\left(\frac{R_{f,2}}{R_{2,1}} + \frac{R_{f,2}}{R_{2,2}}\right) \\
&= -\left[\frac{1\,\text{M}\Omega}{1\,\text{M}\Omega}(-30\,\text{V}) + \frac{1\,\text{M}\Omega}{500\,\text{k}\Omega}(5\,\text{V})\right] \\
&= -[1(-30\,\text{V}) + 2(5\,\text{V})] \\
&= -[-30\,\text{V} + 10\,\text{V}] \\
&= 20\,\text{V} \Leftarrow
\end{aligned}
$$

4.16: In the summing operational amplifier in Fig 4.15, $R_{1,1} = 5\,\text{k}\Omega$. Determine the values of $R_f, R_{1,2}$ and $R_{1,3}$ that will yield an output voltage of $v_o = -(4v_{i1} + 8v_{i2} + 10v_{i3})$.

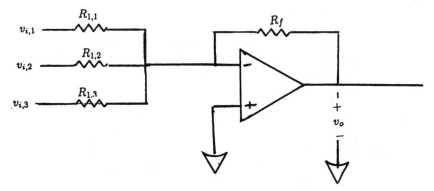

For this summing inverting op-amp, the contribution of the source, v_{i1} to the gain is $-4v_{i1}$. Hence,

$$
v_o = -\frac{R_f}{R_{1,1}} v_{i1} = -4v_{i1}
$$

This establishes R_f

$$-\frac{R_f}{R_{1,1}} = -4 \longrightarrow R_f = 4R_{1,1} = 4(5\,\text{k}\Omega) = 20\,\text{k}\Omega \Leftarrow$$

Then

$$-\frac{R_f}{R_{1,2}} = -8 \longrightarrow R_{1,2} = \frac{R_f}{8} = \frac{20\,\text{k}\Omega}{8} = 2.5\,\text{k}\Omega \Leftarrow$$

and

$$-\frac{R_f}{R_{1,3}} = -10 \longrightarrow R_{1,3} = \frac{R_f}{10} = \frac{20\,\text{k}\Omega}{10} = 2\,\text{k}\Omega \Leftarrow$$

4.17: Determine the current, i_o, and the power drawn by R_o in the ideal operational amplifier circuit of Fig 4.16.

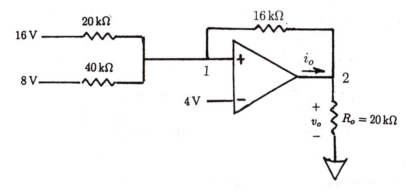

KCL at point-1 yields

$$\frac{v_+ - 16\,\text{V}}{20\,\text{k}\Omega} + \frac{v_+ - 8\,\text{V}}{40\,\text{k}\Omega} + \frac{v_+ - v_o}{16\,\text{k}\Omega} = 0$$

and with $v_+ = v_- = 4\,\text{V}$,

$$\frac{4\,\text{V} - 16\,\text{V}}{20\,\text{k}\Omega} + \frac{4\,\text{V} - 8\,\text{V}}{40\,\text{k}\Omega} + \frac{4\,\text{V} - v_o}{16\,\text{k}\Omega} = 0$$

This equation may be employed to find

$$v_o = -7.20\,\text{V}$$

The use of KCL at point-2 establishes, i_o

$$i_o = \frac{v_o}{20\,\text{k}\Omega} + \frac{v_o - 4\,\text{V}}{16\,\text{k}\Omega}$$

This result may be compared to $v_o = -20\,\text{V}$ for the ideal op amp with infinite gain.

4.20: In problem 4.19, determine the value of the gain, A to make $v_o = -18.5\,\text{V}$.

Reference to Problem 4.19 shows that the relationship between v_o and v_i is

$$\left(R_1 + \frac{R_1 + R_f}{A}\right) v_o = -R_f v_i$$

Solve this for A

$$(R_1 A + R_1 + R_f)v_o = -R_f A v_i$$
$$(R_1 v_o + R_f v_i)A = -(R_1 + R_f)v_o$$

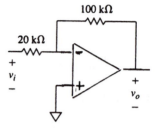

and hence

$$A = -\left[\frac{(R_1 + R_f)v_o}{R_1 v_o + R_f v_i}\right]$$

$$= -\left[\frac{(20\,\text{k}\Omega + 100\,\text{k}\Omega)(-18.5\,\text{V})}{(20\,\text{k}\Omega)(-18.5\,\text{V}) + (100\,\text{k}\Omega)(4\,\text{V})}\right]$$

$$= 74.00 \Leftarrow$$

4.21: Determine the value of R_f required in Fig 4.19 to make the outout voltage, $v_o = 16\,\text{V}$ when $R_1 = 50\,\text{k}\Omega, V_i = -4\,\text{V}$ and the amplifier has an actual gain of $A = 400$.

Refer to Problem 4.20 where it was shown that

$$A = -\left[\frac{(R_1 + R_f)v_o}{R_1 v_o + R_f v_i}\right]$$

This may be solved for R_f

$$R_1 v_o A + R_f v_i A = -R_1 v_o - R_f v_o$$
$$R_f(v_i A + v_o) = -(R_1 v_o A + R_1 v_o)$$

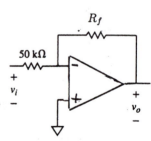

or

$$R_f = -\left[\frac{R_1 v_o(1 + A)}{v_i A + v_o}\right]$$

$$= -\left[\frac{(50\,\text{k}\Omega)(16\,\text{V})(1 + 400)}{(-4\,\text{V})(400) + 16\,\text{V}}\right]$$

$$\approx 202,500\,\Omega \Leftarrow$$

4.22: If the operational amplifier in Fig 4.20 has an actual gain of $A = 200$, determine the value of R_1 required to make $v_o = 18\,\text{V}$ with $v_i = -2\,\text{V}$.

Refer to Problem 4.20 where it was shown that

$$A = -\left[\frac{(R_1 + R_f)v_o}{R_1 v_o + R_f v_i}\right]$$

This may be solved for R_1

$$R_1 v_o A + R_f v_i A = -R_1 v_o - R_f v_o$$
$$R_1(A+1)v_o = -(R_f v_o + R_f v_i A)$$

This gives

$$R_1 = -\left[\frac{R_f(A v_i + v_o)}{(A+1)v_o}\right]$$

$$= -\left[\frac{(100\,\text{k}\Omega)[200(-2\,\text{V}) + 18\,\text{V}]}{(201)(18\,\text{V})}\right]$$

$$\approx 10,560\,\Omega \Leftarrow$$

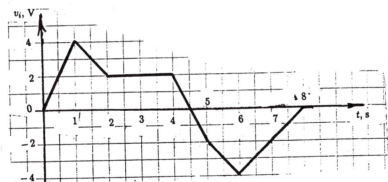

SATURATION EFFECTS

4.23: The input to an ideal operational amplifier in the inverting configuration is shown in Fig 4.21, If $R_f = 200\,\text{k}\Omega$, $R_1 = 50\,\text{k}\Omega$ and $V_{\text{sat}} = 12\,\text{V}$, determine and plot the output characteristic.

For any output voltage, $|v_o| > V_{\text{sat}}$, the operation of the op-amp is outside of the operating region. For $|v_o| = V_{\text{sat}} = 12\,\text{V}$

$$v_i = -\left(\frac{R_1}{R_f}\right)(v_o) = -\left(\frac{50\,\text{k}\Omega}{200\,\text{k}\Omega}\right)(12\,\text{V}) = -3\,\text{V}$$

The op amp is operating in the positive saturation region, Hence, $v_o = 12\,\text{V}$ is is an impossible condition and an application of KCL provides a single node equation at node-1 in the circuit at the right

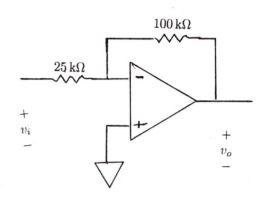

$$\frac{v_+ + 3\,\text{V}}{25\,\text{k}\Omega} + \frac{v_+ - 8\,\text{V}}{100\,\text{k}\Omega} = 0$$

$$\left(\frac{1}{25\,\text{k}\Omega} + \frac{1}{100\,\text{k}\Omega}\right)v_+ = -\frac{3\,\text{V}}{25\,\text{k}\Omega} + \frac{8\,\text{V}}{100\,\text{k}\Omega}$$

$$(5 \times 10^{-5}\,\text{℧})v_+ = -0.12\,\text{mA} + 0.08\,\text{mA}$$

or

$$v_+ = -0.80\,\text{V}$$

and with $v_- = 0$,

$$v_d = v_+ = -0.80\,\text{V}$$

and

$$v_o = 8\,\text{V} \Leftarrow$$

(c) Here

$$v_o = -\frac{R_f}{R_1}v_i = -\left(\frac{100\,\text{k}\Omega}{25\,\text{k}\Omega}\right)(4\,\text{V}) = -4(4\,\text{V}) = -16\,\text{V}$$

and

$$|16\,\text{V}| > V_{\text{sat}}$$

The op amp is operating in the negative saturation region, Hence, $v_o = 12\,\text{V}$ is an impossible condition and an application of KCL provides a single node equation at node-1 in the circuit

$$\frac{v_+ - 4\,\text{V}}{25\,\text{k}\Omega} + \frac{v_+ + 8\,\text{V}}{100\,\text{k}\Omega} = 0$$

$$\left(\frac{1}{25\,\text{k}\Omega} + \frac{1}{100\,\text{k}\Omega}\right)v_+ = \frac{4\,\text{V}}{25\,\text{k}\Omega} - \frac{8\,\text{V}}{100\,\text{k}\Omega}$$

$$(5 \times 10^{-5}\,\text{℧})v_+ = 0.16\,\text{mA} - 0.08\,\text{mA}$$

or

$$v_+ = 1.6\,\text{V}$$

and with $v_- = 0$,

$$v_d = v_+ = -1.6\,\text{V} \qquad \text{and} \qquad v_o = -8\,\text{V} \Leftarrow$$

SELECTION OF OPERATIONAL AMPLIFIER COMPONENTS

4.27: Design an ideal operational amplifier circuit to have the general input-output relationship

$$v_o = -(8v_{a1} + 4v_{a2} + 2v_{a3}) + (3v_{b1} + 6v_{b2} + 9v_{a3})$$

Use Fig 4.22 and set $G_f = 1\,\mho$. Then

$$G_{a1} = 8\,\mho$$
$$G_{a2} = 4\,\mho$$
$$G_{a3} = 2\,\mho$$
$$\sum_k G_{ak} = 14\,\mho$$

and

$$G_{b1} = 3\,\mho$$
$$G_{b2} = 6\,\mho$$
$$G_{b2} = 9\,\mho$$
$$\sum_k G_{bk} = 18\,\mho$$

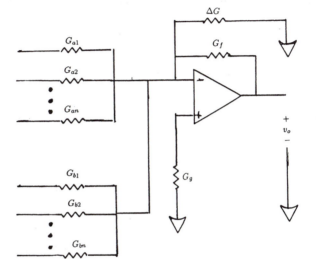

Thus,

$$\delta = \sum_k G_{ak} - \sum_k G_{bk}$$
$$= 14\,\mho - 19\,\mho$$
$$= -5\,\mho$$

Because $\delta < 0$, set $G_g = 1\,\mho$ and

$$\Delta G = |\delta| + G_g = 5\,\mho + 1\,\mho = 6\,\mho$$

Then choose the scaling factors, $K_a = K_b = 10^{-6}$ and obtain

$$G_f = 10^{-6}\,\mho \quad \longrightarrow \quad R_f = 100\,\mathrm{k\Omega}$$
$$\Delta G = 6 \times 10^{-6}\,\mho \quad \longrightarrow \quad \Delta R = 16.67\,\mathrm{k\Omega}$$
$$G_g = 10^{-6}\,\mho \quad \longrightarrow \quad R_g = 100\,\mathrm{k\Omega}$$
$$G_{a1} = 8 \times 10^{-6}\,\mho \quad \longrightarrow \quad R_{a1} = 12.5\,\mathrm{k\Omega}$$
$$G_{a2} = 4 \times 10^{-6}\,\mho \quad \longrightarrow \quad R_{a2} = 25\,\mathrm{k\Omega}$$
$$G_{a3} = 2 \times 10^{-6}\,\mho \quad \longrightarrow \quad R_{a3} = 50\,\mathrm{k\Omega}$$
$$G_{b1} = 3 \times 10^{-6}\,\mho \quad \longrightarrow \quad R_{b1} = 33.33\,\mathrm{k\Omega}$$
$$G_{b2} = 6 \times 10^{-6}\,\mho \quad \longrightarrow \quad R_{b2} = 16.67\,\mathrm{k\Omega}$$
$$G_{b3} = 9 \times 10^{-6}\,\mho \quad \longrightarrow \quad R_{b3} = 11.11\,\mathrm{k\Omega}$$

Because $\delta > 0$, set $G_g = \delta = 8\,\mho$ and $\Delta G = 0$. Then choose the scaling factors, $K_a = K_b = 10^{-6}$ and obtain

$$
\begin{aligned}
G_f &= 10^{-6}\,\mho &\longrightarrow&& R_f &= 100\,\text{k}\Omega \\
G_g &= 8 \times 10^{-6}\,\mho &\longrightarrow&& R_g &= 12.5\,\text{k}\Omega \\
G_{a1} &= 4 \times 10^{-6}\,\mho &\longrightarrow&& R_{a1} &= 25\,\text{k}\Omega \\
G_{a2} &= 2 \times 10^{-6}\,\mho &\longrightarrow&& R_{a2} &= 50\,\text{k}\Omega \\
G_{a3} &= 1 \times 10^{-6}\,\mho &\longrightarrow&& R_{a3} &= 100\,\text{k}\Omega \\
G_{a4} &= 8 \times 10^{-6}\,\mho &\longrightarrow&& R_{a4} &= 12.5\,\text{k}\Omega \\
G_{b1} &= 5 \times 10^{-6}\,\mho &\longrightarrow&& R_{b1} &= 20\,\text{k}\Omega \\
G_{b2} &= 2 \times 10^{-6}\,\mho &\longrightarrow&& R_{b2} &= 50\,\text{k}\Omega
\end{aligned}
$$

An ideal operational amplifier circuit to achieve the given input-output relationship is shown here.

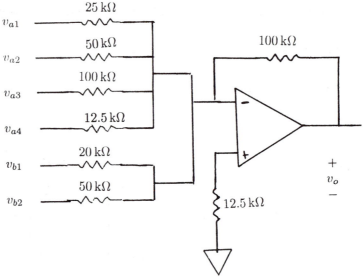

4.30: Use the ideal operational amplifier circuit developed in Problem 4.27 to determine the output voltage if all input voltages are taken at a nominal 1 V.

Here for the inverting input

$$
v_{oa} = \sum_k \left(\frac{R_{f,k}}{R_{1,k}} \right) (v_{a,k})
$$

and with $k = 4$ and $v_{a,k} = 1\,\text{V}$

$$
v_{oa} = - \left(\frac{100\,\text{k}\Omega}{25\,\text{k}\Omega} + \frac{100\,\text{k}\Omega}{50\,\text{k}\Omega} + \frac{100\,\text{k}\Omega}{100\,\text{k}\Omega} + \frac{100\,\text{k}\Omega}{12.5\,\text{k}\Omega} \right)
$$

or

$$v_{oa} = -\left(4\,\mathrm{V} + 2\,\mathrm{V} + 1\,\mathrm{V} + 8\,\mathrm{V}\right) = -15\,\mathrm{V}$$

and for the non-inverting input

$$v_{ob} = \sum_k \left(\frac{R_{f,j}}{R_{1,j}}\right)(v_{b,k})$$

and with $j = 2$ and $v_{b,k} = 1\,\mathrm{V}$

$$v_{ob} = \left(\frac{100\,\mathrm{k\Omega}}{20\,\mathrm{k\Omega}} + \frac{100\,\mathrm{k\Omega}}{50\,\mathrm{k\Omega}}\right)$$

or

$$v_{ob} = \left(5\,\mathrm{V} + 2\,\mathrm{V}\right) = 7\,\mathrm{V}$$

Then

$$v_o = v_{oa} + v_{ob} = -15\,\mathrm{V} + 7\,\mathrm{V} = -8\,\mathrm{V}$$

and if all inputs in the input-output relationship of Problem 4.29

$$v_o = -\left(8v_{a1} + 4v_{a2} + 2v_{a3}\right) + \left(3v_{b1} + 6v_{b2} + 9v_{a3}\right)$$

are taken at $1\,\mathrm{V}$, this is seen to complete the verification of the op amp circuit developed in Problem 4.29.

$$i_1 = a(cv_1) + b(cv_1)^2 = acv_1 + c^2b(v_1)^2 \neq i_1$$

the device is nonlinear $\Leftarrow$

5.5: Consider a diode whose voltage-current relationship is given by

$$i = I(e^{24v} - 2)$$

Is the diode linear?
With

$$i_1 = I(e^{24v_1} - 2) \qquad \text{and} \qquad i_2 = I(e^{24v_2} - 2)$$

if $i_3 = i_1 + i_2$ is due to $v_3 = v_1 + v_2$

$$i_3 = I(e^{24(v_2+v_3)} - 2)$$

which is not equal to

$$i_3 = i_1 + i_2 = I(e^{24v_1} - 2) + I(e^{24v_2} - 2)$$

the diode is nonlinear $\Leftarrow$

PROPORTIONALITY

5.6: In the network of Fig 5.1, the resistance designators correspond to the resistance values. By assuming that $I_4 = 1\,\text{A}$, determine I_3, I_4, V_{12} and V_{14}. (This is a repeat of Problem 2.28).

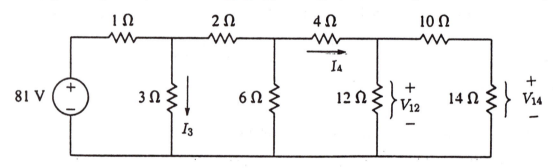

If $I_{14} = 1\,\text{A}$

$$V_{14} = R_{14}I_{14} = (14\,\Omega)(1\,\text{A}) = 14\,\text{V}$$

and because $I_{10} = I_{14} = 1\,\text{A}$

$$V_{10} = R_{10}I_{10} = (10\,\Omega)(1\,\text{A}) = 10\,\text{V}$$

By KVL
$$V_{12} = V_{10} + V_{14} = 10\,\text{V} + 14\,\text{V} = 24\,\text{V}$$

and
$$I_{12} = \frac{V_{12}}{R_{12}} = \frac{24\,\text{V}}{12\,\Omega} = 2\,\text{A}$$

By KCL
$$I_4 = I_{12} + I_{10} = 2\,\text{A} + 1\,\text{A} = 3\,\text{A}$$

so that
$$V_4 = R_4 I_4 = (4\,\Omega)(3\,\text{A}) = 12\,\text{V}$$

and by KVL
$$V_6 = V_4 + V_{12} = 12\,\text{V} + 24\,\text{V} = 36\,\text{V}$$

so that
$$I_6 = \frac{V_6}{R_6} = \frac{36\,\text{V}}{6\,\Omega} = 6\,\text{A}$$

By KCL
$$I_2 = I_6 + I_4 = 6\,\text{A} + 3\,\text{A} = 9\,\text{A}$$

so that
$$V_2 = R_2 I_2 = (2\,\Omega)(9\,\text{A}) = 18\,\text{V}$$

and by KVL
$$V_3 = V_2 + V_6 = 18\,\text{V} + 36\,\text{V} = 54\,\text{V}$$

Finally
$$I_3 = \frac{V_3}{R_3} = \frac{54\,\text{V}}{3\,\Omega} = 18\,\text{A}$$

and by KCL
$$I_1 = I_3 + I_2 = 18\,\text{A} + 9\,\text{A} = 27\,\text{A}$$

Under the assumption that $I_{14} = 1\,\text{A}$, the source voltage would be

$$V_s = R_1 I_1 + V_3 = (1\,\Omega)(27\,\text{A}) + 54\,\text{V} = 81\,\text{V}$$

However, the source voltage is 81 V and no adjustment need be made to the values developed for the currents and voltages. Hence

$$
\begin{aligned}
V_{14} &= 14\,\text{V} \Leftarrow \\
V_{12} &= 24\,\text{V} \Leftarrow \\
I_3 &= 18\,\text{A} \Leftarrow
\end{aligned}
$$

By KVL
$$V_6 = V_2 + V_1 = 2\,\text{V} + 1\,\text{V} = 3\,\text{V}$$

and
$$I_6 = \frac{V_6}{R_6} = \frac{3\,\text{V}}{6\,\Omega} = \frac{1}{2}\,\text{A}$$

KCL gives
$$I_4 = I_6 + I_2 = \frac{1}{2}\,\text{A} + 1\,\text{A} = \frac{3}{2}\,\text{A}$$

so that
$$V_4 = R_4 I_4 = (4\,\Omega)\left(\frac{3}{2}\,\text{A}\right) = 6\,\text{V}$$

and by KVL
$$V_{12} = V_4 + V_6 = 6\,\text{V} + 3\,\text{V} = 9\,\text{V}$$

and
$$I_{12} = \frac{V_{12}}{R_{12}} = \frac{9\,\text{V}}{12\,\Omega} = \frac{3}{4}\,\text{A}$$

Under the assumption that $V_1 = 1\,\text{V}$, the source current would be

$$I_s = I_4 + I_{12} = \frac{3}{2}\,\text{A} + \frac{3}{4}\,\text{V} = \frac{9}{4}\,\text{V}$$

However, the source current is $6\,\text{A}$ so that all values developed for the currents and voltages must be multiplied by the factor
$$F = \frac{6\,\text{A}}{9/4\,\text{A}} = \frac{8}{3}$$

Hence the actual value of I_4 will be

$$I_4 = \left(\frac{8}{3}\right)\left(\frac{3}{2}\,\text{A}\right) = 4\,\text{A} \Leftarrow$$

the actual value of I_6 will be
$$I_6 = \left(\frac{8}{3}\right)\left(\frac{1}{2}\,\text{A}\right) = \frac{4}{3}\,\text{A} \Leftarrow$$

and the actual value of V_2 will be

$$V_2 = \left(\frac{8}{3}\right)(2\,\text{V}) = \frac{16}{3}\,\text{V} \Leftarrow$$

5.9: In the network of Fig 5.4, the resistance designators correspond to the resistance values. By assuming that $I_{10} = 1\,\text{A}$, determine I_2, I_4 and V_8.

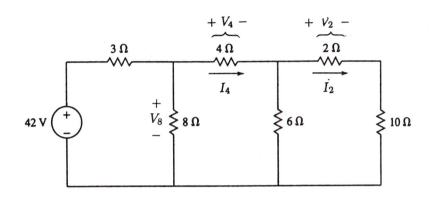

If $I_{10} = I_2 = 1\,\text{A}$

$$V_{10} = R_{10}I_{10} = (10\,\Omega)(1\,\text{A}) = 10\,\text{V}$$

and

$$V_2 = R_2I_2 = (2\,\Omega)(1\,\text{A}) = 2\,\text{V}$$

By KVL

$$V_6 = V_2 + V_{10} = 2\,\text{V} + 10\,\text{V} = 12\,\text{V}$$

and

$$I_6 = \frac{V_6}{R_6} = \frac{12\,\text{V}}{6\,\Omega} = 2\,\text{A}$$

By KCL

$$I_4 = I_6 + I_2 = 2\,\text{A} + 1\,\text{A} = 3\,\text{A}$$

and then

$$V_4 = R_4I_4 = (4\,\Omega)(3\,\text{A}) = 12\,\text{V}$$

and by KVL

$$V_8 = V_4 + V_6 = 12\,\text{V} + 12\,\text{V} = 24\,\text{V}$$

so that

$$I_8 = \frac{V_8}{R_8} = \frac{24\,\text{V}}{8\,\Omega} = 3\,\text{A}$$

Finally by KCL

$$I_3 = I_8 + I_4 = 3\,\text{A} + 3\,\text{A} = 6\,\text{A}$$

and

$$V_3 = R_3I_3 = (3\,\Omega)(6\,\text{A}) = 18\,\text{V}$$

and the actual value of V_7 will be

$$V_7 = \left(\frac{9}{5}\right)(7\,\text{V}) = \frac{63}{5}\,\text{V} \Leftarrow$$

These values check the values of Problem 2.30.

SOURCE TRANSFORMATION PROBLEMS

5.11: Transform the ideal voltage sources shown in Fig 5.6 to ideal current sources.

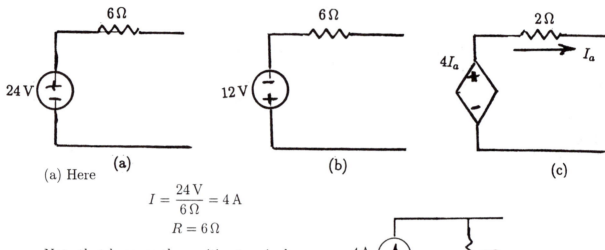

(a) Here

$$I = \frac{24\,\text{V}}{6\,\Omega} = 4\,\text{A}$$

$$R = 6\,\Omega$$

Note that because the positive terminal of the voltage source is at its top, the current leaves the current source in an upward direction.

(b) Here

$$I = \frac{12\,\text{V}}{6\,\Omega} = 2\,\text{A}$$

$$R = 6\,\Omega$$

Note that because the positive terminal of the voltage source is at its bottom, the current leaves the current source in a downward direction.

(c) Here

$$I = \frac{4I_a}{2\,\Omega} = 2I_a\,\text{A}$$

$$R = 2\,\Omega$$

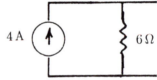

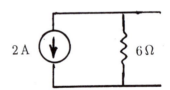

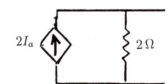

Note that because the positive terminal of the voltage source is at its top, the current leaves the current source in an upward direction.

5.12: Transform the ideal current sources shown in Fig 5.7 to ideal voltage sources.

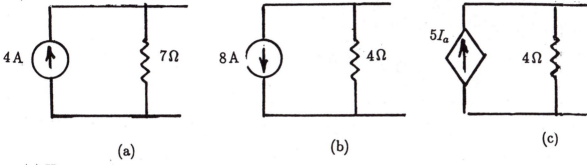

(a)   (b)   (c)

(a) Here

$$V = RI = (7\,\Omega)(4\,\text{A}) = 28\,\text{V}$$

$$R = 7\,\Omega$$

Note that because the current leaves the source in an upward direction, the positive terminal of the voltage source is at its top.

(b) Here

$$V = RI = (4\,\Omega)(8\,\text{A}) = 32\,\text{V}$$

$$R = 4\,\Omega$$

Note that because the current leaves the source in a downward direction, the positive terminal of the voltage source is at its bottom.

(c) Here

$$V = RI = (4\,\Omega)(5I_a) = 20I_a\,\text{V}$$

$$R = 4\,\Omega$$

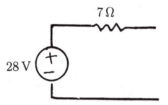

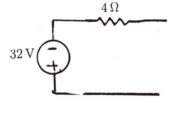

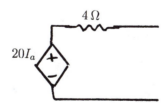

Note that because the current leaves the source in an upward direction, the positive terminal of the voltage source is at its top.

5.13: Transform the network of ideal voltage and current sources in Fig 5.8 to a single ideal voltage source.

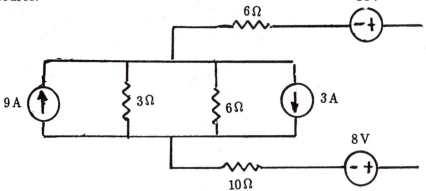

The two current sources are in parallel. Hence they have an equivalent

$$I = I_1 + I_2 = 9\,\text{A} - 3\,\text{A} = 6\,\text{A}$$

with

$$R = \frac{(3\,\Omega)(6\,\Omega)}{3\,\Omega + 6\,\Omega} = \frac{18\,\Omega^2}{9\,\Omega} = 2\,\Omega$$

Thus, the equivalent current source can be transformed to a voltage source

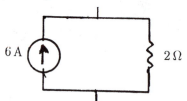 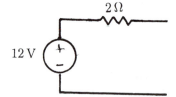

With the foregoing adjustment, the picture is shown at the right. Because the three voltage sources are in series, they have an equivalent

$$
\begin{aligned}
V &= V_1 + V_2 + V_3 \\
&= 14\,\text{V} - 8\,\text{V} + 12\,\text{V} \\
&= 18\,\text{V}
\end{aligned}
$$

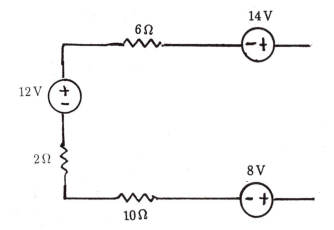

266

and

$$R = R_1 + R_2 + R_3$$
$$= 6\,\Omega + 10\,\Omega + 2\,\Omega$$
$$= 18\,\Omega$$

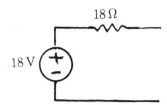

The final result is shown at the right.

5.14: Transform the network of ideal voltage sources, ideal current sources and resistors in Fig 5.9 to a single ideal voltage source.

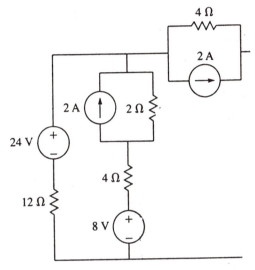

First transform the two current sources to voltage sources, This is shown at the right and involves the transformations.

$$V_1 = RI$$
$$= (2\,\Omega)(2\,\text{A})$$
$$= 4\,\text{V}$$

and

$$V_2 = RI$$
$$= (4\,\Omega)(2\,\text{A})$$
$$= 8\,\text{V}$$

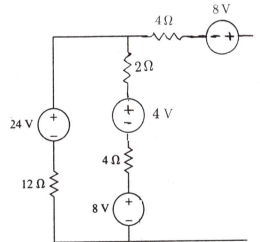

There are three current sources not associated with a voltage source. These may be transformed first.

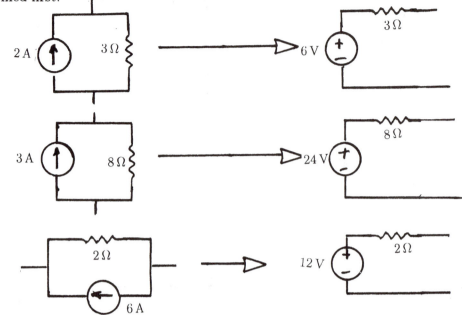

and when these are put into the original network of sources, the result is

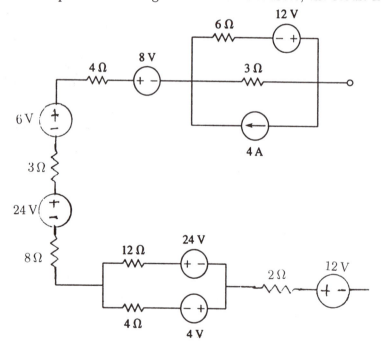

There are three voltage sources which are either in parallel or associated with a current source. These can be transformed to

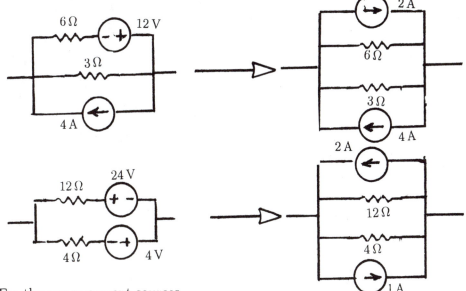

For the upper current sources,

$$I_u = 4\,\text{A} - 2\,\text{A} = 2\,\text{A} \qquad \text{and} \qquad R_u = \frac{(6\,\Omega)(3\,\Omega)}{6\,\Omega + 3\,\Omega} = \frac{18\,\Omega^2}{9\,\Omega} = 2\,\Omega$$

where I_u is from *right to left* and for the lower current sources,

$$I_\ell = 2\,\text{A} - 1\,\text{A} = 1\,\text{A} \qquad \text{and} \qquad R_\ell = \frac{(12\,\Omega)(4\,\Omega)}{12\,\Omega + 4\,\Omega} = \frac{48\,\Omega^2}{16\,\Omega} = 3\,\Omega$$

where I_ℓ is also from right to left. Use of these leads to an equivalent which can then be transformed to a voltage source

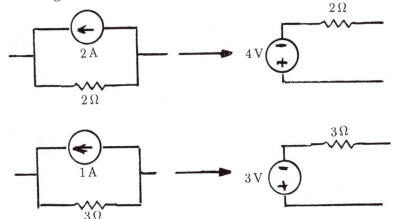

A summary of all of the foregoing along with the final equivalent voltage source is shown.

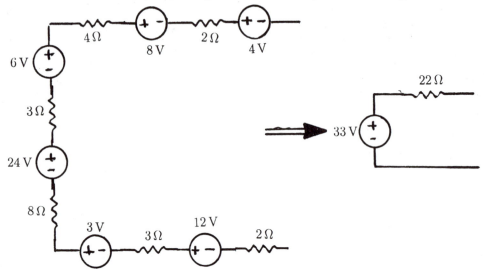

SUPERPOSITION PROBLEMS

5.16: Use superposition to determine the value of the current, I, in the network of Fig 5.11.

Note that the resistance designators correspond to the resistance values. Using superposition, the current, I, will be the sum of two currents

$$I = I_V + I_C$$

where I_V is the current due to the voltage source acting alone and I_C is the current due to the current source acting alone.

To determine I_V, remove the current source and replace it with an open circuit as shown at the right. The equivalent resistance presented to the voltage source will be

$$R_{eq} = R_a + 3\,\Omega$$

where

$$R_a = \frac{(4\,\Omega + 2\,\Omega)(6\,\Omega)}{2\,\Omega + 4\,\Omega + 6\,\Omega} = 3\,\Omega$$

Thus

$$R_{eq} = 3\,\Omega + 3\,\Omega = 6\,\Omega$$

and

$$I_3 = \frac{V_s}{6\,\Omega} = \frac{24\,\text{V}}{6\,\Omega} = 4\,\text{A}$$

By a current divider

$$I_V = \left(\frac{6\,\Omega}{6\,\Omega + 6\,\Omega}\right)(4\,\text{A}) = \left(\frac{1}{2}\right)(4\,\text{A}) = 2\,\text{A}$$

To determine I_C, remove the voltage source and replace it with a short circuit as shown at the right. The resistance of the $2\,\Omega$, $3\,\Omega$ and $4\,\Omega$ resistors can be represented by R_b

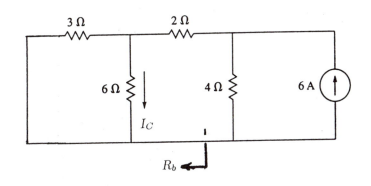

$$\begin{aligned} R_b &= \frac{(6\,\Omega)(3\,\Omega)}{6\,\Omega + 3\,\Omega} + 2\,\Omega \\ &= \frac{18\,\Omega^2}{9\,\Omega} + 2\,\Omega \\ &= 2\,\Omega + 2\,\Omega \\ &= 4\,\Omega \end{aligned}$$

Then by a pair of current dividers

$$I_2 = \left(\frac{4\,\Omega}{4\,\Omega + 4\,\Omega}\right) I_s = \left(\frac{1}{2}\right)(6\,\text{A}) = 3\,\text{A}$$

and

$$I_C = \left(\frac{3\,\Omega}{3\,\Omega + 6\,\Omega}\right)(3\,\text{A}) = \left(\frac{1}{3}\right)(3\,\text{A}) = 1\,\text{A}$$

Hence

$$I = I_V + I_C = 2\,\text{A} + 1\,\text{A} = 3\,\text{A} \Leftarrow$$

5.17: Use superposition to determine the value of the current, I, in the network of Fig 5.12.

Note that the resistance designators correspond to the resistance values. Using superposition, the current, I, will be the sum of two currents

$$I = I_{V1} + I_{V2}$$

where I_{V1} is the current due to the 24 V source acting alone and I_{V2} is the current due to the 16 V source acting alone.

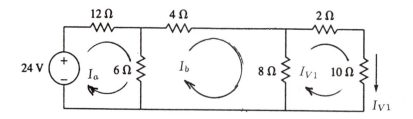

To determine I_{V1}, remove the 16 V source and replace it with a short circuit as shown in the adjusted circuit diagram. Because there are no current sources or controlled sources, the mesh equations for the three mesh currents, I_a, I_b and $I_c = I_{V1}$ can be written by inspection.

$$
\begin{array}{rrrr}
18I_a & -6I_b & & = & 24\,\text{V} \\
-6I_a & +18I_b & -8I_{V1} & = & 0 \\
& -8I_b & +20I_{V1} & = & 0
\end{array}
$$

and these may be written in matrix form as

$$
\begin{bmatrix} 18 & -6 & 0 \\ -6 & 18 & -8 \\ 0 & -8 & 20 \end{bmatrix}
\begin{bmatrix} I_a \\ I_b \\ I_{V1} \end{bmatrix} =
\begin{bmatrix} 24\,\text{V} \\ 0 \\ 0 \end{bmatrix}
$$

A matrix inversion gives

$$
\begin{bmatrix} I_a \\ I_b \\ I_{V1} \end{bmatrix} =
\begin{bmatrix} 18 & -6 & 0 \\ -6 & 18 & -8 \\ 0 & -8 & 20 \end{bmatrix}^{-1}
\begin{bmatrix} 24\,\text{V} \\ 0 \\ 0 \end{bmatrix}
$$

or

$$
\begin{bmatrix} I_a \\ I_b \\ I_{V1} \end{bmatrix} =
\begin{bmatrix} 0.0642 & 0.0260 & 0.0104 \\ 0.0260 & 0.0781 & 0.0313 \\ 0.0104 & 0.0313 & 0.0625 \end{bmatrix}
\begin{bmatrix} 24\,\text{V} \\ 0 \\ 0 \end{bmatrix} =
\begin{bmatrix} 1.542\,\text{A} \\ 0.625\,\text{A} \\ 0.250\,\text{A} \end{bmatrix}
$$

and

$$
I_{V1} = \frac{1}{4}\,\text{A}
$$

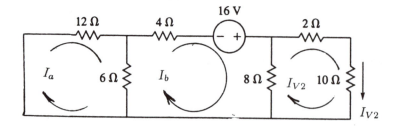

To determine I_{V2}, remove the 24 V source and replace it with a short circuit as shown in the adjusted circuit diagram. The mesh equations here will be the same as those for the case of the 24 V source except that the 16 V appears as the second entry of the column vector containing the voltage sources. In a form ready for a matrix inversion with I_c now equal to I_{V2}

$$\begin{bmatrix} I_a \\ I_b \\ I_{V2} \end{bmatrix} = \begin{bmatrix} 18 & -6 & 0 \\ -6 & 18 & -8 \\ 0 & -8 & 20 \end{bmatrix}^{-1} \begin{bmatrix} 0 \\ 16\text{ V} \\ 0 \end{bmatrix}$$

or

$$\begin{bmatrix} I_a \\ I_b \\ I_{V1} \end{bmatrix} = \begin{bmatrix} 0.0642 & 0.0260 & 0.0104 \\ 0.0260 & 0.0781 & 0.0313 \\ 0,0104 & 0.0313 & 0.0625 \end{bmatrix} \begin{bmatrix} 0 \\ 16\text{ V} \\ 0 \end{bmatrix} = \begin{bmatrix} 0.417\text{ A} \\ 1.250\text{ A} \\ 0.500\text{ A} \end{bmatrix}$$

$$I_{V2} = \frac{1}{2}\text{ A}$$

The current I is equal to

$$I = I_{V1} + I_{V2} = \frac{1}{4}\text{ A} + \frac{1}{2}\text{ A} = \frac{3}{4}\text{ A} \Leftarrow$$

5.18: Use superposition to determine the value of the current, I, in the network of Fig 5.13.

Note that the resistance designators correspond to the resistance values. Using superposition, the current, I, will be the sum of three currents

$$I = I_{C1} + I_{C2} + I_V$$

where I_{C1} is the current due to the 6 A source acting alone, I_{C2} is the current due to the 8 A source acting alone and I_V is the current due to the 24 V source acting alone.

To determine I_{C1}, remove the 8 A source replacing it with an open circuit and remove the 24 V source replacing it with a short circuit as shown in the circuit diagram at the right. Here

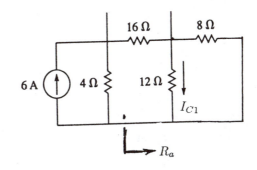

$$\begin{aligned} R_a &= \frac{(8\,\Omega)(12\,\Omega)}{8\,\Omega + 12\,\Omega} + 16\,\Omega \\ &= \frac{96\,\Omega^2}{20\,\Omega} + 16\,\Omega \\ &= 4.8\,\Omega + 16\,\Omega \\ &= 20.8\,\Omega \end{aligned}$$

Then two current dividers give

$$I_a = \left(\frac{4\,\Omega}{4\,\Omega + 20.8\,\Omega}\right)(6\,\text{A}) = (0.161)(6\,\text{A}) = 0.968\,\text{A}$$

and

$$I_{C1} = \left(\frac{8\,\Omega}{8\,\Omega + 12\,\Omega}\right)(0.968\text{A}) = (0.40)(0.968\,\text{A}) = 0.387\,\text{A}$$

To determine I_{C2}, remove the 6 A source replacing it with an open circuit and remove the 24 V source replacing it with short circuit as shown in the adjusted circuit diagram at the right. Here

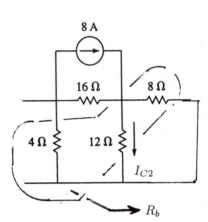

$$
\begin{aligned}
R_b &= \frac{(8\,\Omega)(12\,\Omega)}{8\,\Omega + 12\,\Omega} + 4\,\Omega \\
&= \frac{96\,\Omega^2}{20\,\Omega} + 4\,\Omega \\
&= 4.8\,\Omega + 4\,\Omega \\
&= 8.8\,\Omega
\end{aligned}
$$

and again, two current dividers give

$$I_b = \left(\frac{16\,\Omega}{8.8\,\Omega + 16\,\Omega}\right)(8\,\text{A}) = (0.645)(8\,\text{A}) = 5.161\,\text{A}$$

and

$$I_{C2} = \left(\frac{8\,\Omega}{8\,\Omega + 12\,\Omega}\right)(5.161\text{A}) = (0.40)(5.161\,\text{A}) = 2.065\,\text{A}$$

For I_V, delete both current sources and replace them with open circuits. Here

$$
\begin{aligned}
R_c &= \frac{(16\,\Omega + 4\,\Omega)(12\,\Omega)}{16\,\Omega + 4\,\Omega + 12\,\Omega} \\
&= \frac{240\,\Omega^2}{32\,\Omega} \\
&= 7.5\,\Omega
\end{aligned}
$$

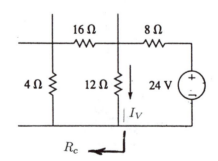

Then, by a voltage divider

$$V_{12} = \left(\frac{7.5\,\Omega}{7.5\,\Omega + 8\,\Omega} \right)(24\,\text{V}) = (0.484)(24\,\text{V}) = 11.613\,\text{V}$$

and

$$I_V = \frac{11.613\,\text{V}}{12\,\Omega} = 0.968\,\text{A}$$

Hence, the current I is equal to

$$\begin{aligned} I &= I_{C1} + I_{C2} + I_V \\ &= 0.387\,\text{A} + 2.065\,\text{A} + 0.968\,\text{A} \\ &= 3.420\,\text{A} \Leftarrow \end{aligned}$$

5.19: Use superposition to determine the value of the current, I, in the network of Fig 5.14.

Note that the network has been redrawn with the current source transformed to a voltage source and that the resistance designators correspond to the resistance values. Using superposition, the current, I, will be the sum of two currents

$$I = I_{V1} + I_{V2}$$

where I_{V1} is the current due to the 16 V source acting alone and, I_{V2} is the current due to the 24 V source acting alone.

To determine I_{V1}, remove the 24 V source and replace it with a short circuit. With

$$2V_a = 2(4I_a) = 8I_a$$

two mesh equations can be written by two applications of KVL

$$6(I_a - I_b) + 4I_a + 8I_a + 8I_a = 0$$

$$6(I_b - I_a) - 16\,\text{V} + 2I_b + 1I_b = 0$$

and after simplification, these can be written in matrix form with $I_b = I_{V1}$

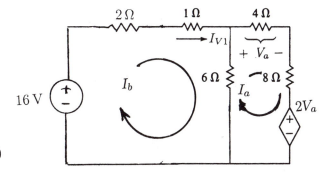

$$\begin{bmatrix} 26 & -6 \\ -6 & 9 \end{bmatrix} \begin{bmatrix} I_a \\ I_{V1} \end{bmatrix} = \begin{bmatrix} 0 \\ 16\,\text{V} \end{bmatrix}$$

The voltage sought will be

$$V = V_{V1} + V_{V2} = 45\,\text{V} + (-\,100\,\text{V}) = 45\,\text{V} - 100\,\text{V} = -55\,\text{V} \Leftarrow$$

5.21: Use superposition to determine the value of the current, I, in the network of Fig 5.16.

Using superposition, the current, I, will be the sum of two currents

$$I = I_{V1} + I_{V2}$$

where V_{V1} is the current due to the 40 V source acting alone and, I_{V2} is the current due to the 16 V source acting alone.

To determine I_{V1}, remove the 16 V source and replace it with a short circuit.

$$R_a = \frac{(16\,\Omega)(48\,\Omega)}{16\,\Omega + 48\,\Omega} = \frac{768\,\Omega^2}{64\,\Omega} = 12\,\Omega$$

$$
\begin{aligned}
R_b &= 12\,\Omega + 8\,\Omega \\
&= 20\,\Omega \\
R_c &= \frac{(10\,\Omega)(20\,\Omega)}{10\,\Omega + 20\,\Omega} \\
&= \frac{200\,\Omega^2}{30\,\Omega} \\
&= \frac{20}{3}\,\Omega
\end{aligned}
$$

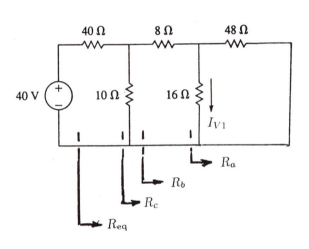

and

$$
\begin{aligned}
R_{\text{eq}} &= 40\,\Omega + \frac{20}{3}\,\Omega \\
&= \frac{140}{3}\,\Omega
\end{aligned}
$$

Then

$$I_{40} = \frac{V_s}{R_{\text{eq}}} = \frac{40\,\text{V}}{140/3\,\Omega} = \frac{6}{7}\,\text{A}$$

and two current dividers give

$$I_8 = \left(\frac{10\,\Omega}{10\,\Omega + 20\,\Omega}\right)\left(\frac{6}{7}\,\text{A}\right) = \left(\frac{1\,\Omega}{3\,\Omega}\right)\left(\frac{6}{7}\,\text{A}\right) = \frac{2}{7}\,\text{A}$$

and

$$I_{V1} = \left(\frac{48\,\Omega}{48\,\Omega + 16\,\Omega}\right) I_8 = \left(\frac{3\,\Omega}{4\,\Omega}\right)\left(\frac{2}{7}\,\text{A}\right) = \frac{3}{14}\,\text{A}$$

To determine I_{V2}, remove the $40\,\text{V}$ source replacing it with a short circuit. Here

$$
\begin{aligned}
R_a &= \frac{(10\,\Omega)(40\,\Omega)}{10\,\Omega + 40\,\Omega} \\
&= \frac{400\,\Omega^2}{50\,\Omega} = 8\,\Omega \\
R_b &= 8\,\Omega + 8\,\Omega = 16\,\Omega \\
R_c &= \left(\frac{1}{2}\right)(16\,\Omega) = 8\,\Omega
\end{aligned}
$$

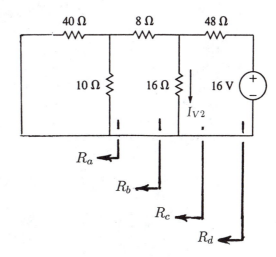

and

$$
\begin{aligned}
R_{eq} &= 8\,\Omega + 48\,\Omega \\
&= 56\,\Omega
\end{aligned}
$$

Then

$$I_{48} = \frac{V_s}{R_{eq}} = \frac{16\,\text{V}}{56\,\Omega} = \frac{2}{7}\,\text{A}$$

and by a current divider

$$I_{V2} = \left(\frac{16\,\Omega}{16\,\Omega + 16\,\Omega}\right) I_{48} = \left(\frac{1}{2}\right)\left(\frac{2}{7}\,\text{A}\right) = \frac{1}{7}\,\text{A}$$

Thus

$$I = I_{V1} + I_{V2} = \frac{3}{14}\,\text{A} + \frac{1}{7}\,\text{A} = \frac{5}{14}\,\text{A} \Leftarrow$$

5.22: Use superposition to determine the value of the voltage, V, in the network of Fig 5.17.

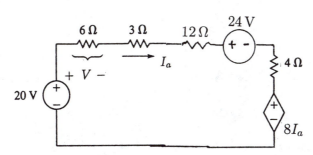

Note that the network has been redrawn with the current source transformed to a voltage source and that the resistance designators correspond to the resistance values. Using superposition, the voltage, V, will be the sum of two voltages

$$V = V_{V1} + V_{V2}$$

where V_{V1} is the current due to the 20 V source acting alone and, V_{V2} is the current due to the 24 V source acting alone.

To determine V_{V1}, remove the 24 V source and replace it with a short circuit. a single application of KVL provides

$$\begin{aligned}
(6 + 3 + 12 + 4)I_a &= 20\,\text{V} - 8I_a \\
33I_a &= 20\,\text{V} \\
I_a &= \frac{20\,\text{V}}{33\,\Omega} \\
&= \frac{20}{33}\,\text{A}
\end{aligned}$$

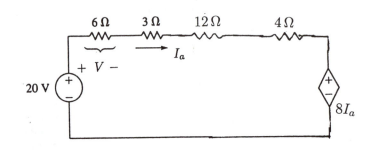

and

$$V_{V1} = R_6 I_a = (6\,\Omega)\left(\frac{20}{33}\,\text{A}\right) = \frac{40}{11}\,\text{V}$$

For V_{V2}, remove the 20 V source and replace it with a short circuit. A single application of KVL provides

$$\begin{aligned}
(6 + 3 + 12 + 4)I_a &= -24\,\text{V} - 8I_a \\
33I_a &= -24\,\text{V} \\
I_a &= -\frac{24\,\text{V}}{33\,\Omega} \\
&= -\frac{8}{11}\,\text{A}
\end{aligned}$$

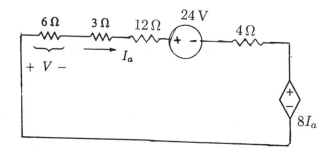

282

and

$$V_{V2} = R_6 I_a = (6\,\Omega)\left(-\frac{8}{11}\,\text{A}\right) = -\frac{48}{11}\,\text{V}$$

Hence

$$
\begin{aligned}
V &= V_{V1} + V_{V2} \\
&= \frac{40}{11}\,\text{V} + \left(-\frac{48}{11}\,\text{V}\right) = -\frac{8}{11}\,\text{V} \Leftarrow
\end{aligned}
$$

5.23: Use superposition to determine the value of the current, I, in the network of Fig 5.18.

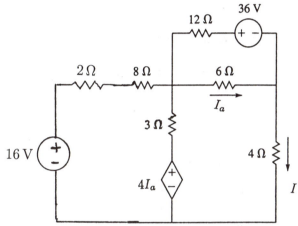

Note that the network has been redrawn with the current source transformed to a voltage source and that the resistance designators correspond to the resistance values. Using superposition, the current, I, will be the sum of two currents

$$I = I_{V1} + I_{V2}$$

where I_{V1} is the current due to the 16 V source acting alone and, I_{V2} is the current due to the 36 V source acting alone.

To determine I_{V1}, remove the 36 V source and replace it with a short circuit. With

$$I_x = I_c - I_b$$

so that

$$4I_x = 4(I_c - I_b)$$

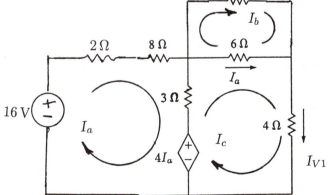

where I_{V1} is the current due to the 48 V source acting alone, I_{V2} is the current due to the 36 V source acting alone and I_V is the current due to the voltage source of unknown magnitude acting alone .

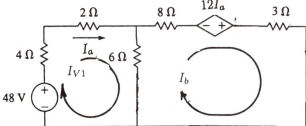

To determine I_{V1}, remove the 36 V source and V_s and replace them with short circuits. Two meshes are evident and two applications of KVL yield

$$-48\,\text{V} + 4I_a + 2I_a + 6(I_a - I_b) = 0$$

$$6(I_b - I_a) + 8I_b - 12I_a + 3I_b = 0$$

and with $I_{V1} = I_a$, these may be simplified and written in matrix form as

$$\begin{bmatrix} 12 & -6 \\ -18 & 17 \end{bmatrix} \begin{bmatrix} I_{V1} \\ I_b \end{bmatrix} = \begin{bmatrix} 48\,\text{V} \\ 0 \end{bmatrix}$$

Solve for the current vector

$$\begin{bmatrix} I_{V1} \\ I_b \end{bmatrix} = \begin{bmatrix} 12 & -6 \\ -18 & 17 \end{bmatrix}^{-1} \begin{bmatrix} 48\,\text{V} \\ 0 \end{bmatrix} = \begin{bmatrix} 0.1771 & 0.0625 \\ 0.1875 & 0.1250 \end{bmatrix} \begin{bmatrix} 48\,\text{V} \\ 0 \end{bmatrix} = \begin{bmatrix} 8.50\,\text{A} \\ 9.00\,\text{A} \end{bmatrix}$$

and

$$I_{V1} = 8.50\,\text{A}$$

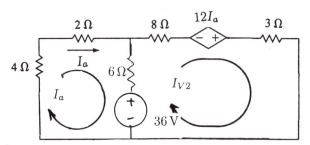

To determine I_{V2}, remove the 48 V source and V_s and replace them with short circuits. Two mesh equations are found from two applications of KVL

$$36\,\text{V} + 4I_a + 2I_a + 6(I_a - I_b) = 0$$

$$-36\,\mathrm{V} + 6(I_b - I_a) + 8I_b - 12I_a + 3I_b = 0$$

and with $I_{V2} = I_a$, these may be simplified and written in matrix form as

$$\begin{bmatrix} 12 & -6 \\ -18 & 17 \end{bmatrix} \begin{bmatrix} I_{V2} \\ I_b \end{bmatrix} = \begin{bmatrix} -36\,\mathrm{V} \\ 36\,\mathrm{V} \end{bmatrix}$$

The coefficient matrix is the same as that for I_{V1} and the current vector is

$$\begin{bmatrix} I_{V2} \\ I_b \end{bmatrix} = \begin{bmatrix} -0.1771 & 0.0625 \\ -0.1875 & 0.1250 \end{bmatrix} \begin{bmatrix} -36\,\mathrm{V} \\ 36\,\mathrm{V} \end{bmatrix} = \begin{bmatrix} -4.125\,\mathrm{A} \\ -2.25\,\mathrm{A} \end{bmatrix}$$

and

$$I_{V2} = -4.125\,\mathrm{A}$$

Now it is required that $I_a = 3.00\,\mathrm{A}$. Thus

$$\begin{aligned} I_V &= I_a - I_{V1} - I_{V2} \\ &= 3.00\,\mathrm{A} - 8.50\,\mathrm{A} - (-4.125\,\mathrm{A}) \\ &= -1.375\,\mathrm{A} \end{aligned}$$

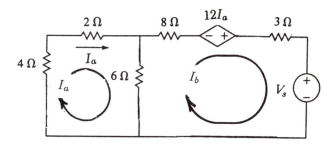

With both the 48 V and 36 V sources removed and replaced with short circuits, two more mesh equations can be written

$$4I_a + 2I_a + 6(I_a - I_b) = 0$$

$$6(I_b - I_a) + 8I_b - 12I_a + 3I_b + V_s = 0$$

or

$$\begin{aligned} 12I_a \quad -6I_b &= \quad 0 \\ -18I_a \quad +17I_b &= \quad -V_s \end{aligned}$$

To determine I_{6A}, remove the $42\,V$ source and replace it with a short circuit and remove the $7\,A$ source and replace it with an open circuit. Here

$$
\begin{aligned}
R_e &= 2\,\Omega + 1\,\Omega \\
&= 3\,\Omega \\
R_f &= \frac{(6\,\Omega)(3\,\Omega)}{6\,\Omega + 3\,\Omega} \\
&= \frac{18\,\Omega^2}{9\,\Omega} \\
&= 2\,\Omega
\end{aligned}
$$

and

$$
\begin{aligned}
R_g &= 2\,\Omega + 8\,\Omega \\
&= 10\,\Omega
\end{aligned}
$$

Then, a simple current divider provides

$$
I_{6A} = -\left(\frac{4\,\Omega}{4\,\Omega + 10\,\Omega}\right)(6\,A) = -\left(\frac{2}{7}\,A\right)(6\,A) = -\frac{12}{7}\,A
$$

and

$$
I = I_{7A} + I_{42V} + I_{6A} = \frac{2}{3}\,A + 1\,A + \left(-\frac{12}{7}\,A\right) = -\frac{1}{21}\,A
$$

The power dissiipated by the $8\,\Omega$ resistor is therefore

$$
P_8 = I^2 R_8 = \left(-\frac{1}{21}\,A\right)^2 (8\,\Omega) = 18.14\,mW \Leftarrow
$$

SUPERPOSITION AND OPERATIONAL AMPLIFIERS

5.26: Figure 5.21 shows an operational amplifier circuit with the op-amp connected as a differential amplifier. Use superposition to evaluate v_+ and v_- in terms of v_{i1} and v_{i2} and find an expression for $v_o = f(v_{i1}, v_{i2})$.

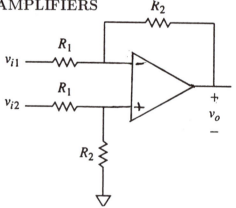

The output is a function of two input voltages, v_{i1} which is obtained by shorting v_{i2} and v_{i2} which is obtained by shorting v_{i1}

$$v_o = v_{o1} + v_{o2} = f(v_{i1}) + f(v_{i2})$$

Thus, with v_{i2} replaced with a short circuit, the op-amp is in the inverting configuration and

$$v_{o1} = -\frac{R_2}{R_1}v_{i1}$$

and with v_{i1} replaced with a short circuit, the op-amp is in the non-inverting configuration and

$$v_{o2} = \left(1 + \frac{R_2}{R_1}\right)v_+$$

But by a voltage divider

$$v_+ = \left(\frac{R_2}{R_1 + R_2}\right)v_{i2}$$

so that

$$v_{o2} = \left(1 + \frac{R_2}{R_1}\right)\left(\frac{R_2}{R_1 + R_2}\right)v_{i2} = \left(\frac{R_1 + R_2}{R_1}\right)\left(\frac{R_2}{R_1 + R_2}\right)v_{i2}$$

or

$$v_{o2} = \frac{R_2}{R_1}v_{i2}$$

Then

$$v_o = v_{o2} + v_{o1} = \frac{R_2}{R_1}v_{i2} - \frac{R_2}{R_1}v_{i1}$$

or

$$v_o = \frac{R_2}{R_1}(v_{i2} - v_{i1}) \Leftarrow$$

5.27: For the operational amplifier arrangement shown in Fig 5.22, if $v_{i2} = 6\,\text{V}$, select the value of R to make $v_o = 24\,\text{V}$.

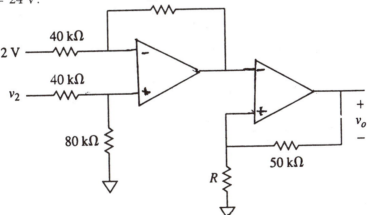

The first op-amp is the differential amplifier of Fig 5.21 considered in Problem 5.26 with

With the specified values of $v_{i1} = 4\,\text{V}, v_{i2} = 8\,\text{V}, R_2 = 80\,\text{k}\Omega$ and $R_4 = 40\,\text{k}\Omega$, let $R_3 = R$ with the constraint that $R_1 = 1.25R_3 = 1.25R$

$$\frac{R_2}{R_1} = \frac{80\,\text{k}\Omega}{1.25R} = \frac{64}{R}$$

where R is in kΩ. Then

$$4 = \left(\frac{1 + 64/R}{1 + R/40}\right)(8\,\text{V}) - \frac{64}{R}(4\,\text{V})$$

$$1 = 2\left(\frac{1 + 64/R}{1 + R/40}\right) - \frac{64}{R}$$

$$1 = \frac{80}{R}\left(\frac{R + 64}{40 + R}\right) - \frac{64}{R}$$

Further algebraic manipulation gives

$$1 = \frac{80(R + 64) - 64(R + 40)}{R(R + 40)}$$

and

$$R^2 + 40R = 80R + 5120 - 64R - 2560$$

This is a quadratic equation

$$R^2 + 24R - 2560 = 0$$

which can be factored to

$$(R - 40)(R + 64) = 0$$

which shows that the two roots are

$$R = 40\,\text{k}\Omega \quad \text{and} \quad R = -64\,\text{k}\Omega$$

The root, $R = -64\,\text{k}\Omega$ is absurd and may discarded. Thus

$$R = R_3 = 40\,\text{k}\Omega \Leftarrow$$

and

$$R_1 = 1.25R = 1.25(40\,\text{k}\Omega) = 50\,\text{k}\Omega \Leftarrow$$

CHAPTER SIX

THEVENIN, NORTON AND
MAXIMUM POWER TRANSFER THEOREMS

THEVENIN AND NORTON THEOREMS FOR
NETWORKS WITHOUT CONTROLLED SOURCES

6.1: Use Thevenin's theorem to find the resistance that must be connected across terminals *a-b* in Fig 6.1 in order for the resistor current to be 3 A.

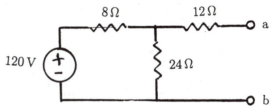

The Thevenin equivalent voltage can be determined from a voltage divider

$$V_T = \left(\frac{24\,\Omega}{24\,\Omega + 8\,\Omega}\right)(120\,\text{V}) = \left(\frac{3}{4}\right)(120\,\text{V}) = 90\,\text{V}$$

and the Thevenin equivalent resistance can be found by removing the 120 V source, replacing it with a short circuit and looking back into terminals a-b

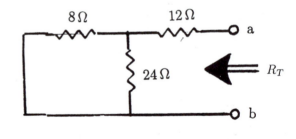

$$
\begin{aligned}
R_T &= \frac{(8\,\Omega)(24\,\Omega)}{8\,\Omega + 24\,\Omega} + 12\,\Omega \\[2mm]
&= \frac{192\,\Omega^2}{32\,\Omega} + 12\,\Omega \\[2mm]
&= 6\,\Omega + 12\,\Omega = 18\,\Omega
\end{aligned}
$$

The Thevenin equivalent circuit is shown with the load, R, connected between terminals a-b at the right. Here

$$I = \frac{V_T}{R_T + R}$$

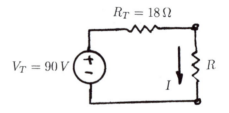

295

$$= \frac{90\,\text{V}}{18\,\Omega + R}$$

$$= 3\,\text{A}$$

This reduces to

$$90\,\text{V} = 54\,\text{V} + (3\,\text{A})R$$

or

$$R = \frac{90\,\text{V} - 54\,\text{V}}{3\,\text{A}} = \frac{36\,\text{V}}{3\,\text{A}} = 12\,\Omega \Leftarrow$$

Notice that if a short circuit is placed between terminals *a-b*, the equivalent resistance will be

$$R_{\text{eq}} = \frac{(12\,\Omega)(24\,\Omega)}{12\,\Omega + 24\,\Omega} + 8\,\Omega$$

$$= \frac{288\,\Omega^2}{36\,\Omega} + 8\,\Omega$$

$$= 8\,\Omega + 8\,\Omega = 16\,\Omega$$

and then

$$I_8 = \frac{120\,\text{V}}{16\,\Omega} = 7.5\,\text{A}$$

and by a current divider

$$I_{SC} = \left(\frac{24\,\Omega}{12\,\Omega + 24\,\Omega}\right) I_8 = \left(\frac{2}{3}\right)(7.5\,\text{A}) = 5\,\text{A}$$

The Norton equivalent circuit with R reconnected is shown at the right and to make $I_R = 3\,\text{A}$, a current divider is used to give

$$I_R = 3\,\text{A}$$

$$= \left(\frac{18\,\Omega}{18\,\Omega + R}\right)(5\,\text{A})$$

$$\frac{18\,\Omega}{18\,\Omega + R} = 0.60$$

$$18\,\Omega = 0.60R + 10.8\,\Omega$$

$$0.6R = 7.2\,\Omega$$

$$R = 12\,\Omega \Leftarrow$$

These can be written in matrix form as

$$\begin{bmatrix} 18 & -6 \\ -6 & 18 \end{bmatrix} \begin{bmatrix} I_a \\ I_b \end{bmatrix} = \begin{bmatrix} 16\,\text{V} \\ 48\,\text{V} \end{bmatrix}$$

and a matrix inversion can be employed to find the current vector

$$\begin{bmatrix} I_a \\ I_b \end{bmatrix} = \begin{bmatrix} 18 & -6 \\ -6 & 18 \end{bmatrix}^{-1} \begin{bmatrix} 16\,\text{V} \\ 48\,\text{V} \end{bmatrix} = \begin{bmatrix} 0.0625 & 0.0208 \\ 0.0208 & 0.0625 \end{bmatrix} \begin{bmatrix} 16\,\text{V} \\ 48\,\text{V} \end{bmatrix} = \begin{bmatrix} 2.000\,\text{A} \\ 3.333\,\text{A} \end{bmatrix}$$

Thus

$$V_T = 4I_a + 2I_b = 4(2.000\,\text{A}) + 2(3.333\,\text{A}) = 8.000\,\text{V} + 6.667\,\text{V} = 14.667\,\text{V}$$

It seems that the most expeditious way to find R_T is by way of I_{sc}. With I_{sc} replacing the $16\,\Omega$ resistor, three clockwise mesh currents are evident and the three mesh equations may be written by inspection

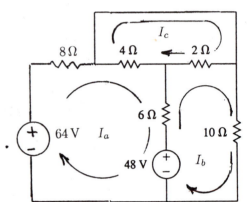

$$(8 + 4 + 6)I_a - 6I_b - 4I_c = 64\,\text{V} - 48\,\text{V}$$
$$-6I_a + (6 + 2 + 10)I_b - 2I_c = 48\,\text{V}$$
$$-4I_a - 2I_b + 6I_c = 0$$

These can be written in matrix form as

$$\begin{bmatrix} 18 & -6 & -4 \\ -6 & 18 & -2 \\ -4 & -2 & 6 \end{bmatrix} \begin{bmatrix} I_a \\ I_b \\ I_c \end{bmatrix} = \begin{bmatrix} 16\,\text{V} \\ 48\,\text{V} \\ 0 \end{bmatrix}$$

and a matrix inversion can be employed to find the current vector

$$\begin{bmatrix} I_a \\ I_b \\ I_c \end{bmatrix} = \begin{bmatrix} 18 & -6 & -4 \\ -6 & 18 & -2 \\ -4 & -2 & 6 \end{bmatrix}^{-1} \begin{bmatrix} 16\,\text{V} \\ 48\,\text{V} \\ 0 \end{bmatrix}$$

or

$$\begin{bmatrix} I_a \\ I_b \\ I_c \end{bmatrix} = \begin{bmatrix} 0.0818 & 0.0346 & 0.0660 \\ 0.0346 & 0.0723 & 0.0472 \\ 0.0660 & 0.0472 & 0.2264 \end{bmatrix} \begin{bmatrix} 16\,\text{V} \\ 48\,\text{V} \\ 0 \end{bmatrix} = \begin{bmatrix} 2.969\,\text{A} \\ 4.025\,\text{A} \\ 3.321\,\text{A} \end{bmatrix}$$

Thus

$$I_{SC} = I_c = 3.321\,\text{A}$$

and

$$R_T = \frac{V_T}{I_{SC}} = \frac{14.667\,\text{V}}{3.321\,\text{A}} = 4.416\,\Omega$$

The Thevenin equivalent, with the $16\,\Omega$ resistor is shown at the right. The current through the $16\,\Omega$ resistor will be

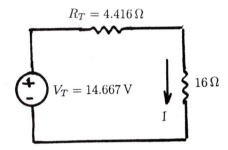

$$
\begin{aligned}
I &= \frac{V_T}{R_T + 16\,\Omega} \\
&= \frac{14.667\,\text{V}}{4.416\,\Omega + 16\,\Omega} \\
&= \frac{14.667\,\text{V}}{20.416\,\Omega} \\
&= 0.718\,\text{A} \Leftarrow
\end{aligned}
$$

6.4: Use Thevenin's theorem to find the current, I, flowing through the $600\,\Omega$ resistor in the network of Fig 6.4.

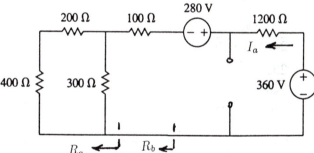

In the revised network diagram, the $600\,\Omega$ resistor has been removed and replaced with an open circuit. Note terminals a and b and that a single loop can be formed using

$$R_a = \frac{(400\,\Omega + 200\,\Omega)(300\,\Omega)}{400\,\Omega + 200\,\Omega + 300\,\Omega} = \frac{180,000\,\Omega^2}{900\,\Omega} = 200\,\Omega$$

and

$$R_b = 200\,\Omega + 100\,\Omega = 300\,\Omega$$

Then

$$I_a = \frac{360\,\text{V} - 280\,\text{V}}{1200\,\Omega + 300\,\Omega} = \frac{80\,\text{V}}{1500\,\Omega} = \frac{8}{150}\,\text{A}$$

Thus

$$R_T = R_N = 400\,\Omega + 200\,\Omega = 600\,\Omega$$

and

$$I_N = \frac{V_T}{R_T} = \frac{57.90\,\text{V}}{600\,\Omega} = 0.0965\,\text{A}$$

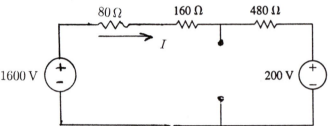

The Norton equivalent circuit is shown at the right.

6.6: Use Thevenin's theorem to find the voltage across the 320 Ω resistor in the network of Fig 6.6.

In the revised network diagram, the 320 Ω resistor is considered as the load resistor and has been removed and replaced with an open circuit. In addition, there has been a current to voltage source transformation. The network reduces to a single loop and

$$I = \frac{1600\,\text{V} - 200\,\text{V}}{80\,\Omega + 160\,\Omega + 480\,\Omega} = \frac{1400\,\text{V}}{720\,\Omega} = 1.944\,\text{A}$$

and

$$
\begin{aligned}
V_T = V_{oc} &= 1600\,\text{V} - 240I \\
&= 1600\,\text{V} + (240\,\Omega)(1.944\,\text{A}) \\
&= 1600\,\text{V} - 467.67\,\text{V} \\
&= 1133.33\,\text{V}
\end{aligned}
$$

Then, with the voltage sources removed and replaced with short circuits

$$
\begin{aligned}
R_T &= \frac{(80\,\Omega + 160\,\Omega)(480\,\Omega)}{80\,\Omega + 160\,\Omega + 480\,\Omega} \\
&= \frac{115,200\,\Omega^2}{720\,\Omega} \\
&= 160\,\Omega
\end{aligned}
$$

The Thevenin equivalent circuit is shown at the right with the $320\,\Omega$ resistor reconnected. By a voltage divider

$$V_{320} = \left(\frac{320\,\Omega}{320\,\Omega + 160\,\Omega}\right)(1133.33\,\text{V})$$

$$= \left(\frac{2}{3}\right)(1133.33\,\text{V})$$

$$= 755.53\,\text{V} \Leftarrow$$

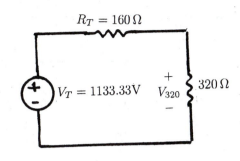

6.7: Use the Thevenin's theorem or the Norton's theorem to determine the value of R that will allow a current of $1\,\text{A}$ to flow through the $2\,\Omega$ resistor in Fig 6.7.

Remove the $2\,\Omega$ resistor and observe that by voltage division

$$V_T = V_{\text{oc}} = \frac{56R}{R + 24\,\Omega}$$

and that the resistance looking into the terminals with the $56\,\text{V}$ source removed and replaced with a short circuit will be

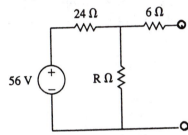

$$R_T = \frac{24R\,\Omega}{24\,\Omega + R} + 6\,\Omega = \frac{6R\,\Omega + 24R\,\Omega + 144\,\Omega^2}{24\,\Omega + R} = \frac{30R\,\Omega + 144\,\Omega^2}{24\,\Omega + R}$$

The Thevenin equivalent with the $2\,\Omega$ load is shown at the right. Here

$$I = \frac{V_T}{R_T + 2\,\Omega} = 1\,\text{A}$$

and

$$I = \frac{V_T}{R_T + 2\,\Omega} = \frac{\dfrac{56R}{R + 24\,\Omega}}{\dfrac{30R + 144\,\Omega}{24\,\Omega + R} + 2\,\Omega} = \frac{56R}{30R\,\Omega + 144\,\Omega^2 + 2R\,\Omega + 48\,\Omega^2}$$

Then

$$I = \frac{56R\,\text{V}}{32R\,\Omega + 192\,\Omega^2} = 1\,\text{A}$$

so that

$$56R\,\Omega = 32R\,\Omega + 192\,\Omega^2$$
$$24R\,\Omega = 192\,\Omega^2$$
$$R = 8\,\Omega \Leftarrow$$

6.8: Use Norton's theorem to determine the current through the $10\,\Omega$ resistor in the network of Fig 6.8.

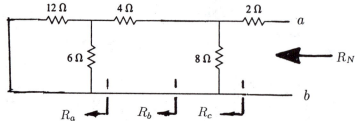

The revamped circuit shows that the $10\,\Omega$ resistor, which will be treated as the load resistor, has been replaced with an open circuit and the Norton equivalent resistance is evaluated by removing both voltage sources and replacing them with short circuits. With respect to terminals a-b

$$R_a = \frac{(6\,\Omega)(12\,\Omega)}{6\,\Omega + 12\,\Omega} = \frac{72\,\Omega^2}{18\,\Omega} = 4\,\Omega$$
$$R_b = 4\,\Omega + 4\,\Omega = 8\,\Omega$$
$$R_c = \left(\frac{1}{2}\right)8\,\Omega = 4\,\Omega$$

and

$$R_N = R_T = 2\,\Omega + 4\,\Omega = 6\,\Omega$$

Then, with a short circuit placed across terminals a-b, there are three clockwise mesh currents whose mesh equations may be written by inspection

$$
\begin{aligned}
(12+6)I_a - 6I_b &= 24\,\text{V} \\
-6I_a + (6+4+8)I_b - 8I_c &= 16\,\text{V} \\
-8I_b + 10I_c &= 0
\end{aligned}
$$

In matrix form, these become

$$
\begin{bmatrix} 18 & -6 & 0 \\ -6 & 18 & -8 \\ 0 & -8 & 10 \end{bmatrix}
\begin{bmatrix} I_a \\ I_b \\ I_c \end{bmatrix} =
\begin{bmatrix} 24\,\text{V} \\ 16\,\text{V} \\ 0 \end{bmatrix}
$$

and a matrix inversion can be employed to find the current vector

$$\begin{bmatrix} I_a \\ I_b \\ I_c \end{bmatrix} = \begin{bmatrix} 18 & -6 & 0 \\ -6 & 18 & -8 \\ 0 & -8 & 10 \end{bmatrix}^{-1} \begin{bmatrix} 24\,\text{V} \\ 16\,\text{V} \\ 0 \end{bmatrix}$$

or

$$\begin{bmatrix} I_a \\ I_b \\ I_c \end{bmatrix} = \begin{bmatrix} 0.0671 & 0.0347 & 0.0278 \\ 0.0347 & 0.1042 & 0.0833 \\ 0.0278 & 0.0833 & 0.1667 \end{bmatrix} \begin{bmatrix} 24\,\text{V} \\ 16\,\text{V} \\ 0 \end{bmatrix} = \begin{bmatrix} 2.167\,\text{A} \\ 2.500\,\text{A} \\ 2.000\,\text{A} \end{bmatrix}$$

Thus

$$I_{\text{sc}} = I_c = 2\,\text{A}$$

and in the Norton equivalent circuit shown at the right with the $10\,\Omega$ resistor reconnected, a current divider gives

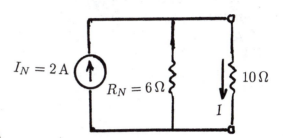

$$I = \left(\frac{6\,\Omega}{6\,\Omega + 10\,\Omega}\right)(2\,\text{A}) = \left(\frac{3}{8}\right)(2\,\text{A}) = \frac{3}{4}\,\text{A}$$

THEVENIN AND NORTON THEOREMS FOR NETWORKS WITH CONTROLLED SOURCES

6.9: Use Thevenin's theorem to determine the power dissipated by the $12\,\Omega$ resistor in Fig 6.9.

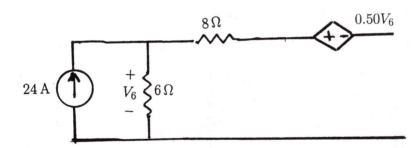

The original network in Fig 6.9 is adjusted to show the removal the $12\,\Omega$ resistor and its replacement by an open circuit. The voltage controlled current source has been transformed to a voltage controlled voltage source. Then

$$V_{\text{oc}} = V_T = V_6 - 0.50V_6 = 0.50V_6$$

With

$$V_6 = (6\,\Omega)(24\,\text{A}) = 144\,\text{V}$$

The Thevenin equivalent voltage will be

$$V_T = 0.50V_6 = 0.50(144\,\text{V}) = 72\,\text{V}$$

Then, with a fictitious 1 V source added to to open-circuited terminals and the removal of the 24 A source with its replacement by an open circuit, a single loop results with a loop current, I_o

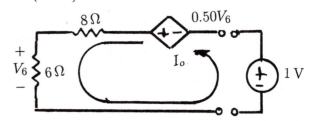

$$V_6 = 6I_o$$

and KVL gives

$$6I_o - 0.50(6I_o) + 8I_o - 1\,\text{V} = 0$$

or

$$11I_o = 1\,\text{V}$$

and

$$R_T = \left|\frac{1\,\text{V}}{I_o}\right| = 11\,\Omega$$

Thus, the Thevenin equivalent circuit with the 12 Ω resistor reconnected is shown at the right and by a voltage divider

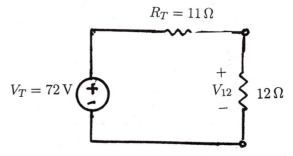

$$\begin{aligned}
V_{12} &= \left(\frac{12\,\Omega}{12\,\Omega + 11\,\Omega}\right)(72\,\text{V}) \\
&= (0.522)(72\,\text{V}) = 37.58\,\text{V}
\end{aligned}$$

and

$$P_{12} = \frac{V_{12}^2}{R_{12}} = \frac{(37.58\,\text{V})^2}{12\,\Omega} = 117.6\,\text{W} \Leftarrow$$

6.10: Determine the Norton equivalent for terminals a-b in the network of Fig 6.10.

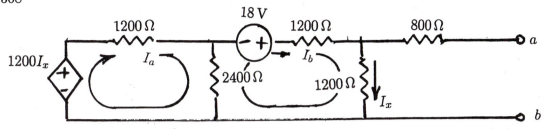

The network is shown here with two clockwise mesh currents inserted. The strategy here is to determine the Thevenin equivalent and, from it, the Norton equivalent. The open circuit voltage is

$$V_{oc} = V_T = 1200 I_b$$

and with $I_x = I_b$, two applications of KCL give the two mesh equations

$$-1200 I_b + 1200 I_a + 2400(I_a - I_b) = 0$$

$$2400(I_b - I_a) - 18\,\text{V} + 1200 I_b + 1200 I_b = 0$$

or

$$
\begin{aligned}
3600 I_a \quad - 3600 I_b &= \quad 0 \\
-2400 I_a \quad + 4800 I_b &= \quad 18\,\text{V}
\end{aligned}
$$

Here it is observed in the first equation that $I_a = I_b$ so that in the second equation

$$-2400 I_a + 4800 I_b = -2400 I_b + 4800 I_b = 2400 I_b = 18\,\text{V}$$

and

$$I_b = \frac{18\,\text{V}}{2400\,\Omega} = 7.5\,\text{mA}$$

The value of the Thevenin equivalent voltage will be

$$V_{oc} = V_T = (1200\,\Omega)(0.0075\,\text{A}) = 9\,\text{V}$$

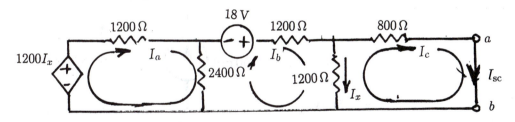

A similar process can be employed to find I_{sc}. Here, a short circuit has been placed across the terminals and three clockwise mesh currents are noted. With $I_x = I_b - I_c$, the three mesh equations derive from three applications of KVL

6.12: Use Norton's theorem to determine the current flowing through the right-hand $4\,\Omega$ resistor in Fig 6.12.

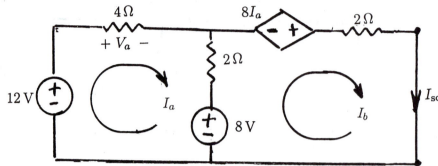

To obtain $I_{sc} = I_N$, remove the right-hand $4\,\Omega$ resistor and replace it with a short circuit. With the $4\,A$ current source transformed to an $8\,V$ voltage source, the network is seen to contain two mesh currents which can be found from two applications of KVL. With

$$V_a = 4I_a \qquad \text{so that} \qquad 2V_a = 8I_a$$

the two mesh equations are

$$-12\,V + 4I_a + 2(I_a - I_b) + 8\,V = 0$$

$$-8\,V + 2(I_b - I_a) - 8I_a + 2I_b = 0$$

These may be simplified to

$$
\begin{aligned}
6I_a \quad -2I_b &= \quad 4\,V \\
-10I_a \quad +4I_b &= \quad 8\,V
\end{aligned}
$$

The first equation may be multiplied by 2 so that the pair of equations becomes

$$
\begin{aligned}
12I_a \quad -4I_b &= \quad 8\,V \\
-10I_a \quad +4I_b &= \quad 8\,V
\end{aligned}
$$

Then, a simple addition gives

$$
\begin{aligned}
2I_a &= \quad 16\,V \\
I_a &= \quad 8\,A
\end{aligned}
$$

and from the first equation

$$
\begin{aligned}
6I_a - 2I_b &= \quad 4\,V \\
6(8\,A) - 2I_b &= \quad 4\,V \\
-2I_b &= \quad -44\,V
\end{aligned}
$$

and

$$I_b = I_{sc} = 22 \text{ A}$$

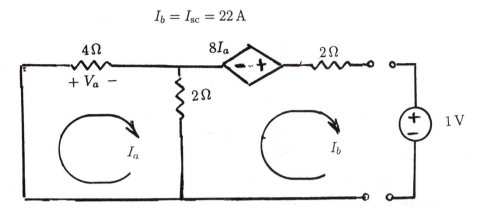

A fictitious voltage source of 1 V may be applied at the terminals as shown in the figure. Note that the both ideal voltage sources have been removed and replaced with short circuits. This time the mesh equations are

$$4I_a + 2(I_a - I_b) = 0$$

$$2(I_b - I_a) - 8I_a + 2I_b - 1 \text{ V} = 0$$

and these may be simplified to

$$\begin{aligned} 6I_a \quad -2I_b &= \quad 0 \\ -10I_a \quad +4I_b &= \quad 1 \text{ V} \end{aligned}$$

Solution of these gives

$$I_b = \frac{3}{2} \text{ A}$$

and

$$R_N = R_T = \left| \frac{1 \text{ V}}{I_b} \right| = \frac{2}{3} \, \Omega$$

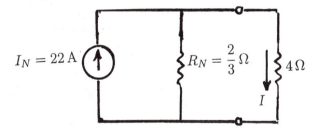

The Norton equivalent is shown at the right with the right-hand 4 Ω resistor connected and the current through the 4 Ω resistor, designated by I, will be

$$I = \left[\frac{2/3}{(2/3) + 4} \right] (22 \text{ A}) = \left(\frac{2}{14} \right) (22 \text{ A}) = \frac{22}{7} \text{ A} \Leftarrow$$

6.13: Use Thevenin's theorem to determine the current flowing through the 16 Ω resistor in Fig 6.13.

or

$$V_T = -14,400(0.010\,\text{A})$$
$$= -144\,\text{V}$$

Note that the positive terminal of V_T is at point-b.

To determine the Thevenin equivalent
resistance, remove the 12 V source and
replace it with a short circuit. In this
case, $I_a = 0$ and the resistance across
the terminals a-b is just $600\,\Omega$. The
Thevenin equivalent circuit with the
$300\,\Omega$ resistor reconnected is shown
at the right. Here

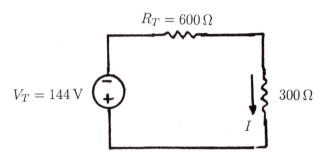

$$|I| = \frac{144\,\text{V}}{600\,\Omega + 300\,\Omega} = \frac{144\,\text{V}}{900\,\Omega} = 160\,\text{mA}$$

and the power dissipated by the $300\,\Omega$ resistor will be

$$P = (|I|)^2 R = (0.160\,\text{A})^2(300\,\Omega) = 7.68\,\text{W} \Leftarrow$$

6.15: Use Thevenin's theorem to determine the current through the $6\,\Omega$ resistor in the network
of Fig 6.15.

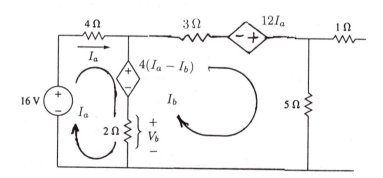

The revamped circuit has several features. First, the $6\,\Omega$ resistor, which is treated as the
load resistor has been removed and replaced with an open circuit. The Thevenin equivalent
voltage is this open circuit voltage and, in terms of the two mesh currents will be

$$V_T = 5I_b$$

Second, the voltage controlled current source (VCVS) is controlled by the voltage across the $2\,\Omega$ resistor

$$V_b = 2(I_a - I_b)$$

so that

$$2V_b = 4(I_a - I_b)$$

Finally, the the current controlled current source with strength $4I_a$ can be transformed to a current controlled voltage source of strength $12I_a$ and the two clockwise mesh currents can be found from two applications of KVL

$$-16\,\mathrm{V} + 4I_a + 4(I_a - I_b) + 2(I_a - I_b) = 0$$

$$3I_b - 12I_a + 5I_b + 2(I_b - I_a) - 4(I_a - I_b) = 0$$

and, in matrix form, these may be simplified to

$$\begin{bmatrix} 10 & -6 \\ -18 & 14 \end{bmatrix} \begin{bmatrix} I_a \\ I_b \end{bmatrix} = \begin{bmatrix} 16\,\mathrm{V} \\ 0 \end{bmatrix}$$

A matrix inversion gives

$$\begin{bmatrix} I_a \\ I_b \end{bmatrix} = \begin{bmatrix} 10 & -6 \\ -18 & 14 \end{bmatrix}^{-1} \begin{bmatrix} 16\,\mathrm{V} \\ 0 \end{bmatrix} = \begin{bmatrix} 0.4375 & 0.1875 \\ 0.5625 & 0.3125 \end{bmatrix} \begin{bmatrix} 16\,\mathrm{V} \\ 0 \end{bmatrix} = \begin{bmatrix} 7.000\,\mathrm{A} \\ 9.000\,\mathrm{A} \end{bmatrix}$$

and

$$V_T = 5I_b = (5\,\Omega)(9.00\,\mathrm{A}) = 45\,\mathrm{V}$$

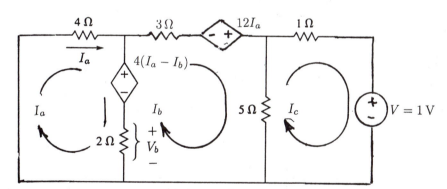

To find R_T, apply a 1 V fictitious source at the network terminals after replacing the 16 V source with a short circuit. Three clockwise mesh currents are evident and the mesh currents can be found from three applications of KVL.

$$4I_a + 4(I_a - I_b) + 2(I_a - I_b) = 0$$

so that the first can be used to give V_b

$$8(-2.25V_b) + 7V_b = 288\text{ V}$$
$$-18V_b + 7V_b = 288\text{ V}$$
$$-11V_b = 288\text{ V}$$

and

$$V_{24} = V_b = -\frac{288}{11}\text{ V}$$

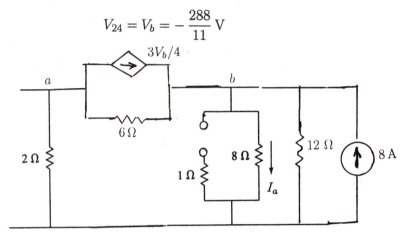

To obtain V_8, remove the 24 A source and replace it with an open circuit. This time, the two node equations are

$$\frac{1}{2}V_a + \frac{V_a - V_b}{6} + \frac{3}{4}V_b = 0$$

$$\frac{V_b - V_a}{6} + \frac{1}{8}V_b + \frac{1}{12}V_b - 8\text{ A} - \frac{3}{4}V_b = 0$$

or

$$\frac{2}{3}V_a + \frac{7}{12}V_b = 0$$

$$-\frac{1}{6}V_a - \frac{3}{8}V_b = 8\text{ A}$$

Simplification provides

$$8V_a + 7V_b = 0$$
$$-4V_a - 9V_b = 192\text{ A}$$

The first of these shows that

$$V_a = -\frac{7}{8}V_b$$

so that the second can be used to give V_b

$$-4\left(-\frac{7}{8}\right)V_b - 9V_b = 192\,\text{V}$$

$$-\left(\frac{11}{2}\right)V_b = 192\,\text{V}$$

and

$$V_8 = V_b = -\frac{384}{11}\,\text{V}$$

The open circuit voltage (the Thevenin equivalent voltage) will be

$$V_{\text{oc}} = V_{24} + V_8 = -\left(\frac{288}{11}\,\text{V}\right) + \left(-\frac{384}{11}\,\text{V}\right) = -\frac{672}{11}\,\text{V}$$

Observe that superposition was required by the statement of the problem. It is more direct (and easier) to just employ a nodal analysis with both sources in which case, the node equations are

$$\frac{2}{3}V_a + \frac{7}{12}V_b = 24\,\text{A}$$

$$-\frac{1}{6}V_a - \frac{3}{8}V_b = 8\,\text{A}$$

or

$$8V_a + 7V_b = 288\,\text{A}$$

$$-4V_a - 9V_b = 192\,\text{A}$$

and with these in matrix form

$$\begin{bmatrix} 8 & 7 \\ -4 & -9 \end{bmatrix} \begin{bmatrix} V_a \\ V_b \end{bmatrix} = \begin{bmatrix} 288\,\text{A} \\ 192\,\text{A} \end{bmatrix}$$

A matrix inversion can be employed to find the voltage vector

$$\begin{bmatrix} V_a \\ V_b \end{bmatrix} = \begin{bmatrix} 8 & 7 \\ -4 & -9 \end{bmatrix}^{-1} \begin{bmatrix} 288\,\text{A} \\ 192\,\text{A} \end{bmatrix} = -\frac{1}{11}\begin{bmatrix} -9 & -7 \\ 4 & 8 \end{bmatrix} \begin{bmatrix} 72\,\text{A} \\ 48\,\text{A} \end{bmatrix} = \frac{1}{11}\begin{bmatrix} 984\,\text{V} \\ -672\,\text{V} \end{bmatrix}$$

and

$$V_{\text{oc}} = V_T = -\frac{672}{11}\,\text{V}$$

and a current divider may be used to show that

$$I_a = \left(\frac{4\,\Omega}{4\,\Omega + 8\,\Omega}\right)(3\,\text{A}) = \left(\frac{1}{3}\right)(3\,\text{A}) = 1\,\text{A}$$

and

$$V_{3\text{A}} = (6\,\Omega)(1\,\text{A}) = 6\,\text{V}$$

The Thevenin equivalent voltage is therefore

$$V_T = V_{12\text{V}} + V_{3A} = 6\,\text{V} + 6\,\text{V} = 12\,\text{V}$$

The Thevenin equivalent resistance is determined by removing both sources, replacing the 12 V source with a short circuit and replacing the 3 A source with an open circuit. The result is shown at the right and it is noted that the $6\,\Omega$ resistor is in parallel with a resistance of $4\,\Omega + 2\,\Omega = 6\,\Omega$, Hence

$$R_T = \left(\frac{1}{2}\right)(6\,\Omega) = 3\,\Omega$$

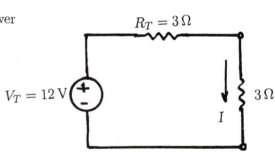

The resistance required for maximum power transfer is

$$R = R_T = 3\,\Omega \Leftarrow$$

and this is shown at the right in a Thevenin equivalent circuit. The current, I, through the $3\,\Omega$ resistor is

$$I = \frac{V_T}{2R_T} = \frac{12\,\text{V}}{2(3\,\Omega)} = 2\,\text{A}$$

and the power drawn by this resistor is

$$P = I^2 R = (2\,\text{A})^2(3\,\Omega) = 12\,\text{W} \Leftarrow$$

6.18: In the network of Fig 6.18, determine the value of the load to be placed across terminals a-b in order for the load to draw maximum power. Then, determine the value of this power.

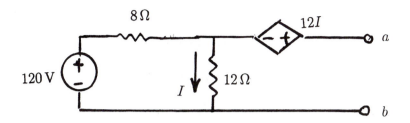

Here,

$$I = \frac{120\,\text{V}}{8\,\Omega + 12\,\Omega} = \frac{120\,\text{V}}{20\,\Omega} = 6\,\text{A}$$

and the current controlled voltage source will have a strength of

$$12I = 12(6\,\text{A}) = 72\,\text{V}$$

The Thevenin equivalent voltage is

$$V_{\text{oc}} = V_T = 12I + 12I = 24I = 24(6\,\text{A}) = 144\,\text{V}$$

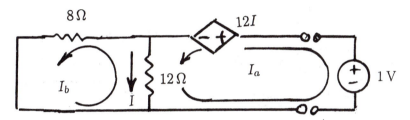

With the 120 V source replaced by a short circuit, a fictitious 1 V source may be placed across the terminals and two counterclockwise mesh currents can be inserted. These mesh equations can be written from two applications of KVL. With $I = I_a - I_b$,

$$12(I_a - I_b) + 12(I_a - I_b) - 1\,\text{V} = 0$$

$$12(I_b - I_a) + 8I_b = 0$$

and these can be simplified to

$$\begin{aligned} 24I_a & \quad -24I_b & = & \quad 1\,\text{V} \\ -12I_a & \quad +20I_b & = & \quad 0 \end{aligned}$$

The second of these shows that

$$I_a = \frac{.5}{3}I_b$$

and the power dissipated by the load will be

$$P = I^2 R = \left(\frac{14}{5}\,\mathrm{A}\right)^2 \left(\frac{30}{7}\right)\,\Omega) = 33.6\,\mathrm{W} \Leftarrow$$

6.20: In the network of Fig 6.20, determine the value of the load to be placed across terminals a-b in order for the load to draw maximum power. Then, determine the value of this power.

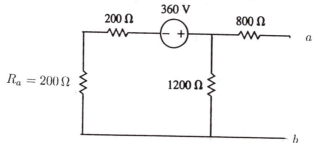

The network has been altered by replacing the combination of the $500\,\Omega, 300\,\Omega$ and $100\,\Omega$ resistors at the extreme left by a single resistance

$$R_a = \frac{(500\,\Omega + 100\,\Omega)(300\,\Omega)}{500\,\Omega + 100\,\Omega + 300\,\Omega} = \frac{18,000\,\Omega^2}{900\,\Omega} = 200\,\Omega$$

Thus, there is a single loop and a voltage division gives

$$V_{\mathrm{oc}} = V_T = \left(\frac{1200\,\Omega}{1200\,\Omega + 200\,\Omega + 200\,\Omega}\right)(360\,\mathrm{V}) = \left(\frac{3}{4}\right)(360\,\mathrm{V}) = 270\,\mathrm{V}$$

With the 360 V source removed and replaced by a short circuit, the resistance looking back into terminals a-b is shown in the figure. Here, R_a is as before

$$R_a = 200\,\Omega$$

$$R_b = 200\,\Omega + 200\,\Omega = 400\,\Omega$$

$$R_c = \frac{(400\,\Omega)(1200\,\Omega)}{400\,\Omega + 1200\,\Omega} = \frac{480,000\,\Omega^2}{1600\,\Omega} = 300\,\Omega$$

and

$$R_T = 300\,\Omega + 800\,\Omega = 1100\,\Omega$$

For maximum power transfer, make
$R_L = R_T$

$$R_L = R_T = 1100\,\Omega \Leftarrow$$

The Thevenin equivalent circuit with the load resistor reconnected is shown at the right. Here, the current, I, through the load is

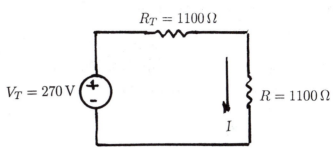

$$I = \frac{V_T}{2R_T} = \frac{270\,\text{V}}{2(1100\,\Omega)} = 0.123\,\text{A}$$

and the power dissipated by the load will be

$$P = I^2R = (0.123\,\text{A})^2(1100\,\Omega) = 16.57\,\text{W} \Leftarrow$$

6.21: In the network of Fig 6.21, determine the value of the load to be placed across terminals a-b in order for the load to draw maximum power. Then, determine the value of this power.

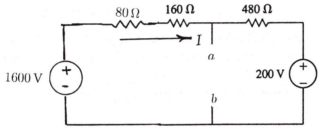

With the transformation of the current source on the left to a voltage source, the current in the single loop is

$$I = \frac{1600\,\text{V} - 200\,\text{V}}{80\,\Omega + 160\,\Omega + 480\,\Omega} = \frac{1400\,\text{V}}{720\,\Omega} = 1.944\,\text{A}$$

Then with $V_{\text{oc}} = V_T$

$$
\begin{aligned}
V_T &= 1600\,\text{V} - (80\,\Omega + 160\,\Omega)I \\
&= 1600\,\text{V} - (240\,\Omega)(1.944\,\text{A}) \\
&= 1600\,\text{V} - 466.67\,\text{V} \\
&= 1133.33\,\text{V}
\end{aligned}
$$

The Thevenin equivalent resistance is obtained by removing both sources and looking back into the terminals. With both voltage sources removed and replaced by short circuits, the picture is as shown at the right.

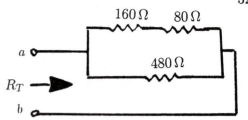

$$R_T = \frac{(160\,\Omega + 80\,\Omega)(480\,\Omega)}{160\,\Omega + 80\,\Omega + 480\,\Omega} = \frac{115,200\,\Omega^2}{720\,\Omega} = 160\,\Omega$$

For maximum power transfer, make $R_L = R_T$

$$R_L = R_T = 160\,\Omega \Leftarrow$$

The Thevenin equivalent circuit with the load resistor reconnected is shown at the right. Here, the current, I, through the load is

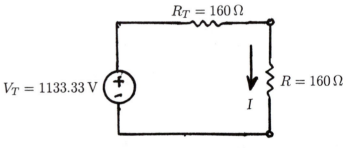

$$I = \frac{V_T}{2R_T} = \frac{1133.33\,\text{V}}{2(160\,\Omega)} = 3.542\,\text{A}$$

and the power dissipated by the load will be

$$P = I^2 R = (3.542\,\text{A})^2 (160\,\Omega) \approx 2007\,\text{W} \Leftarrow$$

6.22: In the network of Fig 6.22, determine the value of the load to be placed across terminals a-b in order for the load to draw maximum power. Then, determine the value of this power.

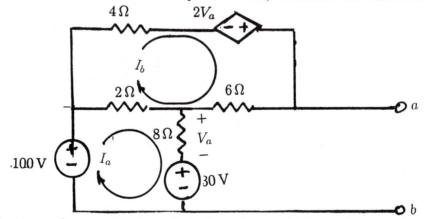

Two clockwise mesh currents, I_a and I_b, have been added to the original figure. Two applications of KVL give two mesh equations with $V_a = 8I_a$ and $2V_a = 16I_a$

$$-100\,\text{V} + 2(I_a - I_b) + 8I_a + 30\,\text{V} = 0$$

$$4I_b - 16I_a + 6I_b + 2(I_b - I_a) = 0$$

These can be simplified to

$$
\begin{aligned}
10I_a \quad - 2I_b &= 70\,\mathrm{V} \\
-18I_a \quad + 12I_b &= 0
\end{aligned}
$$

The second of these equations shows that

$$12I_b = 18I_a$$

or

$$I_b = \frac{3}{2}I_a$$

and with this in the first equation

$$10I_a - 2\left(\frac{3}{2}I_a\right) = 70\,\mathrm{V}$$

$$7I_a = 70\,\mathrm{V}$$

$$I_a = 10\,\mathrm{A}$$

and then

$$I_b = \left(\frac{3}{2}\right)I_a = \left(\frac{3}{2}\right)(10\,\mathrm{A}) = 15\,\mathrm{A}$$

Thus

$$
\begin{aligned}
V_{\mathrm{oc}} = V_T &= 30\,\mathrm{V} + 8I_a + 6I_b \\
&= 30\,\mathrm{V} + 8(10\,\mathrm{A}) + 6(15\,\mathrm{A}) \\
&= 30\,\mathrm{V} + 80\,\mathrm{V} + 90\,\mathrm{V} \\
&= 200\,\mathrm{V}
\end{aligned}
$$

For the Thevenin equivalent resistance place a short circuit across the terminals as shown in the figure at the right and note three meshes carrying clockwise mesh currents I_a, I_b and I_c. With

$$V_a = 8(I_a - I_c)$$

so that

$$2V_a = 16(I_a - I_c)$$

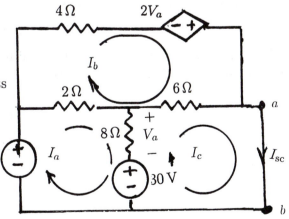

CHAPTER SEVEN
INDUCTORS, CAPACITORS AND DUALITY

INDUCTORS

7.1: Determine the voltage induced in a 250 mH inductor when the current changes at the rate of 50A/s.

Here,

$$v = L\frac{di}{dt} = (0.250\,\text{H})(50\,\text{A/s}) = 12.5\,\text{V} \Leftarrow$$

7.2: Two inductors of 300 mH and 600 mH are connected in parallel. Determine the equivalent inductance.

Here,

$$L_{\text{eq}} = \frac{L_1 L_2}{L_1 + L_2} = \frac{(0.300\,\text{H})(0.600\,\text{H})}{0.300\,\text{H} + 0.600\,\text{H}} = \frac{0.180\,\text{H}^2}{0.900\,\text{H}} = 0.200\,\text{H} \Leftarrow$$

7.3: If the current through an 80 mH inductor is

$$i = 20\sqrt{2}\sin 400t\,\text{A}$$

determine the power drawn by the inductor.

Here

$$
\begin{aligned}
v &= L\frac{di}{dt} \\[2mm]
&= (0.080\,\text{H})\frac{d}{dt}(20\sqrt{2}\sin 400t\,\text{A}) \\[2mm]
&= (0.080\,\text{H})[20\sqrt{2}(400)\cos 400t\,\text{A}]
\end{aligned}
$$

or

$$v = 640\sqrt{2}\cos 400t\,\text{V}$$

Then

$$p = vi = (640\sqrt{2}\cos 400t\,\text{V})(20\sqrt{2}\sin 400t\,\text{A})$$

or

$$p = 25{,}600\cos 400t\sin 400t\,\text{W}$$

Recall the trigonometric identity

$$\sin 2x = 2\cos x \sin x$$

With $x = 400t$

$$p = 12,800 \sin 800t \text{ W} \Leftarrow$$

7.4: The current through a 400 mH inductor is given by

$$i = 4e^{-t} - 4e^{-2t} \text{ A}$$

Determine
(a) the energy stored in the inductor at $t = 0\,s$,
(b) the energy stored in the inductor between $t = 0\,s$ and $t = 0.5\,s$,
(c) the voltage at $t = 0.5\,s$ and
(d) the instantaneous power at $t = 0.5\,s$

(a) With

$$w(t) = \frac{1}{2}Li^2 = \frac{1}{2}(0.400\,\text{H})(4e^{-t} - 4e^{-2t})^2$$

or

$$w(t) = 3.2(e^{-2t} - 2e^{-3t} + e^{-4t})\,\text{J}$$

at $t = 0\,s$

$$w(t = 0\,s) = 3.2(1 - 2 + 1) = 3.2(0) = 0\,\text{J} \Leftarrow$$

(b) At $t = 0.5\,s$

$$
\begin{aligned}
w(t = 0.5\,s) &= 3.2(e^{-1} - 2e^{-1.5} + e^{-2}) \\
&= 3.2[0.3679 - 2(0.2231) + 0.1353] \\
&= 3.2(0.0570) \\
&= 0.1823\,\text{J} \Leftarrow
\end{aligned}
$$

(c) With

$$v = L\frac{di}{dt} = (0.400\,\text{H})\frac{d}{dt}(4e^{-t} - 4e^{-2t}) = 1.600(-e^{-t} + 2e^{-2t})$$

and at $t = 0.5\,s$

$$
\begin{aligned}
v &= 1.600(-e^{-0.5} + 2e^{-1}) \\
&= 1,600(-0.6065 + 0.7358) \\
&= 1.600(0.1292) \\
&= 0.2068\,\text{V} \Leftarrow
\end{aligned}
$$

(a) Here with

$$i = 4\sin 400t\ \text{A}$$

and

$$v = L\frac{di}{dt} = (0.240\,\text{H})\frac{d}{dt}(4\sin 400t) = 384\cos 400t\ \text{V}$$

The power will be

$$p = vi = (384\cos 400t)(4\sin 400t) = 1536\cos 400t\sin 400t\ \text{W}$$

and at $t = \pi/200\,\text{s}$

$$p(t = \pi/200\,\text{s}) = 1536(\cos 2\pi)(\sin 2\pi) = 1536(1)(0) = 0\,\text{W}\ \Leftarrow$$

(b) The energy stored is the integral of the power

$$W(0 \to \pi/200\,\text{s}) = \int_0^{\pi/200} 1536\cos 400t\sin 400t\ dt$$

The form of the power is somewhat cumbersome. Making use of the trigonometric identity

$$\sin 2x = 2\cos x\sin x$$

the energy stored can be written as

$$W(0 \to \pi/200\,\text{s}) = \int_0^{\pi/200} 768\sin 800t\ dt$$

and hence

$$W(0 \to \pi/200\,\text{s}) = -0.96\cos 800t\Big|_0^{\pi/200} = -0.96(\cos 4\pi - \cos 0) == 0(1-1) = 0\,\text{J}\ \Leftarrow$$

7.8: If the current through a 135 mH inductor is given by $i = 8t^2 + 4t + 2\ \text{A}$, how much energy is stored or removed in the inductor during the period, $0 \le t \le 1.2\,\text{s}$?

Here

$$W = \frac{1}{2}L(i_f^2 - i_o^2)$$

With

$$i_f = 8(1.2\,\text{s})^2 + 4(1.2\,\text{s}) + 2 = 11.52\,\text{A} + 4.8\,\text{A} + 2\,\text{A} = 18.32\,\text{A}$$

and

$$i_o = 8(0\,\text{s})^2 + 4(0\,\text{s}) + 2 = 2\,\text{A}$$

$$W(t = 1.2\,\text{s}) = \frac{1}{2}L(i_f^2 - i_o^2)$$

$$= \frac{1}{2}(0.135\,\text{H})[(18.32\,\text{A})^2 - (2\,\text{A})^2]$$

$$= \frac{1}{2}(0.135\,\text{H})[335.62\,\text{A}^2 - 4\,\text{A}^2]$$

$$= 22.38\,\text{J} \Leftarrow$$

7.9: In Fig 7.2, the current source is supplying 288 W at $t = 2\,\text{s}$. Determine the value of R

Here

$$i = t^2 + 4\,\text{A}$$

and

$$\frac{di}{dt} = 2t\ \text{A/s}$$

At $t = 2\,\text{s}$

$$i = (2\ \text{s})^2 + 4 = 8\,\text{A}$$

and

$$\left.\frac{di}{dt}\right|_{t=2\,\text{s}} = 2(2\,\text{s}) = 4\,\text{A/s}$$

Then

$$p_L = vi = \left(L\frac{di}{dt}\right)i = (1.0\,\text{H})(2t)(t^2 + 4)\ \text{W}$$

and at $t = 2\,\text{s}$

$$p_L = (1.0\,\text{H})(4\,\text{A/s})(8\,\text{A})\ \text{W} = 32\,\text{W}$$

and because energy must be conserved

$$p_R = p_s - p_L = 288\,\text{W} - 32\,\text{W} = 256\,\text{W}$$

Because $p_R = i^2 R$

$$R = \frac{p_R}{I^2}$$

and at $t = 2\,\text{s}$

$$R = \frac{256\,\text{W}}{(8\,\text{A})^2} = 4\,\Omega \Leftarrow$$

7.10: Two inductors of 0.080 H and 0.120 H are connected in parallel. If the current at a particular instant of time is 12.5 A, determine the currents in each inductor.

CAPACITORS

7.13: Determine the current through and the charge on a $50\,\mu\text{F}$ capacitor subjected to a voltage of

$$v = 120 \sin 250t\ \text{V}$$

Here,

$$
\begin{aligned}
i &= C\frac{dv}{dt} \\
&= (50\,\mu\text{F})\frac{d}{dt}(120\sin 250t\ \text{V}) \\
&= (50\,\mu\text{F})(120)(250)(\cos 250t\ \text{V/s}) \\
&= 1.50\cos 250t\ \text{A} \Leftarrow
\end{aligned}
$$

With

$$v = \frac{1}{C}\int i\,dt \quad\text{and}\quad q = \int i\,dt$$

it is seen that

$$q = Cv = (50\,\mu\text{F})(120\sin 250t\ \text{V}) = 6\sin 250t\ \text{mC} \Leftarrow$$

7.14: An $80\,\mu\text{F}$ capacitor is charged so that it stores 0.40 J. If an uncharged $120\,\mu\text{F}$ capacitor is hooked up across the terminals of the $80\,\mu\text{F}$ capacitor, determine the final energy in the system.

The initial charge on the $80\,\mu\text{F}$ capacitor can be obtained from

$$W = \frac{1}{2}\frac{Q^2}{C}$$

Thus

$$Q = (2CW)^{1/2} = [2(80\,\mu\text{F})(0.40\,\text{J})]^{1/2} = (64\times 10^{-6})^{1/2} = 8\times 10^{-3}\ \text{C}$$

When the two capacitors are connected in parallel, the voltage across the terminals of each capacitor will be

$$V = \frac{Q_1 + Q_2}{C_1 + C_2} = \frac{8\times 10^{-3}\ \text{C}}{80\,\mu\text{F} + 120\,\mu\text{F}} = 40\ \text{V}$$

After the connection, the energy stored in of C_1 which is the original capacitor will be

$$W_1 = \frac{1}{2}C_1 V^2 = \frac{1}{2}(80\,\mu\text{F})(40\,\text{V})^2 = 0.064\ \text{J}$$

and the energy stored in C_2 which is the added capacitor will be

$$W_2 = \frac{1}{2}C_2 V^2 = \frac{1}{2}(120\,\mu\text{F})(40\,\text{V})^2 = 0.096\ \text{J}$$

and the total energy will be

$$W = W_1 + W_2 = 0.064\,\text{J} + 0.096\,\text{J} = 0.160\,\text{J} \Leftarrow$$

7.15: If an $80\,\mu\text{F}$ capacitor is charged to $600\,\mu\text{C}$ and then connected across the terminals of an uncharged $160\,\mu\text{F}$ capacitor, determine the charge transferred from the $80\,\mu\text{F}$ capacitor to the $160\,\mu\text{F}$ capacitor.

Because, in general,

$$V = \frac{Q}{C}$$

for the two capacitors at the same voltage

$$\frac{Q_2}{Q_1} = \frac{C_2}{C_1}$$

and here with $C_1 = 80\,\mu\text{F}$ and $C_2 = 160\,\mu\text{F}$

$$Q_2 = \frac{C_2}{C_1}Q_1 = \frac{160\,\mu\text{F}}{80\,\mu\text{F}}Q_1 = 2Q_1$$

Then because $Q_1 + Q_2 = 600\,\mu\text{C}$

$$\begin{aligned}
Q_1 + 2Q_1 &= 600\,\mu\text{C} \\
3Q_1 &= 600\,\mu\text{C} \\
Q_1 &= 200\,\mu\text{C} \Leftarrow
\end{aligned}$$

and

$$\begin{aligned}
Q_2 &= 2Q_1 \\
&= 400\,\mu\text{C} \Leftarrow
\end{aligned}$$

7.16: Two capacitors of $100\,\mu\text{F}$ and $400\,\mu\text{F}$ are connected in series. Determine the equivalent capacitance.

Here,

$$C_{\text{eq}} = \frac{C_1 C_2}{C_1 + C_2} = \frac{(100\,\mu\text{F})(400\,\mu\text{F})}{100\,\mu\text{F} + 400\,\mu\text{F}} = \frac{40{,}000\,\mu\text{F}^2}{500\,\mu\text{F}} = 80\,\mu\text{F} \Leftarrow$$

7.17: If the voltage across a $200\,\mu\text{F}$ capacitor is

$$v = 120\sin 400t\;\text{V}$$

determine the power provided to the capacitor.

$$= \int_0^4 (0.20t^2 + 1.60t + 3.2\,\text{A})\,dt$$

$$= \left[\frac{1}{15}t^3 + 0.80t^2 + 3.2t\right]_0^4$$

$$= \left[\frac{64}{15} + 12.80 + 12.80\right]$$

$$= 29.87\,\text{J} \Leftarrow$$

(e) For the capacitor between $0 \le t \le 4\,\text{s}$

$$W_C = \int_0^4 p_C\,dt$$

$$= \int_0^4 (t + 4)\,dt$$

$$= \left[\frac{1}{2}t^2 + 4t\right]_0^4$$

$$= (8 + 16)\,\text{J}$$

$$= 24\,\text{J} \Leftarrow$$

(f) For the source between $0 \le t \le 4\,\text{s}$

$$W_s = W_R + W_C$$

$$= 29.87\,\text{J} + 24.00\,\text{W}$$

$$= 53.87\,\text{A} \Leftarrow$$

7.20: The voltage across a $40\,\mu\text{F}$ capacitor is given by $v = 4e^{-t} - 2^{-2t}\,\text{V}$. Determine how much energy will be stored or removed during the period $0 \le t \le 2\,\text{s}$.

Here with v_o designating the voltage at $t = 0\,\text{s}$ and v_f designating the voltage at $t = 2\,\text{s}$

$$v_o = 4\,\text{V} - 2\,\text{V} = 2\,\text{V}$$

and

$$v_f = 4e^{-2} - 2e^{-4}\,\text{V}$$

$$= 4(0.1353) - 2(0.0183)\,\text{V}$$

$$= 0.5047\,\text{V}$$

Then

$$W_C = \frac{1}{2}C(v_f^2 - v_o^2)$$
$$= \frac{1}{2}(40\,\mu\text{F})[(0.5047\,\text{V})^2 - (2\,\text{V})^2]$$
$$= -74.91\,\mu\text{J} \Leftarrow$$

This energy is being discharged by (removed from) the capacitor.

7.21: How much energy is stored in an initially uncharged $125\,\mu\text{F}$ capacitor between 0 and 8 s when a voltage of $18e^{-t/10}\,\text{V}$ is placed across its terminals.

Here with v_o designating the voltage at $t = 0\,\text{s}$ and v_f designating the voltage at $t = 8\,\text{s}$, $v_o = 0\,\text{V}$ because the capacitor is initially uncharged and

$$v_f = 18e^{-8\,\text{s}/10}\,\text{V} = 18e^{-0.8}\,\text{V} = 18(0.4493)\,\text{V} = 8.0879\,\text{V}$$

Then

$$W_C = \frac{1}{2}C(v_f^2 - v_o^2)$$
$$= \frac{1}{2}(125\,\mu\text{F})[(8.0879)^2 - 0^2]$$
$$= 4.088\,\text{mJ} \Leftarrow$$

7.22: Each pair of two pairs of capacitors are connected in parallel. The first pair has $C_1 = 60\,\mu\text{F}$ and $C_2 = 30\,\mu\text{F}$ and the second pair has $C_3 = 120\,\mu\text{F}$ with C_4 unknown. The two pairs of capacitors are connected in series and if the equivalent capacitance of the entire arrangement is $C_{\text{eq}} = 60\,\mu\text{F}$, determine the value of the unknown capacitor, C.

Here, the equivalent capacitance of each pair of capacitors in parallel is

$$C_{\text{eq},1} = C_1 + C_2 = 60\,\mu\text{F} + 30\,\mu\text{F} = 90\,\mu\text{F}$$

and

$$C_{\text{eq},2} = C_3 + C = 120\,\mu\text{F} + C$$

The equivalent capacitance for the entire arrangement is

$$C_{\text{eq}} = 60\,\mu\text{F}$$
$$= \frac{(C_{\text{eq},1})(C_{\text{eq},2})}{C_{\text{eq},1} + C_{\text{eq},2}}$$
$$= \frac{(90\,\mu\text{F})(120\,\mu\text{F} + C)}{90\,\mu\text{F} + 120\,\mu\text{F} + C}$$

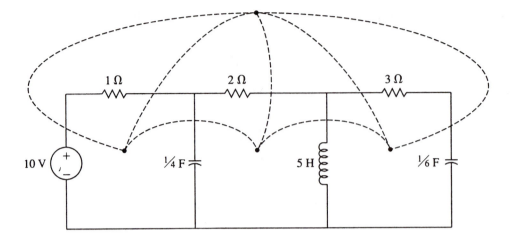

In Fig 7.4, three mesh currents, i_1, i_2 and i_3 have been inserted and, by a repeated application of KVL, three intego-differential mesh equations can be written noting four node-points including a "ground node."

$$i_1 + 4 \int i_1 \, dt - 4 \int i_2 \, dt = 10 \, \text{V}$$

$$-4 \int i_1 \, dt + 5 \frac{di_2}{dt} + 2i_2 + 4 \int i_2 \, dt - 5 \frac{di_3}{dt} = 0$$

$$-5 \frac{di_2}{dt} + 5 \frac{di_3}{dt} + 3i_3 + 6 \int i_3 \, dt = 0$$

The mechanism for constructing the dual network is also shown in the foregoing figure and the dual network is shown on the next page. In the dual network, there are four nodes indicated by numerals within circles. The datum node is node-4 and by three applications of KCL, three integro-differential equations can be written

$$v_1 + 4 \int v_1 \, dt - 4 \int v_2 \, dt = 10 \, \text{A}$$

$$-4 \int v_1 \, dt + 5 \frac{dv_2}{dt} + 2v_2 + 4 \int v_2 \, dt - 5 \frac{dv_3}{dt} = 0$$

$$-5 \frac{dv_2}{dt} + 5 \frac{dv_3}{dt} + 3v_3 + 6 \int v_3 \, dt = 0$$

350

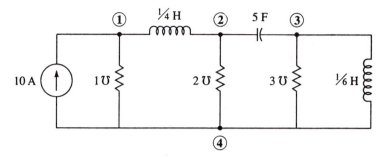

Observe that the mesh equations and the node equations match.

7.26: Construct the dual network for the network in Fig 7.5 and then verify that the node equations for the dual network are in the same form as the mesh equations for the network in Fig 7.5

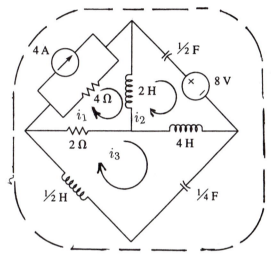

In an adjustment to Fig 7.5, three meshes containing mesh currents, i_1, i_2 and i_3 and an outer mesh have been inserted and, after a transformation of the 4 A current source to a voltage source, by a repeated application of KVL, three intego-differential mesh equations can be written.

$$6i_1 + 2\frac{di_1}{dt} - 2\frac{di_2}{dt} - 2i_3 = 16 \text{ V}$$

$$-2\frac{di_1}{dt} + 6\frac{di_2}{dt} + 2\int i_2\,dt - 4\frac{di_3}{dt} = -8 \text{ V}$$

$$-2i_1 - 4\frac{di_2}{dt} + 2i_3 + \frac{9}{2}\frac{di_3}{dt} + 4\int i_3\,dt = 0$$

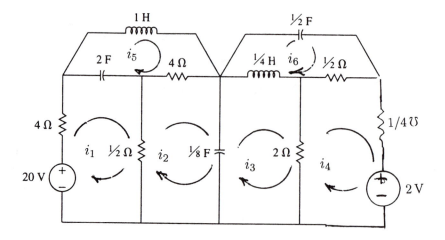

A rework of Fig 7.7 shows six meshes containing mesh currents, i_1 through i_6 and the 8 A current source has been transformed to a voltage source. Mesh-0 represents the outer mesh and, by a repeated application of KVL, six mesh equations can be written

$$\frac{9}{2}i_1 - \frac{1}{2}i_2 + \frac{1}{2}\int i_1\,dt - \frac{1}{2}\int i_5\,dt = 20\,\text{V}$$

$$-\frac{1}{2}i_1 + \frac{9}{2}i_2 + 8\int i_2\,dt - 8\int i_3\,dt - 4i_5 = 0$$

$$-8\int i_2\,dt + 2i_3 + 8\int i_3\,dt + \frac{1}{4}\frac{di_3}{dt} - 2i_4 - \frac{1}{4}\frac{di_6}{dt} = 0$$

$$-2i_3 + \frac{11}{4}i_4 - \frac{1}{2}i_6 = -2\,\text{V}$$

$$-\frac{1}{2}\int i_1\,dt - 4i_2 + \frac{1}{2}\int i_5\,dt + 4i_5 + \frac{di_5}{dt} = 0$$

$$-\frac{1}{4}\frac{di_3}{dt} - \frac{1}{2}i_4 + \frac{1}{4}\frac{di_3}{dt} + \frac{1}{2}i_6 + 2\int i_6\,dt = 0$$

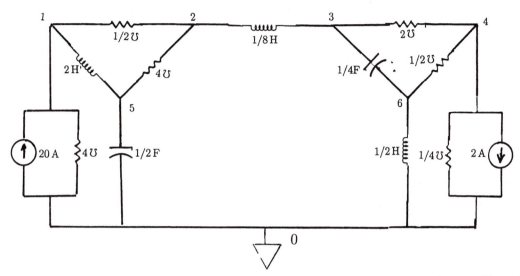

The dual network has seven nodes, one of which is the datum or "ground" node and the node equations are

$$\frac{9}{2}v_1 - \frac{1}{2}v_2 + \frac{1}{2}\int v_1\, dt - \frac{1}{2}\int v_5\, dt = 20\,\text{A}$$

$$-\frac{1}{2}v_1 + \frac{9}{2}v_2 + 8\int v_2\, dt - 8\int v_3\, dt - 4v_5 = 0$$

$$-8\int v_2\, dt + 2v_3 + 8\int v_3\, dt + \frac{1}{4}\frac{dv_3}{dt} - 2v_4 - \frac{1}{4}\frac{dv_6}{dt} = 0$$

$$-2v_3 + \frac{11}{4}v_4 - \frac{1}{2}v_6 = -2\,\text{A}$$

$$-\frac{1}{2}\int v_1\, dt - 4v_2 + \frac{1}{2}\int v_5\, dt + 4v_5 + \frac{dv_5}{dt} = 0$$

$$-\frac{1}{4}\frac{dv_3}{dt} - \frac{1}{2}v_4 + \frac{1}{4}\frac{dv_3}{dt} + \frac{1}{2}v_6 + 2\int v_6\, dt = 0$$

Observe that the mesh equations and the node equations match.

With

$$\frac{1}{T} = \frac{1}{0.2\,\text{s}} = 5\,\text{s}^{-1}$$

and use of the condition at $0.12\,\text{s}$ permits writing

$$v(t = 0.12\,\text{s}) = 658.574\,\text{V} = Ve^{-(5\,\text{s}^{-1})(0.12\,\text{s})} = Ve^{-0.60}$$

Thus, the initial value of the voltage is

$$V = \frac{658.574\,\text{V}}{e^{-0.60}} = \frac{658.574\,\text{V}}{0.5488} = 1200\,\text{V} \Leftarrow$$

The voltage is represented by

$$v(t) = (1200\,\text{V})e^{-5t}$$

where t is in s and at $t = 0.0325\,\text{s}$ $(32.5\,\text{ms})$

$$v(t = 0.0325\,\text{s}) = (1200\,\text{V})e^{-(5\,\text{s}^{-1})(0.0325\,\text{s})}$$

$$v(t = 0.0325\,\text{s}) = (1200\,\text{V})e^{-0.1625} = (1200\,\text{V})(0.8500) = 1020\,\text{V} \Leftarrow$$

8.3: If an exponentially decaying current has a time constant of $100\,\text{ms}$ and if the current has a value of $12.42\,\text{A}$ at $t = 165\,\text{ms}$, determine the value of the current at $t = 0$ and the time at which the current has a value of $4.571\,\text{A}$.

With

$$i(t) = Ie^{-t/T}$$

and with the time constant, T, given as $100\,\text{ms}$ and $i(t = 165\,\text{ms}) = 12.42\,\text{A}$,

$$
\begin{aligned}
i(t = 0.165\,\text{s}) &= Ie^{-0.165\,\text{s}/0.100\,\text{s}} \\
12.42\,\text{A} &= Ie^{-1.65} \\
&= I(0.1920) \\
I &= \frac{12.42\,\text{A}}{0.1920} \\
i(t = 0\,\text{s}) &= 64.67\,\text{A} \Leftarrow
\end{aligned}
$$

Thus,

$$i(t) = (64.67\,\text{A})e^{-t/100\,\text{ms}}$$

and

$$4.571\,\text{A} = (64.67\,\text{A})e^{-t/100\,\text{ms}}$$

$$e^{-t/100\,\text{ms}} = \frac{4.571\,\text{A}}{64.671\,\text{A}}$$

$$-\frac{t}{100\,\text{ms}} = \ln(0.0707)$$

$$-\frac{t}{100\,\text{ms}} = -2.65$$

$$t = 100(2.65\,\text{ms}) = 265\,\text{ms} \Leftarrow$$

8.4: An exponentially decaying voltage has a time constant of 400 ms and an initial value of 1.2 kV. What is the value of the voltage at $t = 125$ ms and what is the value of the voltage five time constants later.

With

$$v(t) = Ve^{-t/T}$$

and with the time constant, T, given as 400 ms and $V = 1200$ V

$$v(t) = (1200\,\text{V})e^{-t/400\,\text{ms}}$$

At $t = 125$ ms

$$v(t = 125\,\text{ms}) = (1200\,\text{V})e^{-125\,\text{ms}/400\,\text{ms}} = (1200\,\text{V})e^{-0.3125} = (1200\,\text{V})(0.7316) = 877.9\,\text{V} \Leftarrow$$

Five time constants later fixes the time at

$$t = 0.125\,\text{s} + 5(0.400\,\text{s}) = 2.125\,\text{s}$$

and

$$v(t = 2.125\,\text{s}) = (1200\,\text{V})e^{-2125\,\text{ms}/400\,\text{ms}} = (1200\,\text{V})e^{-5.3125}$$

$$v(t = 2.125\,\text{s}) = (1200\,\text{V})(4.930 \times 10^{-3}) = 5.92\,\text{V} \Leftarrow$$

8.5 Measurements of an exponentially decaying current yield the following data:

Time, s	Current, A
0.250	294.30
0.275	266.30

Determine the initial value of the current and the time constant.

Here, the data may be used in the decaying exponential of the form

$$i(t) = Ie^{-t/T} \text{ A}$$

With t and T in s

$$294.30 = Ie^{-0.250/T}$$
$$266.20 = Ie^{-0.275/T}$$

If the first is divided by the second, the result is

$$1.1056 = e^{0.025/T}$$

and this may be solved for the time constant, T

$$\ln 1.1056 = 0.025/T$$
$$0.1000 = 0.025/T$$
$$T = \frac{0.025}{0.1000}$$
$$= 0.25 \text{ s} \Leftarrow$$

With

$$\frac{1}{T} = \frac{1}{0.25 \text{ s}} = 4 \text{ s}^{-1}$$

use of the condition at 0.25 s permits writing

$$i(t = 0.25 \text{ s}) = 294.30 \text{ A} = Ie^{-(4 \text{ s}^{-1})(0.25 \text{ s})}$$

$$294.30 \text{ A} = Ie^{-1.00}$$

Thus, the initial value of the current is

$$I = \frac{294.30 \text{ A}}{e^{-1.00}} = \frac{294.30 \text{ A}}{0.3679} = 800 \text{ A} \Leftarrow$$

8.6 Measurements of an exponentially decaying voltage yield the following data:

Time, ms	Voltage, A
10	53.375
25	25.214

and

$$R_T = 300\,\Omega + 800\,\Omega = 1100\,\Omega$$

For maximum power transfer, make
$R_L = R_T$

$$R_L = R_T = 1100\,\Omega \Leftarrow$$

The Thevenin equivalent circuit with the load resistor reconnected is shown at the right. Here, the current, I, through the load is

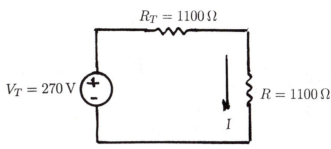

$$I = \frac{V_T}{2R_T} = \frac{270\,\text{V}}{2(1100\,\Omega)} = 0.123\,\text{A}$$

and the power dissipated by the load will be

$$P = I^2 R = (0.123\,\text{A})^2 (1100\,\Omega) = 16.57\,\text{W} \Leftarrow$$

6.21: In the network of Fig 6.21, determine the value of the load to be placed across terminals a-b in order for the load to draw maximum power. Then, determine the value of this power.

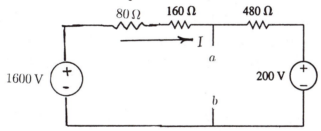

With the transformation of the current source on the left to a voltage source, the current in the single loop is

$$I = \frac{1600\,\text{V} - 200\,\text{V}}{80\,\Omega + 160\,\Omega + 480\,\Omega} = \frac{1400\,\text{V}}{720\,\Omega} = 1.944\,\text{A}$$

Then with $V_{\text{oc}} = V_T$

$$
\begin{aligned}
V_T &= 1600\,\text{V} - (80\,\Omega + 160\,\Omega)I \\
&= 1600\,\text{V} - (240\,\Omega)(1.944\,\text{A}) \\
&= 1600\,\text{V} - 466.67\,\text{V} \\
&= 1133.33\,\text{V}
\end{aligned}
$$

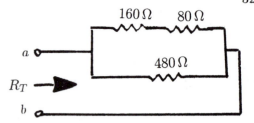

The Thevenin equivalent resistance is obtained by removing both sources and looking back into the terminals. With both voltage sources removed and replaced by short circuits, the picture is as shown at the right.

$$R_T = \frac{(160\,\Omega + 80\,\Omega)(480\,\Omega)}{160\,\Omega + 80\,\Omega + 480\,\Omega} = \frac{115,200\,\Omega^2}{720\,\Omega} = 160\,\Omega$$

For maximum power transfer, make $R_L = R_T$

$$R_L = R_T = 160\,\Omega \Leftarrow$$

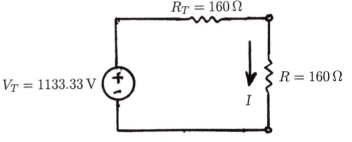

The Thevenin equivalent circuit with the load resistor reconnected is shown at the right. Here, the current, I, through the load is

$$I = \frac{V_T}{2R_T} = \frac{1133.33\,\text{V}}{2(160\,\Omega)} = 3.542\,\text{A}$$

and the power dissipated by the load will be

$$P = I^2 R = (3.542\,\text{A})^2(160\,\Omega) \approx 2007\,\text{W} \Leftarrow$$

6.22: In the network of Fig 6.22, determine the value of the load to be placed across terminals a-b in order for the load to draw maximum power. Then, determine the value of this power.

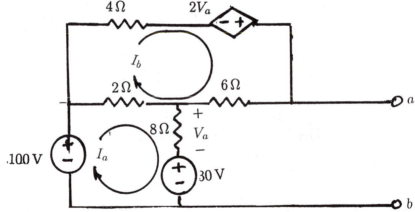

Two clockwise mesh currents, I_a and I_b, have been added to the original figure. Two applications of KVL give two mesh equations with $V_a = 8I_a$ and $2V_a = 16I_a$

$$-100\,\text{V} + 2(I_a - I_b) + 8I_a + 30\,\text{V} = 0$$

$$4I_b - 16I_a + 6I_b + 2(I_b - I_a) = 0$$

These can be simplified to

$$
\begin{aligned}
10I_a \quad - 2I_b &= 70\,\text{V} \\
-18I_a \quad + 12I_b &= 0
\end{aligned}
$$

The second of these equations shows that

$$12I_b = 18I_a$$

or

$$I_b = \frac{3}{2}I_a$$

and with this in the first equation

$$10I_a - 2\left(\frac{3}{2}I_a\right) = 70\,\text{V}$$

$$7I_a = 70\,\text{V}$$

$$I_a = 10\,\text{A}$$

and then

$$I_b = \left(\frac{3}{2}\right)I_a = \left(\frac{3}{2}\right)(10\,\text{A}) = 15\,\text{A}$$

Thus

$$
\begin{aligned}
V_{\text{oc}} = V_T &= 30\,\text{V} + 8I_a + 6I_b \\
&= 30\,\text{V} + 8(10\,\text{A}) + 6(15\,\text{A}) \\
&= 30\,\text{V} + 80\,\text{V} + 90\,\text{V} \\
&= 200\,\text{V}
\end{aligned}
$$

For the Thevenin equivalent resistance place a short circuit across the terminals as shown in the figure at the right and note three meshes carrying clockwise mesh currents I_a, I_b and I_c. With

$$V_a = 8(I_a - I_c)$$

so that

$$2V_a = 16(I_a - I_c)$$

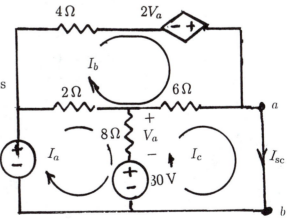

three applications of KVL give

$$-100\,\text{V} + 2(I_a - I_b) + 8(I_a - I_c) + 30\,\text{V} = 0$$
$$4I_b - 16(I_a - I_c) + 6(I_b - I_c) + 2(I_b - I_a) = 0$$
$$6(I_c - I_b) - 30\,\text{V} + 8(I_c - I_a) = 0$$

and these can be simplified to

$$
\begin{aligned}
10I_a &\quad -2I_b &\quad -8I_c &= 70\,\text{V} \\
-18I_a &\quad +12I_b &\quad +10I_c &= 0 \\
-8I_a &\quad -6I_b &\quad +14I_c &= 30\,\text{V}
\end{aligned}
$$

When written in matrix form, the set of equations becomes

$$
\begin{bmatrix} 10 & -2 & -8 \\ -18 & 12 & 10 \\ -8 & -6 & 14 \end{bmatrix}
\begin{bmatrix} I_a \\ I_b \\ I_c \end{bmatrix}
=
\begin{bmatrix} 70\,\text{V} \\ 0 \\ 30\,\text{V} \end{bmatrix}
$$

and a matrix inversion can be employed to find the current vector

$$
\begin{bmatrix} I_a \\ I_b \\ I_c \end{bmatrix}
=
\begin{bmatrix} 10 & -2 & -8 \\ -18 & 12 & 10 \\ -8 & -6 & 14 \end{bmatrix}^{-1}
\begin{bmatrix} 70\,\text{V} \\ 0 \\ 30\,\text{V} \end{bmatrix}
$$

or

$$
\begin{bmatrix} I_a \\ I_b \\ I_c \end{bmatrix}
=
\begin{bmatrix} 0.7500 & 0.2500 & 0.2500 \\ 0.5658 & 0.2500 & 0.1447 \\ 0.6711 & 0.2500 & 0.2763 \end{bmatrix}
\begin{bmatrix} 70\,\text{V} \\ 0 \\ 30\,\text{V} \end{bmatrix}
=
\begin{bmatrix} 60.000\,\text{A} \\ 43.947\,\text{A} \\ 55.263\,\text{A} \end{bmatrix}
$$

With $I_c = I_{sc} = 55.263\,\text{A}$ the Thevenin equivalent resistance is

$$R_T = \left| \frac{V_T}{I_{sc}} \right| = \frac{200\,\text{V}}{55.263\,\text{A}} = 3.619\,\Omega$$

For maximum power transfer, make $R_L = R_T$

$$R_L = R_T = 3.619\,\Omega$$

The Thevenin equivalent circuit with the load resistor reconnected is shown on the next page. Here, the current, I, through the load is

$$
\begin{aligned}
I &= \frac{V_T}{2R_T} \\[2mm]
&= \frac{200\,\text{V}}{2(3.619\,\Omega)} \\[2mm]
&= 27.63\,\text{A}
\end{aligned}
$$

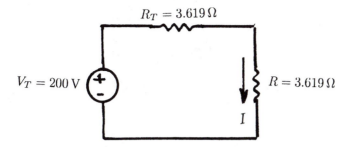

and the power dissipated by the load will be

$$P = I^2 R = (27.63 \, \text{A})^2 (3.619 \, \Omega) \approx 2763 \, \text{W}$$

CHAPTER SEVEN
INDUCTORS, CAPACITORS AND DUALITY

INDUCTORS

7.1: Determine the voltage induced in a 250 mH inductor when the current changes at the rate of 50 A/s.

Here,

$$v = L\frac{di}{dt} = (0.250\,\text{H})(50\,\text{A/s}) = 12.5\,\text{V} \Leftarrow$$

7.2: Two inductors of 300 mH and 600 mH are connected in parallel. Determine the equivalent inductance.

Here,

$$L_{\text{eq}} = \frac{L_1 L_2}{L_1 + L_2} = \frac{(0.300\,\text{H})(0.600\,\text{H})}{0.300\,\text{H} + 0.600\,\text{H}} = \frac{0.180\,\text{H}^2}{0.900\,\text{H}} = 0.200\,\text{H} \Leftarrow$$

7.3: If the current through an 80 mH inductor is

$$i = 20\sqrt{2}\sin 400t\,\text{A}$$

determine the power drawn by the inductor.

Here

$$\begin{aligned} v &= L\frac{di}{dt} \\[2mm] &= (0.080\,\text{H})\frac{d}{dt}(20\sqrt{2}\sin 400t\,\text{A}) \\[2mm] &= (0.080\,\text{H})[20\sqrt{2}(400)\cos 400t\,\text{A}] \end{aligned}$$

or

$$v = 640\sqrt{2}\cos 400t\,\text{V}$$

Then

$$p = vi = (640\sqrt{2}\cos 400t\,\text{V})(20\sqrt{2}\sin 400t\,\text{A})$$

or

$$p = 25,600\cos 400t\sin 400t\,\text{W}$$

Recall the trigonometric identity

$$\sin 2x = 2\cos x\sin x$$

With $x = 400t$

$$p = 12,800 \sin 800t \, \text{W} \Leftarrow$$

7.4: The current through a 400 mH inductor is given by

$$i = 4e^{-t} - 4e^{-2t} \, \text{A}$$

Determine
(a) the energy stored in the inductor at $t = 0\,s$,
(b) the energy stored in the inductor between $t = 0\,s$ and $t = 0.5\,s$,
(c) the voltage at $t = 0.5\,s$ and
(d) the instantaneous power at $t = 0.5\,s$

(a) With

$$w(t) = \frac{1}{2}Li^2 = \frac{1}{2}(0.400 \, \text{H})(4e^{-t} - 4e^{-2t})^2$$

or

$$w(t) = 3.2(e^{-2t} - 2e^{-3t} + e^{-4t}) \, \text{J}$$

at $t = 0\,s$

$$w(t = 0\,s) = 3.2(1 - 2 + 1) = 3.2(0) = 0 \, \text{J} \Leftarrow$$

(b) At $t = 0.5\,s$

$$
\begin{aligned}
w(t = 0.5\,s) &= 3.2(e^{-1} - 2e^{-1.5} + e^{-2}) \\
&= 3.2[0.3679 - 2(0.2231) + 0.1353] \\
&= 3.2(0.0570) \\
&= 0.1823 \, \text{J} \Leftarrow
\end{aligned}
$$

(c) With

$$v = L\frac{di}{dt} = (0.400 \, \text{H})\frac{d}{dt}(4e^{-t} - 4e^{-2t}) = 1.600(-e^{-t} + 2e^{-2t})$$

and at $t = 0.5\,s$

$$
\begin{aligned}
v &= 1.600(-e^{-0.5} + 2e^{-1}) \\
&= 1,600(-0.6065 + 0.7358) \\
&= 1.600(0.1292) \\
&= 0.2068 \, \text{V} \Leftarrow
\end{aligned}
$$

(d) With $p = vi$

$$\begin{aligned} p &= [1.600(2e^{-2t} - e^{-t})][4e^{-t} - 4e^{-2t}] \\ &= 1.600(12e^{-3t} - 4e^{-2t} - 8e^{-4t}) \end{aligned}$$

At $t = 0.5\,\text{s}$

$$\begin{aligned} p &= 1.600(12e^{-1.5} - 4e^{-1} - 8e^{-2}) \\ &= 1.600(2.6776 - 1.4715 - 1.0827) \\ &= 1.600(0.1234) \\ &= 0.1974\,\text{W} \Leftarrow \end{aligned}$$

7.5: A parallel combination of two inductors ($L_1 = 1.2\,\text{H}$ and $L_2 = 0.60\,\text{H}$) is placed in series with two more inductors ($L_3 = 1.0\,\text{H}$ and $L_2 = 0.80\,\text{H}$). Determine the equivalent inductance.

Here

$$\begin{aligned} L_{eq} &= \frac{L_1 L_2}{L_1 + L_2} + L_3 + L_4 \\ &= \frac{(1.20\,\text{H})(0.60\,\text{H})}{1.20\,\text{H} + 0.60\,\text{H}} + 1.00\,\text{H} + 0.80\,\text{H} \\ &= \frac{0.72\,\text{H}^2}{1.80\,\text{H}} + 1.80\,\text{H} \\ &= 0.40\,\text{H} + 1.80\,\text{H} \\ &= 2.20\,\text{H} \Leftarrow \end{aligned}$$

7.6: The simple network shown in Fig 7.1 is connected to a current source at $t = 0$. The current provided by the source is $i = 4t\,\text{A}$.

At $t = 5\,\text{s}$, find
(a) the instantaneous power drawn by the resistor,
(b) the instantaneous power drawn by the inductor,
(c) the instantaneous power delivered by the source,

and between $0 \le t \le 4\,\text{s}$,

(d) the energy dissipated by the resistor
(e) the energy stored by the inductor
(f) the energy delivered by the current source.

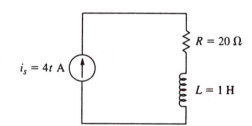

(a) For the resistor
$$p_R = i^2 R = (4t)^2 (20\,\Omega) = 320t^2\,\text{W}$$

and at $t = 5\,\text{s}$
$$p_R(t = 5\,\text{s}) = 320(5\,\text{s})^2 = 8000\,\text{W} \Leftarrow$$

(b) For the inductor
$$v = L\frac{di}{dt} = (1\,\text{H})\frac{d}{dt}(4t) = 4\,\text{V}$$

and
$$p_L = vi = (4\,\text{V})(4t) = 16t\,\text{W}$$

and at $t = 5\,\text{s}$
$$p_L(t = 5\,\text{s}) = 16(5\,\text{s}) = 80\,\text{W} \Leftarrow$$

(c) For the source
$$
\begin{aligned}
p(t = 5\,\text{s}) &= p_R(t = 5\,\text{s}) + p_L(t = 5\,\text{s}) \\
&= 8000\,\text{W} + 80\,\text{W} \\
&= 8080\,\text{W} \Leftarrow
\end{aligned}
$$

(d) For the resistor between $0 \le t \le 4\,\text{s}$
$$W_R = \int_0^4 p_R\,dt = \int_0^4 320t^2\,dt = \frac{320}{3}t^3\Big|_0^4 = \frac{320}{3}(4\,\text{s})^3 = 6826.67\,\text{J} \Leftarrow$$

(e) For the inductor between $0 \le t \le 4\,\text{s}$ with i_f and i_o designating the final and initial values of the current respectively. With $i = 4t\,\text{A}$, $i_f = 4(4\,\text{s}) = 16\,\text{A}$ and $i_o = 0\,\text{A}$
$$W_L = \frac{1}{2}L(i_f^2 - i_o^2) = \frac{1}{2}(1.00\,\text{H})[(16\,\text{A})^2 - (0\,\text{A})^2] = 128\,\text{J} \Leftarrow$$

(f) For the source between $0 \le t \le 4\,\text{s}$
$$
\begin{aligned}
W_s &= W_R + W_L \\
&= 6826.7\,\text{J} + 128\,\text{J} \\
&= 6954.67\,\text{J} \Leftarrow
\end{aligned}
$$

7.7: The current through a 240 mH inductor is given by $i = 4\sin 400t\,\text{A}$. Determine

(a) the instantaneous power at $t = \pi/200\,\text{s}$ and
(b) the energy stored during the period $0 \le t \le \pi/200\,\text{s}$.

(a) Here with

$$i = 4\sin 400t \text{ A}$$

and

$$v = L\frac{di}{dt} = (0.240\,\text{H})\frac{d}{dt}(4\sin 400t) = 384\cos 400t \text{ V}$$

The power will be

$$p = vi = (384\cos 400t)(4\sin 400t) = 1536\cos 400t \sin 400t \text{ W}$$

and at $t = \pi/200\,\text{s}$

$$p(t = \pi/200\,\text{s}) = 1536(\cos 2\pi)(\sin 2\pi) = 1536(1)(0) = 0\,\text{W} \Leftarrow$$

(b) The energy stored is the integral of the power

$$W(0 \to \pi/200\,\text{s}) = \int_0^{\pi/200} 1536\cos 400t \sin 400t\, dt$$

The form of the power is somewhat cumbersome. Making use of the trigonometric identity

$$\sin 2x = 2\cos x \sin x$$

the energy stored can be written as

$$W(0 \to \pi/200\,\text{s}) = \int_0^{\pi/200} 768\sin 800t\, dt$$

and hence

$$W(0 \to \pi/200\,\text{s}) = -0.96\cos 800t\Big|_0^{\pi/200} = -0.96(\cos 4\pi - \cos 0) == 0(1-1) = 0\,\text{J} \Leftarrow$$

7.8: If the current through a 135 mH inductor is given by $i = 8t^2 + 4t + 2$ A, how much energy is stored or removed in the inductor during the period, $0 \le t \le 1.2\,\text{s}$?

Here

$$W = \frac{1}{2}L(i_f^2 - i_o^2)$$

With

$$i_f = 8(1.2\,\text{s})^2 + 4(1.2\,\text{s}) + 2 = 11.52\,\text{A} + 4.8\,\text{A} + 2\,\text{A} = 18.32\,\text{A}$$

and

$$i_o = 8(0\,\text{s})^2 + 4(0\,\text{s}) + 2 = 2\,\text{A}$$

$$W(t = 1.2\,\text{s}) = \frac{1}{2}L(i_f^2 - i_o^2)$$

$$= \frac{1}{2}(0.135\,\text{H})[(18.32\,\text{A})^2 - (2\,\text{A})^2]$$

$$= \frac{1}{2}(0.135\,\text{H})[335.62\,\text{A}^2 - 4\,\text{A}^2]$$

$$= 22.38\,\text{J} \Leftarrow$$

7.9: In Fig 7.2, the current source is supplying $288\,\text{W}$ at $t = 2\,\text{s}$. Determine the value of R

Here

$$i = t^2 + 4\,\text{A}$$

and

$$\frac{di}{dt} = 2t\ \text{A/s}$$

At $t = 2\,\text{s}$

$$i = (2\ \text{s})^2 + 4 = 8\,\text{A}$$

and

$$\left.\frac{di}{dt}\right|_{t=2\,\text{s}} = 2(2\,\text{s}) = 4\,\text{A/s}$$

Then

$$p_L = vi = \left(L\frac{di}{dt}\right)i = (1.0\,\text{H})(2t)(t^2 + 4)\ \text{W}$$

and at $t = 2\,\text{s}$

$$p_L = (1.0\,\text{H})(4\,\text{A/s})(8\,\text{A})\ \text{W} = 32\,\text{W}$$

and because energy must be conserved

$$p_R = p_s - p_L = 288\,\text{W} - 32\,\text{W} = 256\,\text{W}$$

Because $p_R = i^2 R$

$$R = \frac{p_R}{I^2}$$

and at $t = 2\,\text{s}$

$$R = \frac{256\,\text{W}}{(8\,\text{A})^2} = 4\,\Omega \Leftarrow$$

7.10: Two inductors of $0.080\,\text{H}$ and $0.120\,\text{H}$ are connected in parallel. If the current at a particular instant of time is $12.5\,\text{A}$, determine the currents in each inductor.

CAPACITORS

7.13: Determine the current through and the charge on a $50\,\mu\text{F}$ capacitor subjected to a voltage of

$$v = 120 \sin 250t \text{ V}$$

Here,

$$
\begin{aligned}
i &= C\frac{dv}{dt} \\[2mm]
&= (50\,\mu\text{F})\frac{d}{dt}(120 \sin 250t \text{ V}) \\[2mm]
&= (50\,\mu\text{F})(120)(250)(\cos 250t \text{ V/s}) \\[2mm]
&= 1.50 \cos 250t \text{ A} \Leftarrow
\end{aligned}
$$

With

$$v = \frac{1}{C}\int i\,dt \quad \text{and} \quad q = \int i\,dt$$

it is seen that

$$q = Cv = (50\,\mu\text{F})(120 \sin 250t \text{ V}) = 6 \sin 250t \text{ mC} \Leftarrow$$

7.14: An $80\,\mu\text{F}$ capacitor is charged so that it stores 0.40 J. If an uncharged $120\,\mu\text{F}$ capacitor is hooked up across the terminals of the $80\,\mu\text{F}$ capacitor, determine the final energy in the system.

The initial charge on the $80\,\mu\text{F}$ capacitor can be obtained from

$$W = \frac{1}{2}\frac{Q^2}{C}$$

Thus

$$Q = (2CW)^{1/2} = [2(80\,\mu\text{F})(0.40\,\text{J})]^{1/2} = (64 \times 10^{-6})^{1/2} = 8 \times 10^{-3}\,\text{C}$$

When the two capacitors are connected in parallel, the voltage across the terminals of each capacitor will be

$$V = \frac{Q_1 + Q_2}{C_1 + C_2} = \frac{8 \times 10^{-3}\,\text{C}}{80\,\mu\text{F} + 120\,\mu\text{F}} = 40\,\text{V}$$

After the connection, the energy stored in of C_1 which is the original capacitor will be

$$W_1 = \frac{1}{2}C_1 V^2 = \frac{1}{2}(80\,\mu\text{F})(40\,\text{V})^2 = 0.064\,\text{J}$$

and the energy stored in C_2 which is the added capacitor will be

$$W_2 = \frac{1}{2}C_2 V^2 = \frac{1}{2}(120\,\mu\text{F})(40\,\text{V})^2 = 0.096\,\text{J}$$

and the total energy will be

$$W = W_1 + W_2 = 0.064\,\text{J} + 0.096\,\text{J} = 0.160\,\text{J} \Leftarrow$$

7.15: If an $80\,\mu\text{F}$ capacitor is charged to $600\,\mu\text{C}$ and then connected across the terminals of an uncharged $160\,\mu\text{F}$ capacitor, determine the charge transferred from the $80\,\mu\text{F}$ capacitor to the $160\,\mu\text{F}$ capacitor.

Because, in general,

$$V = \frac{Q}{C}$$

for the two capacitors at the same voltage

$$\frac{Q_2}{Q_1} = \frac{C_2}{C_1}$$

and here with $C_1 = 80\,\mu\text{F}$ and $C_2 = 160\,\mu\text{F}$

$$Q_2 = \frac{C_2}{C_1}Q_1 = \frac{160\,\mu\text{F}}{80\,\mu\text{F}}Q_1 = 2Q_1$$

Then because $Q_1 + Q_2 = 600\,\mu\text{C}$

$$
\begin{aligned}
Q_1 + 2Q_1 &= 600\,\mu\text{C} \\
3Q_1 &= 600\,\mu\text{C} \\
Q_1 &= 200\,\mu\text{C} \Leftarrow
\end{aligned}
$$

and

$$
\begin{aligned}
Q_2 &= 2Q_1 \\
&= 400\,\mu\text{C} \Leftarrow
\end{aligned}
$$

7.16: Two capacitors of $100\,\mu\text{F}$ and $400\,\mu\text{F}$ are connected in series. Determine the equivalent capacitance.

Here,

$$C_{\text{eq}} = \frac{C_1 C_2}{C_1 + C_2} = \frac{(100\,\mu\text{F})(400\,\mu\text{F})}{100\,\mu\text{F} + 400\,\mu\text{F}} = \frac{40{,}000\,\mu\text{F}^2}{500\,\mu\text{F}} = 80\,\mu\text{F} \Leftarrow$$

7.17: If the voltage across a $200\,\mu\text{F}$ capacitor is

$$v = 120\sin 400t\ \text{V}$$

determine the power provided to the capacitor.

Here

$$i = C\frac{dv}{dt}$$

$$= (200\,\mu\text{F})\frac{d}{dt}(120\sin 400t\,\text{V})$$

$$= (200\,\mu\text{F})(120)(400)\cos 400t\,\text{V}$$

or

$$i = 9.60\cos 400t\,\text{A}$$

Then

$$p = vi = (120\sin 400t\,\text{V})(9.60\cos 400t\,\text{A})$$

or

$$p = 1152\sin 400t\cos 400t\,\text{W}$$

Recall the trigonometric identity

$$\sin 2x = 2\cos x\sin x$$

With $x = 400t$

$$p = 576\sin 800t\,\text{W} \Leftarrow$$

7.18: A parallel combination of two capacitors ($C_1 = 80\,\mu\text{F}$ and $C_2 = 120\,\mu\text{F}$) is placed in parallel with a series combination of two more capacitors ($C_3 = 40\,\mu\text{F}$ and $C_4 = 160\,\mu\text{F}$). Determine the equivalent capacitance.

Here

$$C_{\text{eq}} = C_1 + C_2 + \frac{C_3 C_4}{C_3 + C_4}$$

$$= 80\,\mu\text{F} + 120\,\mu\text{F} + \frac{(40\,\mu\text{F})(160\,\mu\text{F})}{40\,\mu\text{F} + 160\,\mu\text{F}}$$

$$= 200\,\mu\text{F} + \frac{6400\,\mu\text{F}^2}{200\,\mu\text{F}}$$

$$= 200\,\mu\text{F} + 32\,\mu\text{F}$$

$$= 232\,\mu\text{F} \Leftarrow$$

7.19: The simple network shown in Fig 7.3 is connected to a voltage source at $t = 0$. The voltage provided by the source is $v = 2t + 8\,\text{V}$.

At $t = 4\,\text{s}$, find
(a) the instantaneous power drawn by the resistor,
(b) the instantaneous power drawn by the capacitor,
(c) the instantaneous power delivered by the source,

and between $0 \le t \le 4\,\text{s}$,

(d) the energy dissipated by the resistor
(e) the energy stored by the capacitor
(f) the energy delivered by the current source.

(a) For the resistor with

$$i_R = \frac{v}{R} = \frac{2t + 8\,\text{V}}{20\,\Omega} = 0.1t + 0.40\,\text{A}$$

$$p_R = vi_R = (2t + 8\,\text{V})(0.1t + 0.4\,\text{A}) = 0.20t^2 + 1.60t + 3.2\,\text{W}$$

and at $t = 4\,\text{s}$

$$
\begin{aligned}
p_R(t = 4\,\text{s}) &= 0.20(4\,\text{s})^2 + 1.60(4\,\text{s}) + 3.20\,\text{W} \\
&= (3.20 + 6.40 + 3.20)\,\text{W} \\
&= 12.80\,\text{W} \Leftarrow
\end{aligned}
$$

(b) For the capacitor

$$i_C = C\frac{dv}{dt} = (0.250\,\text{F})\frac{d}{dt}(2t + 8\,\text{V}) = 0.50\,\text{A}$$

and

$$p_C = vi_C = (2t + 8\,\text{V})(0.50\,\text{A}) = t + 4\,\text{W}$$

and at $t = 4\,\text{s}$

$$p_C(t = 4\,\text{s}) = 4\,\text{W} + 4\,\text{W} = 8\,\text{W} \Leftarrow$$

(c) For the source

$$
\begin{aligned}
p(t = 4\,\text{s}) &= p_R(t = 4\,\text{s}) + p_C(t = 4\,\text{s}) \\
&= 12.80\,\text{W} + 8.00\,\text{W} \\
&= 20.80\,\text{W} \Leftarrow
\end{aligned}
$$

(d) For the resistor between $0 \le t \le 4\,\text{s}$

$$W_R = \int_0^4 p_R\,dt$$

$$= \int_0^4 (0.20t^2 + 1.60t + 3.2 \, \text{A}) \, dt$$

$$= \left[\frac{1}{15}t^3 + 0.80t^2 + 3.2t \right]_0^4$$

$$= \left[\frac{64}{15} + 12.80 + 12.80 \right]$$

$$= 29.87 \, \text{J} \Leftarrow$$

(e) For the capacitor between $0 \leq t \leq 4 \, \text{s}$

$$W_C = \int_0^4 p_C \, dt$$

$$= \int_0^4 (t + 4) \, dt$$

$$= \left[\frac{1}{2}t^2 + 4t \right]_0^4$$

$$= (8 + 16) \, \text{J}$$

$$= 24 \, \text{J} \Leftarrow$$

(f) For the source between $0 \leq t \leq 4 \, \text{s}$

$$W_s = W_R + W_C$$

$$= 29.87 \, \text{J} + 24.00 \, \text{W}$$

$$= 53.87 \, \text{A} \Leftarrow$$

7.20: The voltage across a $40 \, \mu\text{F}$ capacitor is given by $v = 4e^{-t} - 2^{-2t} \, \text{V}$. Determine how much energy will be stored or removed during the period $0 \leq t \leq 2 \, \text{s}$.

Here with v_o designating the voltage at $t = 0 \, \text{s}$ and v_f designating the voltage at $t = 2 \, \text{s}$

$$v_o = 4 \, \text{V} - 2 \, \text{V} = 2 \, \text{V}$$

and

$$v_f = 4e^{-2} - 2e^{-4} \, \text{V}$$

$$= 4(0.1353) - 2(0.0183) \, \text{V}$$

$$= 0.5047 \, \text{V}$$

Then

$$W_C = \frac{1}{2}C(v_f^2 - v_o^2)$$

$$= \frac{1}{2}(40\,\mu\text{F})[(0.5047\,\text{V})^2 - (2\,\text{V})^2]$$

$$= -74.91\,\mu\text{J} \Leftarrow$$

This energy is being discharged by (removed from) the capacitor.

7.21: How much energy is stored in an initially uncharged $125\,\mu\text{F}$ capacitor between 0 and 8 s when a voltage of $18e^{-t/10}\,\text{V}$ is placed across its terminals.

Here with v_o designating the voltage at $t = 0\,\text{s}$ and v_f designating the voltage at $t = 8\,\text{s}$, $v_o = 0\,\text{V}$ because the capacitor is initially uncharged and

$$v_f = 18e^{-8\,\text{s}/10}\,\text{V} = 18e^{-0.8}\,\text{V} = 18(0.4493)\,\text{V} = 8.0879\,\text{V}$$

Then

$$W_C = \frac{1}{2}C(v_f^2 - v_o^2)$$

$$= \frac{1}{2}(125\,\mu\text{F})[(8.0879)^2 - 0^2]$$

$$= 4.088\,\text{mJ} \Leftarrow$$

7.22: Each pair of two pairs of capacitors are connected in parallel. The first pair has $C_1 = 60\,\mu\text{F}$ and $C_2 = 30\,\mu\text{F}$ and the second pair has $C_3 = 120\,\mu\text{F}$ with C_4 unknown. The two pairs of capacitors are connected in series and if the equivalent capacitance of the entire arrangement is $C_{\text{eq}} = 60\,\mu\text{F}$, determine the value of the unknown capacitor, C.

Here, the equivalent capacitance of each pair of capacitors in parallel is

$$C_{\text{eq},1} = C_1 + C_2 = 60\,\mu\text{F} + 30\,\mu\text{F} = 90\,\mu\text{F}$$

and

$$C_{\text{eq},2} = C_3 + C = 120\,\mu\text{F} + C$$

The equivalent capacitance for the entire arrangement is

$$C_{\text{eq}} = 60\,\mu\text{F}$$

$$= \frac{(C_{\text{eq},1})(C_{\text{eq},2})}{C_{\text{eq},1} + C_{\text{eq},2}}$$

$$= \frac{(90\,\mu\text{F})(120\,\mu\text{F} + C)}{90\,\mu\text{F} + 120\,\mu\text{F} + C}$$

$$60\,\mu\text{F} = \frac{10,800\,\mu\text{F}^2 + 90C\,\mu\text{F}}{210\,\mu\text{F} + C}$$

$$12,600\,\mu\text{F}^2 + 60C\,\mu\text{F} = 10,800\,\mu\text{F}^2 + 90C\,\mu\text{F}$$

$$30C\,\mu\text{F} = 1800\,\mu\text{F}^2$$

$$C = 60\,\mu\text{F} \Leftarrow$$

7.23: A $100\,\mu\text{F}$ capacitor is connected in series to a $300\,\mu\text{F}$ capacitor. At a certain instant of time, the total voltage across the two capacitors is $20\,\text{V}$. Determine the voltage distribution across the two capacitors.

Here

$$v_1 = \frac{1}{C_1}\int i\,dt \quad \text{and} \quad v_2 = \frac{1}{C_2}\int i\,dt$$

and

$$\int dt = v_1 C_1 = v_2 C_2$$

or

$$v_1 = \frac{300\,\mu\text{F}}{100\,\mu\text{F}} v_2 = 3v_2$$

Thus

$$v_1 + v_2 = 20\,\text{V}$$

$$3v_2 + v_2 = 20\,\text{V}$$

$$4v_2 = 20\,\text{V}$$

$$v_2 = 5\,\text{V} \Leftarrow$$

and

$$v_1 = 15\,\text{V} \Leftarrow$$

7.24: The voltage across a $1\,\text{F}$ capacitor during a period of $10\,\text{s}$ is described by the waveform

$$v = \begin{cases} t\,\text{V}; & 0 < t < 2\,\text{s} \\ 3 - 0.5t\,\text{V}; & 2\,\text{s} < t < 4\,\text{s} \\ 1.5t - 5\,\text{V}; & 4\,\text{s} < t < 6\,\text{s} \\ 22 - 3t\,\text{V}; & 6\,\text{s} < t < 7\,\text{s} \\ 0.5t - 2.5\,\text{V}; & 7\,\text{s} < t < 9\,\text{s} \\ t - 7\,\text{V}; & 9\,\text{s} < t < 10\,\text{s} \end{cases}$$

Determine the current through the capacitor and sketch the waveform.

348

The given waveform consists of several "ramp" functions and because

$$i = C\frac{dv}{dt}$$

all of the currents will have constant values. Note that $C = 1\,\text{F}$ so that

$$i = \frac{dv}{dt}$$

For $t < 0\,\text{s}$, $i = 0\,\text{V}$

For $0\,\text{s} < t < 2\,\text{s}$, $i = 1\,\text{A}$
For $2\,\text{s} < t < 4\,\text{s}$, $i = -0.5\,\text{A}$
For $4\,\text{s} < t < 6\,\text{s}$, $i = 1.5\,\text{A}$
For $6\,\text{s} < t < 7\,\text{s}$, $i = -3\,\text{A}$
For $7\,\text{s} < t < 9\,\text{s}$, $i = 0.5\,\text{A}$,
For $9\,\text{s} < t < 10\,\text{s}$, $i = 1\,\text{A}$

and the required sketch of the current waveform is shown below.

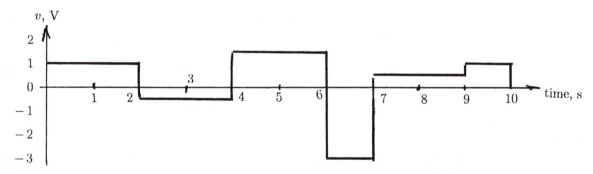

DUAL NETWORKS

7.25: Construct the dual network for the network in Fig 7.4 and then verify that the node equations for the dual network are in the same form as the mesh equations for the network in Fig 7.4.

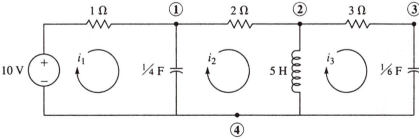

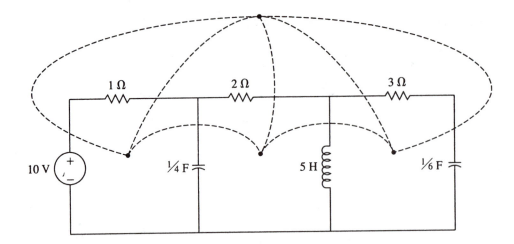

In Fig 7.4, three mesh currents, i_1, i_2 and i_3 have been inserted and, by a repeated application of KVL, three intego-differential mesh equations can be written noting four node-points including a "ground node."

$$i_1 + 4 \int i_1 \, dt - 4 \int i_2 \, dt = 10 \, \text{V}$$

$$-4 \int i_1 \, dt + 5\frac{di_2}{dt} + 2i_2 + 4 \int i_2 \, dt - 5\frac{di_3}{dt} = 0$$

$$-5\frac{di_2}{dt} + 5\frac{di_3}{dt} + 3i_3 + 6 \int i_3 \, dt = 0$$

The mechanism for constructing the dual network is also shown in the foregoing figure and the dual network is shown on the next page. In the dual network, there are four nodes indicated by numerals within circles. The datum node is node-4 and by three applications of KCL, three integro-differential equations can be written

$$v_1 + 4 \int v_1 \, dt - 4 \int v_2 \, dt = 10 \, \text{A}$$

$$-4 \int v_1 \, dt + 5\frac{dv_2}{dt} + 2v_2 + 4 \int v_2 \, dt - 5\frac{dv_3}{dt} = 0$$

$$-5\frac{dv_2}{dt} + 5\frac{dv_3}{dt} + 3v_3 + 6 \int v_3 \, dt = 0$$

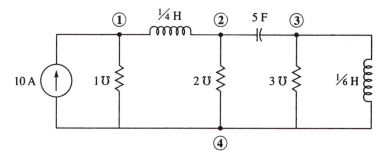

Observe that the mesh equations and the node equations match.

7.26: Construct the dual network for the network in Fig 7.5 and then verify that the node equations for the dual network are in the same form as the mesh equations for the network in Fig 7.5

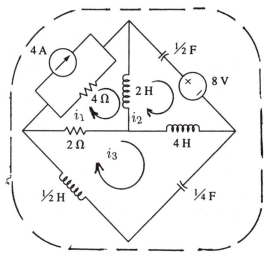

In an adjustment to Fig 7.5, three meshes containing mesh currents, i_1, i_2 and i_3 and an outer mesh have been inserted and, after a transformation of the 4 A current source to a voltage source, by a repeated application of KVL, three intego-differential mesh equations can be written.

$$6i_1 + 2\frac{di_1}{dt} - 2\frac{di_2}{dt} - 2i_3 = 16 \text{ V}$$

$$-2\frac{di_1}{dt} + 6\frac{di_2}{dt} + 2\int i_2 \, dt - 4\frac{di_3}{dt} = -8 \text{ V}$$

$$-2i_1 - 4\frac{di_2}{dt} + 2i_3 + \frac{9}{2}\frac{di_3}{dt} + 4\int i_3 \, dt = 0$$

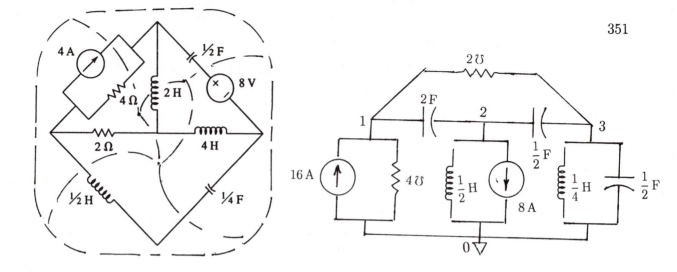

Noting four node-points including a "ground node," the mechanism for constructing the dual network is indicated and the dual network is is shown at the right. In the dual network, there are four nodes indicated by numerals within circles. The datum node is node-4 and by three applications of KCL, three integro-differential equations can be written

$$6v_1 + 2\frac{dv_1}{dt} - 2\frac{dv_2}{dt} - 2v_3 = 16 \text{ A}$$

$$-2\frac{dv_1}{dt} + 6\frac{dv_2}{dt} + 2\int v_2\,dt - 4\frac{dv_3}{dt} = -8\,\text{A}$$

$$-2v_1 - 4\frac{dv_2}{dt} + 2v_3 + \frac{9}{2}\frac{dv_3}{dt} + 4\int v_3\,dt = 0$$

Observe that the mesh equations and the node equations match.

7.27: Construct the dual network for the network in Fig 7.6 and then verify that the node equations for the dual network are in the same form as the mesh equations for the network in Fig 7.6.

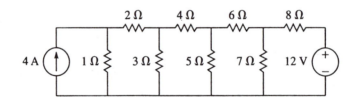

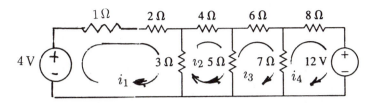

In an adjustment to Fig 7.6, four meshes containing mesh currents, i_1 through i_4 have been inserted and mesh-0 is the outer node. After a transformation of the 4 A current source to a voltage source, by a repeated application of KVL, four mesh equations can be written

$$
\begin{array}{rrrrl}
6i_1 & -3i_2 & & & = & 4\,\text{V} \\
-3i_1 & +12i_2 & -5i_3 & & = & 0 \\
& -5i_2 & +18i_3 & -7i_4 & = & 0 \\
& & -7i_3 & +15i_4 & = & -12\,\text{V}
\end{array}
$$

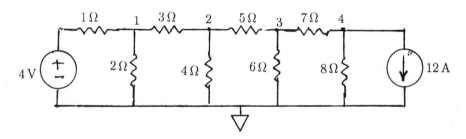

The dual network has five nodes, one of which is the datum or "ground" node and the node equations are

$$
\begin{array}{rrrrl}
6v_1 & -3v_2 & & & = & 4\,\text{A} \\
-3v_1 & +12v_2 & -5v_3 & & = & 0 \\
& -5v_2 & +18v_3 & -7v_4 & = & 0 \\
& & -7v_3 & +15v_4 & = & -12\,\text{A}
\end{array}
$$

Observe that the mesh equations and the node equations match.

7.28: Construct the dual network for the network in Fig 7.7 and then verify that the node equations for the dual network are in the same form as the mesh equations for the network in Fig 7.7.

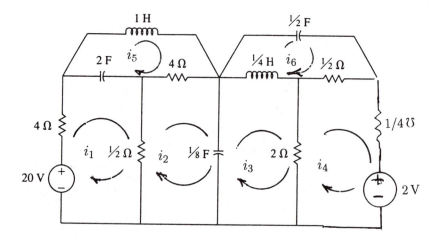

A rework of Fig 7.7 shows six meshes containing mesh currents, i_1 through i_6 and the 8 A current source has been transformed to a voltage source. Mesh-0 represents the outer mesh and, by a repeated application of KVL, six mesh equations can be written

$$\frac{9}{2}i_1 - \frac{1}{2}i_2 + \frac{1}{2}\int i_1\,dt - \frac{1}{2}\int i_5\,dt = 20\,\text{V}$$

$$-\frac{1}{2}i_1 + \frac{9}{2}i_2 + 8\int i_2\,dt - 8\int i_3\,dt - 4i_5 = 0$$

$$-8\int i_2\,dt + 2i_3 + 8\int i_3\,dt + \frac{1}{4}\frac{di_3}{dt} - 2i_4 - \frac{1}{4}\frac{di_6}{dt} = 0$$

$$-2i_3 + \frac{11}{4}i_4 - \frac{1}{2}i_6 = -2\,\text{V}$$

$$-\frac{1}{2}\int i_1\,dt - 4i_2 + \frac{1}{2}\int i_5\,dt + 4i_5 + \frac{di_5}{dt} = 0$$

$$-\frac{1}{4}\frac{di_3}{dt} - \frac{1}{2}i_4 + \frac{1}{4}\frac{di_3}{dt} + \frac{1}{2}i_6 + 2\int i_6\,dt = 0$$

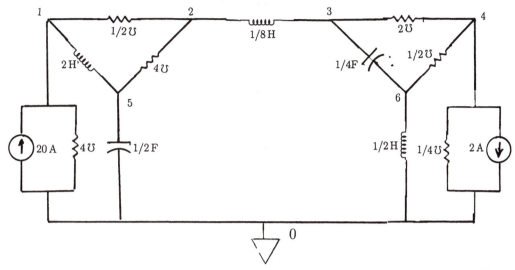

The dual network has seven nodes, one of which is the datum or "ground" node and the node equations are

$$\frac{9}{2}v_1 - \frac{1}{2}v_2 + \frac{1}{2}\int v_1\,dt - \frac{1}{2}\int v_5\,dt = 20\,\text{A}$$

$$-\frac{1}{2}v_1 + \frac{9}{2}v_2 + 8\int v_2\,dt - 8\int v_3\,dt - 4v_5 = 0$$

$$-8\int v_2\,dt + 2v_3 + 8\int v_3\,dt + \frac{1}{4}\frac{dv_3}{dt} - 2v_4 - \frac{1}{4}\frac{dv_6}{dt} = 0$$

$$-2v_3 + \frac{11}{4}v_4 - \frac{1}{2}v_6 = -2\,\text{A}$$

$$-\frac{1}{2}\int v_1\,dt - 4v_2 + \frac{1}{2}\int v_5\,dt + 4v_5 + \frac{dv_5}{dt} = 0$$

$$-\frac{1}{4}\frac{dv_3}{dt} - \frac{1}{2}v_4 + \frac{1}{4}\frac{dv_3}{dt} + \frac{1}{2}v_6 + 2\int v_6\,dt = 0$$

Observe that the mesh equations and the node equations match.

With

$$\frac{1}{T} = \frac{1}{0.2\,\text{s}} = 5\,\text{s}^{-1}$$

and use of the condition at 0.12 s permits writing

$$v(t = 0.12\,\text{s}) = 658.574\,\text{V} = Ve^{-(5\,\text{s}^{-1})(0.12\,\text{s})} = Ve^{-0.60}$$

Thus, the initial value of the voltage is

$$V = \frac{658.574\,\text{V}}{e^{-0.60}} = \frac{658.574\,\text{V}}{0.5488} = 1200\,\text{V} \Leftarrow$$

The voltage is represented by

$$v(t) = (1200\,\text{V})e^{-5t}$$

where t is in s and at $t = 0.0325\,\text{s}$ (32.5 ms)

$$v(t = 0.0325\,\text{s}) = (1200\,\text{V})e^{-(5\,\text{s}^{-1})(0.0325\,\text{s})}$$

$$v(t = 0.0325\,\text{s}) = (1200\,\text{V})e^{-0.1625} = (1200\,\text{V})(0.8500) = 1020\,\text{V} \Leftarrow$$

8.3: If an exponentially decaying current has a time constant of 100 ms and if the current has a value of 12.42 A at $t = 165$ ms, determine the value of the current at $t = 0$ and the time at which the current has a value of 4.571 A.

With

$$i(t) = Ie^{-t/T}$$

and with the time constant, T, given as 100 ms and $i(t = 165\,\text{ms}) = 12.42\,\text{A}$,

$$
\begin{aligned}
i(t = 0.165\,\text{s}) &= Ie^{-0.165\,\text{s}/0.100\,\text{s}} \\
12.42\,\text{A} &= Ie^{-1.65} \\
&= I(0.1920) \\
I &= \frac{12.42\,\text{A}}{0.1920} \\
i(t = 0\,\text{s}) &= 64.67\,\text{A} \Leftarrow
\end{aligned}
$$

Thus,

$$i(t) = (64.67\,\text{A})e^{-t/100\,\text{ms}}$$

and

$$4.571\,\text{A} = (64.67\,\text{A})e^{-t/100\,\text{ms}}$$
$$e^{-t/100\,\text{ms}} = \frac{4.571\,\text{A}}{64.671\,\text{A}}$$
$$-\frac{t}{100\,\text{ms}} = \ln(0.0707)$$
$$-\frac{t}{100\,\text{ms}} = -2.65$$
$$t = 100(2.65\,\text{ms}) = 265\,\text{ms} \Leftarrow$$

8.4: An exponentially decaying voltage has a time constant of 400 ms and an initial value of 1.2 kV. What is the value of the voltage at $t = 125$ ms and what is the value of the voltage five time constants later.

With
$$v(t) = Ve^{-t/T}$$

and with the time constant, T, given as 400 ms and $V = 1200$ V

$$v(t) = (1200\,\text{V})e^{-t/400\,\text{ms}}$$

At $t = 125$ ms

$$v(t = 125\,\text{ms}) = (1200\,\text{V})e^{-125\,\text{ms}/400\,\text{ms}} = (1200\,\text{V})e^{-0.3125} = (1200\,\text{V})(0.7316) = 877.9\,\text{V} \Leftarrow$$

Five time constants later fixes the time at

$$t = 0.125\,\text{s} + 5(0.400\,\text{s}) = 2.125\,\text{s}$$

and

$$v(t = 2.125\,\text{s}) = (1200\,\text{V})e^{-2125\,\text{ms}/400\,\text{ms}} = (1200\,\text{V})e^{-5.3125}$$

$$v(t = 2.125\,\text{s}) = (1200\,\text{V})(4.930 \times 10^{-3}) = 5.92\,\text{V} \Leftarrow$$

8.5 Measurements of an exponentially decaying current yield the following data:

Time, s	Current, A
0.250	294.30
0.275	266.30

Determine the initial value of the current and the time constant.

Here, the data may be used in the decaying exponential of the form

$$i(t) = Ie^{-t/T} \text{ A}$$

With t and T in s

$$294.30 = Ie^{-0.250/T}$$
$$266.20 = Ie^{-0.275/T}$$

If the first is divided by the second, the result is

$$1.1056 = e^{0.025/T}$$

and this may be solved for the time constant, T

$$\ln 1.1056 = 0.025/T$$
$$0.1000 = 0.025/T$$
$$T = \frac{0.025}{0.1000}$$
$$= 0.25 \text{ s} \Leftarrow$$

With

$$\frac{1}{T} = \frac{1}{0.25 \text{ s}} = 4 \text{ s}^{-1}$$

use of the condition at $0.25\,$s permits writing

$$i(t = 0.25 \text{ s}) = 294.30 \text{ A} = Ie^{-(4 \text{ s}^{-1})(0.25 \text{ s})}$$

$$294.30 \text{ A} = Ie^{-1.00}$$

Thus, the initial value of the current is

$$I = \frac{294.30 \text{ A}}{e^{-1.00}} = \frac{294.30 \text{ A}}{0.3679} = 800 \text{ A} \Leftarrow$$

8.6 Measurements of an exponentially decaying voltage yield the following data:

Time, ms	Voltage, A
10	53.375
25	25.214

$$= 40 + (0 - 40)e^{-(15/16)t} \text{ V}$$
$$= 40[1 - e^{-(15/16)t}] \text{ V}$$

and the current current leaving the parallel combination of capacitors is

$$i_C = C_{eq} \frac{dv_c}{dt} = \left(\frac{4}{150} \text{ F}\right)\left[\left(-\frac{15}{16} \text{ s}^{-1}\right)\left(-40e^{-(15/16)t} \text{ V}\right)\right] = e^{-(15/16)t} \text{ A}$$

and by a current divider

$$i_1(t) = \frac{1}{3}e^{-(15/16)t} \text{ A} \Leftarrow$$

and

$$i_2(t) = \frac{2}{3}e^{-(15/16)t} \text{ A} \Leftarrow$$

This confirms the result of Problem 8.19.

8.27: Provide an alternate solution to Problem 8.20 that is based on initial and final values of the response and the time constant.

In Problem 8.20, it was shown that the time constant is

$$T = \frac{L_{eq}}{R_{eq}} = \frac{4 \text{ H}}{12 \text{ }\Omega} = \frac{1}{3} \text{ s}$$

and from Fig 8.14, shown at the right it is seen that at $t = 0$, no current is flowing anywhere in the network

$$i(0^+) = 0 \text{ A}$$

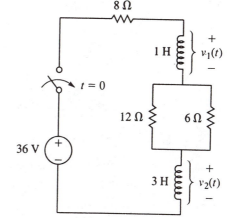

Moreover, at $t \to \infty$, the inductors are acting as short circuits so that

$$i(\infty) = \frac{v_s}{R_T} = \frac{36 \text{ V}}{12 \text{ }\Omega} = 3 \text{ A}$$

Thus,

$$i(t) = i(\infty) + [i(0^+) - i(\infty)]e^{-t/T}$$

so that

$$= 3 + (0 - 3)e^{-3t} \text{ A}$$
$$= 3(1 - e^{-3t}) \text{ A} \Leftarrow$$

This confirms the result of Problem 8.20.

8.28: Provide an alternate solution to Problem 8.23 that is based on initial and final values of the response and the time constant.

In Problem 8.23, a Thevenin equivalent network was formed with $R_T = 13.454\,\Omega$ and $v_T = 31.272\,\text{V}$. This network is shown at the right with the capacitor reconnected across its terminals. With the polarity of the voltage source noted, the time constant is

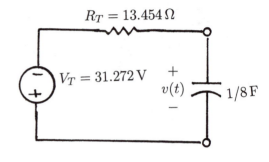

$$T = R_T C = (13.454\,\Omega)\left(\frac{1}{8}\right)\,\text{F} = 1.682\,\text{s}$$

The capacitor is initially uncharged so that

$$v(0^+) = 0\,\text{V}$$

and the final value of the voltage will occur when the network is at steady state (when the capacitor is acting as an open circuit)

$$v(\infty) = -31.272\,\text{V}$$

Thus,

$$v(t) \;=\; v(\infty) + [v(0^+) - v(\infty)]e^{-t/T}$$

so that

$$
\begin{aligned}
&= -31.272 + [0 - (-31.272)]e^{-0.595t}\,\text{V} \\
&= 31.272(e^{-0.595t} - 1)\,\text{V} \Leftarrow
\end{aligned}
$$

This confirms the result of Problem 8.23.

8.29: Provide an alternate solution to Problem 8.24 that is based on initial and final values of the response and the time constant.

In Problem 8.24, a Thevenin equivalent network was formed with $R_T = 196/17\,\Omega$ and $v_T = 600/17\,\text{V}$. This network is shown at the right with the inductor reconnected across its terminals. The time constant is

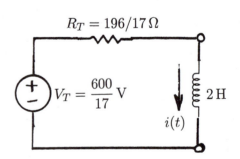

$$T = \frac{L}{R_{\text{eq}}} = \frac{2\,\text{H}}{196/17\,\Omega} = \frac{17}{98}\,\text{s}$$

The initial inductor current was seen to be

$$i(0^+) = \frac{75}{31}\,\text{A}$$

and the final value of the current will oocur when the network is at steady state (when the inductor is acting as a short circuit)

$$i(\infty) = \frac{v_T}{R_T} = \frac{600/17\,\text{V}}{196/17\,\Omega} = \frac{150}{49}\,\text{A}$$

Thus,

$$i(t) \;=\; i(\infty) + [i(0^+) - i(\infty)]e^{-t/T}$$

so that

$$= \frac{150}{49} + \left(\frac{150}{49} - \frac{150}{49}\right)e^{-(98/17)t}\,\text{A}$$

$$= \frac{150}{49} + \left(\frac{150}{49} - \frac{975}{1519}\right)e^{-(98/17)t}\,\text{A} \Leftarrow$$

or

$$= 3.061 - 0.642e^{-5.675t}\,\text{A} \Leftarrow$$

This confirms the result of Problem 8.24.

8.30: Provide an alternate solution to Problem 8.25 that is based on initial and final values of the response and the time constant.

In Problem 8.25, a Thevenin equivalent network was formed with $R_T = 196/17\,\Omega$ and $v_T = 600/17\,\text{V}$. This network is shown at the right with the inductor reconnected across its terminals. The time constant is

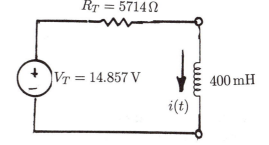

$$T = \frac{L}{R_{\text{eq}}} = \frac{0.400\,\text{H}}{5.714\,\text{k}\Omega} = 70\,\mu\text{s}$$

The initial inductor current was seen to be

$$i(0^+) = 7.778\,\text{mA}$$

Here, for the first op-amp which is a differentiator,

$$v_{o1} = R_1 C_1 \frac{dv_1}{dt} = (50\,\mu\text{F})(100\,\text{k}\Omega)(400\,\text{rad/s})(12\cos 400t) = 24\cos 400t \text{ V}$$

and for the second op-amp which is an integrator,

$$v_{o2} = -\frac{1}{R_2 C_2}\int_o^t v_{i2}\,dz = \frac{1}{(100\,\mu\text{F})(50\,\text{k}\Omega)}\int_0^t (24\cos 400z\,dz + 12\sin 400t)\,dz$$

or

$$v_{o2} = v_o = \frac{1}{5\,\text{s}}\left[\frac{24}{400}\sin 400z - \frac{12}{400}\cos 400z\right]_{z=0}^{z=t} = 12\sin 400t - 6(\cos 400t - 1)\,\text{mV} \Leftarrow$$

8.34: The set of simultaneous differential equations

$$\frac{dv_1}{dt} + 5v_1 - 2v_2 = 18$$

and

$$-2v_1 + 2\frac{dv_2}{dt} + 2v_2 = 0$$

is to be programmed for an analog computer. Draw a circuit diagram using ideal operational amplifiers and specifying values for all R's and C's and considering no initial voltages on any of the capacitors to accomplish this program.

Here, let $p = d/dt$ Then

$$pv_1 + 5v_1 - 2v_2 = 18$$

and

$$-2v_1 + 2pv_2 + 2v_2 = 0$$

or

$$pv_1 = -5v_1 + 2v_2 + 18$$

and

$$pv_2 = v_1 - v_2 = 0$$

These equations lead to the operational amplifier configuration shown on the next page.

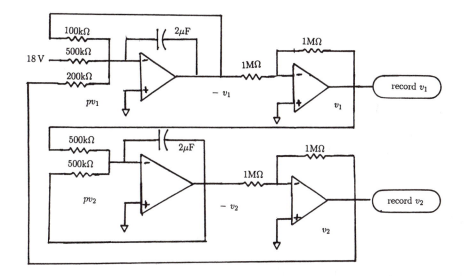

8.36: The differential equation

$$\frac{d^2i}{dt^2} + 10\frac{di}{dt} + 16i = 40t$$

is to be programmed for an analog computer. Draw a circuit diagram using ideal operational amplifiers and specifying values for all R's and C's, a separate function generator for the $40t$ input and considering no initial voltages on any of the capacitors to accomplish this program.

Here, with $p = d/dt$ Then

$$p^2 i + 10pi + 16i = 40t$$

or

$$p^2 i = -10pi - 16i + 40t$$

These equations lead to the operational amplifier configuration shown below.

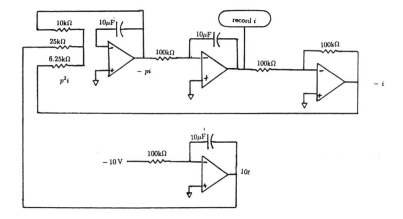

As a check

$$
\begin{aligned}
i(t = 4 \times 10^{-6}\,\text{s}) &= (0.040\,\text{A})\cos(0.20\pi - 0.20\pi) \\
&= (0.040)\cos(0)\,\text{A} \\
&= 0.040\,\text{A}\checkmark
\end{aligned}
$$

9.3: A sinusoidally varying charge reaches a maximum of $20\,\mu\text{C}$ at $t = 8\,\text{ms}$ and the next negative maximum of at $t = 16\,\text{ms}$. Find Q, ω and ϕ in

$$
q(t) = Q\cos(\omega t + \phi)
$$

The time to proceed from a positive maximum to a negative maximum is one-half a period. Thus

$$
\frac{1}{2}T = 16\,\text{ms} - 8\,\text{ms} = 8\,\text{ms}
$$

so that

$$
T = 16\,\text{ms}
$$

The frequency is the reciprocal of the period

$$
f = \frac{1}{T} = \frac{1}{16 \times 10^{-3}\,\text{s}} = 62.5\,\text{Hz}
$$

and hence

$$
\omega = 2\pi f = 2\pi(62.5\,\text{Hz}) = 125\pi\,\text{rad/s} \Longleftarrow
$$

so that with Q specified as $Q = 20\,\mu\text{C} \Longleftarrow$

$$
q(t) = 20\cos(125\pi t + \phi)\,\mu\text{C}
$$

and at $t = 8\,\text{ms}$,

$$
i(t = 0.008\,\text{s}) = 20\cos[(125\pi)(0.008) + \phi] = 20\cos(\pi + \phi) = 20\,\mu\text{C}
$$

Thus

$$
\cos(\pi + \phi) = 1
$$

and

$$
\pi + \phi = 0
$$

or

$$
\phi = \pm\pi\,\text{rad} \Longleftarrow
$$

9.4: A sinusoidally varying voltage has a positive maximum of 208 V at $t = 0$ and decreases to a value of 120 V at $t = 0.125$ ms. Find V, ω and ϕ in

$$v(t) = V \cos(\omega t + \phi) \, \text{V}$$

Here, V is specified as $V = 208$ V $\Leftarrow$

$$v(t) = 208 \cos(\omega t + \phi) \, \text{V}$$

At $t = 0$,
$$v(t = 0) = 208 \cos \phi = 208 \, \text{V}$$

so that
$$\cos \phi = 1$$

Thus
$$\phi = 0 \, \text{rad} \quad \text{or} \quad \pi \, \text{rad}$$

For $t = 0$ and $\phi = 0$ rad
$$v(t) = 120 \cos(0) \, \text{V} = 120 \, \text{V}$$

This corresponds to the given data and as a result, $\phi = \pi$ rad can be discarded. Hence

$$\phi = 0 \, \text{rad} \Leftarrow$$

and at $t = 0.125$ ms
$$v(t) = 208 \cos \left[(1.25 \times 10^{-4})\omega + 0 \right] = 120 \, \text{V}$$

or
$$\cos(1.25 \times 10^{-4}\omega) = \frac{120 \, \text{V}}{208 \, \text{V}} = 0.5769$$

and
$$1.25 \times 10^{-4}\omega = 0.9558 \, \text{rad}$$

or
$$\omega = \frac{0.9558 \, \text{rad}}{1.25 \times 10^{-4} \, \text{s}} = 7646.7 \, \text{rad/s} \Leftarrow$$

9.5 Determine the frequency and the period of a sinusoidally varying voltage that has a value of 60 V at $t = 0$ and reaches its first maximum of 120 V at $t = 2.5$ ms.

Here
$$v(t) = 120 \cos(2\pi f t + \phi) \, \text{V}$$

and at $t = 0$
$$60 = 120 \cos \phi \, \text{V}$$

If $R = 8\,\Omega$ and $C = 1/100\,\text{F}$, determine the value of L to make the response underdamped with a damped natural frequency of $6\,\text{rad/s}$. Then, determine the natural frequency and the damping factor.

Here

$$\sigma = \frac{R}{2L} = \frac{8\,\Omega}{2L} = \frac{4}{L}$$

and

$$\omega_n^2 = \frac{1}{LC} = \frac{100}{L}$$

With ω_d specified as $6\,\text{rad/s}$

$$\omega_d = \left(\omega_n^2 - \sigma^2\right)^{1/2} = \left(\frac{100}{L} - \frac{16}{L^2}\right)^{1/2} = 6\,\text{rad/s}$$

so that

$$\frac{100}{L} - \frac{16}{L^2} = 36$$

or

$$36L^2 - 100L + 16 = 0$$

and there are two values of inductance

$$L = 0.171\,\text{H} \Leftarrow \qquad \text{and} \qquad 2.607\,\text{H} \Leftarrow$$

both of which are valid.

For this underdamped case where $\zeta < 1$,

$$\zeta = \frac{\sigma}{\omega_n} = \frac{R}{2}\sqrt{\frac{C}{L}}$$

or

$$\zeta = \frac{4}{10\sqrt{L}} = \frac{0.40}{\sqrt{L}}$$

Thus, for $L = 0.171\,H$,

$$\zeta = \frac{0.40}{\sqrt{L}} = \frac{0.40}{\sqrt{0.171\,\text{H}}} = 0.967 \Leftarrow$$

and

$$\omega_n = \frac{10}{\sqrt{L}} = \frac{10}{\sqrt{0.171\,\text{H}}} = 24.18\,\text{rad/s} \Leftarrow$$

and for $L = 2.607\,\text{H}$,

$$\zeta = \frac{0.40}{\sqrt{L}} = \frac{0.40}{\sqrt{2.607\,\text{H}}} = 0.248 \Leftarrow$$

and

$$\omega_n = \frac{10}{\sqrt{L}} = \frac{10}{\sqrt{2.607\,\text{H}}} = 6.19\,\text{rad/s} \Leftarrow$$

9.9: The voltage in an RLC parallel network is governed by the differential equation

$$C\frac{d^2v}{dt^2} + \frac{1}{R}\frac{dv}{dt} + \frac{1}{L}v = 0$$

If $R = 4\,\Omega$ and $L = 1/8\,\text{H}$, determine the value of C to make the response overdamped with a damping factor of 2.0.

Here, for the overdamped condition where $\zeta > 1$

$$\sigma = \frac{1}{2RC} = \frac{1}{8C}$$

$$\omega_n^2 = \frac{1}{LC} = \frac{8}{C}$$

and

$$\zeta = \frac{\sigma}{\omega_n} = \frac{1}{2R}\sqrt{\frac{L}{C}} = \frac{1}{8}\sqrt{\frac{1}{8C}}$$

If $\zeta = 2.0$, then

$$\frac{1}{8}\sqrt{\frac{1}{8C}} = 2$$

$$\sqrt{\frac{1}{8C}} = 16$$

$$8C = \left(\frac{1}{16}\right)^2$$

or

$$C = \frac{1}{2048}\,\text{F} \quad \text{or} \quad 488.3\,\mu\text{F} \Leftarrow$$

9.10: The voltage in an RLC parallel network is governed by the differential equation

$$C\frac{d^2v}{dt^2} + \frac{1}{R}\frac{dv}{dt} + \frac{1}{L}v = 0$$

If $R = 4\,\Omega$ and $L = 70.82\,\text{mH}$, determine the value of C required to yield a damped natural frequency of 36 rad/s. What is the damping factor?

Critical damping requires that

$$\sigma^2 = \omega_n^2$$

Thus

$$\frac{36}{C_{\text{eq}}} = (72)^2 = 5134$$

and

$$C_{\text{eq}} = \frac{36}{5134} = \frac{3}{432} \, \text{F}$$

Because C_{eq} consists of a parallel arrangement of two capacitors

$$C_2 = C_{\text{eq}} - C_1 = \frac{3}{432} \, \text{F} - \frac{1}{2304} \, \text{F} = \frac{15}{2304} \, \text{F} = 0.00651 \, \text{F} \Leftarrow$$

UNDRIVEN SECOND ORDER NETWORKS

9.13: In the network of Fig 9.3, the opens instantaneously at $t = 0$. Find the current response for $t \geq 0$.

The initial conditions are derived from the condition with the switch closed. The capacitor in the right hand leg blocks the flow of current so that the current through the inductor is $i_L(0^-) = 0\,\text{A}$. Moreover, the capacitor voltage appears across the $6\,\Omega$ resistor so that, by a voltage divider,

$$v_C(0^-) = \left(\frac{6\,\Omega}{6\,\Omega + 2\,\Omega}\right)(24\,\text{V}) = 18\,\text{V}$$

Hence, for the right hand leg of the parallel combination, by KVL

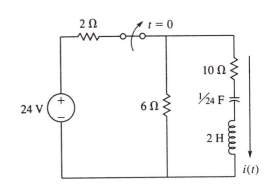

$$
\begin{aligned}
v_C(0^-) + v_L(0^-) + v_R(0^-) &= 0 \\
18\,\text{V} + v_L(0^-) + 0 &= 0 \\
v_L(0^-) &= -18\,\text{V} \\
L\frac{di}{dt}\bigg|_{t=0^-} &= -18\,\text{V} \\
\frac{di}{dt}\bigg|_{t=0^-} &= \frac{-18\,\text{V}}{L} \\
\frac{di}{dt}\bigg|_{t=0^-} &= -\frac{18\,\text{V}}{2\,\text{H}} = -9\,\text{A/s}
\end{aligned}
$$

When the switch moves to the open position at $t = 0$, the configuration is as as shown at the right. Here, KVL gives for all time, t

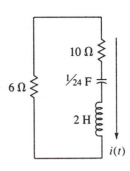

$$v_L + v_R + v_C = 0$$
$$L\frac{di}{dt} + Ri + \frac{1}{C}\int i\,dt = 0$$

and via a single differentiation

$$\frac{d^2i}{dt^2} + \frac{R}{L}\frac{di}{dt} + \frac{1}{LC}i = 0$$

With $R = 6\,\Omega + 10\,\Omega = 16\,\Omega$, $L = 2\,\text{H}$ and $C = 1/24\,\text{F}$, the second order differential equation to be solved is

$$\frac{d^2i}{dt^2} + 8\frac{di}{dt} + 12i = 0$$

and this is subject to the initial conditions obtained from the continuity of stored energy principle

$$i(0^+) = 0\,\text{A} \qquad \text{and} \qquad i'(0^+) = -9\,\text{A/s}$$

Use of the trial solution $i = Ie^{st}$ gives

$$s^2 + 8s + 12 = 0$$

and with $\sigma = 8/2 = 4$ and $\omega_n^2 = 12$, it is observed that

$$\sigma^2 > \omega_n^2$$

so that the response is overdamped. Thus

$$s^2 + 8s + 12 = 0 = (s + 2)(s + 6)$$

and with $s = -2$ and $s = -6$

$$i(t) = I_1e^{-2t} + I_2e^{-6t}$$

This is the general solution.

The arbitrary constants, I_1 and I_2 are evaluated from the initial conditions. With

$$\frac{di}{dt} = -2I_1e^{-2t} - 6I_1e^{-6t}$$

so that

$$i(t) = e^{-12t} \left(A \cos 8t + B \sin 8t \right)$$

is the general solution.

The arbitrary constants, I_1 and I_2 are evaluated from the initial conditions. First,

$$i(0^+) = e^0 \left(A \cos 0 + B \sin 0 \right) = A = 0$$

With $A = 0$, this leaves

$$i(t) = Be^{-12t} \sin 8t$$

with the derivative

$$i'(t) = -12Be^{-12t} \sin 8t + 8Be^{-12t} \cos 8t$$

and at $t = 0^+$ where $i'(t = 0^+) = -192 \, \text{A/s}$

$$i'(t = 0^+) = -12Be^0 \sin 0 + 8Be^0 \cos 0 = -192 \, \text{A/s}$$

or

$$B = \frac{-192 \, \text{A/s}}{8} = -24 \, \text{A/s}$$

The particular solution for the current response will be

$$i(t) = -24e^{-12t} \sin 8t \, \text{A}$$

and for the voltage across the inductor

$$
\begin{aligned}
v_L(t) &= L\frac{di}{dt} \\
&= (1 \, \text{H})\frac{d}{dt}\left(-24e^{-12t} \sin 8t \right) \\
&= 288e^{-12t} \sin 8t - 192e^{-12t} \cos 8t
\end{aligned}
$$

or

$$v_L(t) = 96e^{-12t} \left[3 \sin 8t - 2 \cos 8t \right] \, \text{V} \Leftarrow$$

9.15: In the network of Fig 9.5, the switch moves instantaneously from position-1 to position-2 at $t = 0$. Find the voltage response for $t \geq 0$.

The initial conditions are derived from the condition with switch in position-1. The capacitor, in parallel with the $100/11 \, \Omega$ resistor blocks the flow of current so that the current through the capacitor is $i_C(0^-) = 0\,\text{A}$. Moreover, the capacitor voltage, the voltage across the $100/11\,\Omega$ resistor, is $v_C(0^-) = 11\,\text{V}$.

By KCL, with the switch in position-2

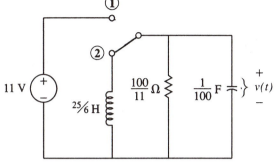

$$i_L + i_R + i_C = 0$$

$$C\frac{dv}{dt} + \frac{1}{R}v + \frac{1}{L}\int v\,dt = 0$$

and via a single differentiation

$$\frac{d^2v}{dt^2} + \frac{1}{RC}\frac{dv}{dt} + \frac{1}{LC}v = 0$$

With $R = 100/11\,\Omega$, $L = 25/6\,\text{H}$ and $C = 1/100\,\text{F}$, the second order differential equation to be solved is

$$\frac{d^2v}{dt^2} + 11\frac{dv}{dt} + 24v = 0$$

The continuity of stored energy principle may be used to develop the initial conditions. First

$$v(0^+) = v_C(0^+) = v(0^-) = 11\,\text{V}$$

and then by KCL with $i_L(0^+) = i_L(0^-) = 0\,\text{A}$

$$i_C(0^+) + i_L(0^+) + i_R(0^+) = 0$$

$$Cv'(0^+) + 0 + \frac{v(0^+)}{R} = 0$$

$$v'(0^+) = -\frac{v(0^+)}{RC}$$

$$= -\frac{11\,\text{V}}{(100/11\,\Omega)(1/100\,\text{F})}$$

$$= -121\,\text{V/s}$$

Use of the trial solution $v = Ve^{st}$ gives

$$s^2 + 11s + 24 = 0$$

and with $\sigma = 11/2$ and $\omega_n^2 = 24$, it is observed that

$$\sigma^2 > \omega_n^2$$

so that the response is underdamped. Thus

$$s^2 + 400s + 200,000 = 0 = [s - (-200 + j400)][s + (-200 - j400)]$$

so that

$$v(t) = e^{-200t}(A\cos 400t + B\sin 400t) \text{ V}$$

is the general solution.

The arbitrary constants, V_1 and V_2 are evaluated from the initial conditions. First,

$$v(0^+) = e^0(A\cos 0 + B\sin 0) = A = 0 \text{ V}$$

With $A = 0$, this leaves

$$v(t) = Be^{-200t}\sin 400t \text{ V}$$

with the derivative

$$v'(t) = -200Be^{-200t}\sin 400t + 400Be^{-200t}\cos 400t$$

and at $t = 0^+$ where $v'(t = 0^+) = -18 \times 10^4 \text{ V/s}$

$$v'(t = 0^+) = -200Be^0\sin 0 + 400Be^0\cos 0 = -18 \times 10^4 \text{ V/s}$$

or

$$B = \frac{-18 \times 10^4 \text{ V/s}}{400} = -450 \text{ V/s}$$

The particular solution for the voltage response will be

$$v(t) = -450e^{-200t}\sin 400t \text{ V}$$

and for the current through the inductor

$$i_L(t) = -i_C(t) - i_R(t)$$

$$
\begin{aligned}
i_C(t) &= C\frac{dv}{dt} \\
&= (2\,\mu\text{F})\frac{d}{dt}\left(-450e^{-200t}\sin 400t\right) \\
&= -9 \times 10^{-4}e^{-200t}[-200\sin 400t + 400\cos 400t] \\
&= e^{-200t}[0.180\sin 400t - 0.360\cos 400t] \text{ A}
\end{aligned}
$$

and

$$i_R(t) \quad = \quad \frac{v(t)}{1250\,\Omega}$$

$$= \quad -0.360e^{-200t}\sin 400t \text{ A}$$

so that

$$i_L(t) \quad = \quad -i_C(t) - i_R(t)$$

$$= \quad e^{-200t}\left[-0.180\sin 400t + 0.360\cos 400t - (-0.360\sin 400t)\right]$$

$$= \quad e^{-200t}\left[0.180\sin 400t + 0.360\cos 400t\right] \text{ A} \Leftarrow$$

9.17: In the network of Fig 9.7 the switch goes from position-1 to position-2 instantaneously at $t = 0$. Determine the voltage across the capacitor for $t \geq 0$ if the capacitor is initially uncharged.

In this problem, the strategy is to first solve for the current, $i(t)$, and then obtain $v(t) = v_C(t)$.

The initial conditions are derived from the condition with the switch in position -1. The inductor is acting as a short circuit and the inductor current will be equal to

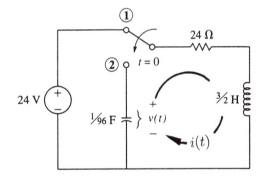

$$i_L(0^-) = \frac{24\,\text{V}}{24\,\Omega} = 1\,\text{A}$$

The capacitor is specified as having no initial charge so that

$$v_C(0^-) = 0\,\text{V}$$

and when the switch is thrown to position-2, the continuity of stored energy principle demands that

$$v_C(0^+) = 0\,\text{V} \qquad \text{and} \qquad i_L(0^+) = 1\,\text{A}$$

Then, by KVL

$$v_C(0^+) + v_L(0^+) + v_R(0^+) \quad = \quad 0$$

$$0 + L\frac{di}{dt}\bigg|_{t=0+} + 24i(0^+) \quad = \quad 0$$

9.18: In the network of Fig 9.8 the switch goes from position-1 to position-2 instantaneously at $t = 0$. Determine the voltage across the capacitor for $t \geq 0$ if the capacitor is initially uncharged.

The initial conditions are derived from the condition with the switch in position -1. The inductor is acting as a short circuit and the inductor current will be equal to

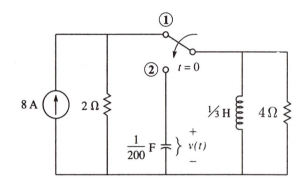

$$i_L(0^-) = 8\,\text{A}$$

and, because the capacitor is uncharged

$$v_C(0^-) = 0\,\text{V}$$

When the switch is thrown to position-2, the continuity of stored energy principle demands that

$$v_C(0^+) = 0\,\text{V} \qquad \text{and} \qquad i_L(0^+) = 8\,\text{A}.$$

Then, by KCL

$$
\begin{aligned}
i_C(0^+) + i_R(0^+) + i_L(0^+) &= 0 \\
C\frac{dv}{dt}\bigg|_{t=0+} + \frac{v(0^+)}{R} + i_L(0+) &= 0 \\
\frac{dv}{dt}\bigg|_{t=0+} &= -\frac{0}{RC} - \frac{8\,\text{A}}{C} \\
\frac{dv}{dt}\bigg|_{t=0+} &= -\frac{8\,\text{A}}{1/200\text{F}} \\
\frac{dv}{dt}\bigg|_{t=0+} &= -1600\,\text{V/s}
\end{aligned}
$$

With the switch in position-2, an application of KCL gives at the single node gives

$$
\begin{aligned}
i_C + i_R + i_L &= 0 \\
C\frac{dv}{dt} + \frac{v}{R} + \frac{1}{L}\int v\,dt &= 0
\end{aligned}
$$

and via a single differentiation

$$\frac{d^2v}{dt^2} + \frac{1}{RC}\frac{dv}{dt} + \frac{1}{LC}v = 0$$

and with $C = 1/200\,\mathrm{F}$, $R = 4\,\Omega$ and $L = 1/3\,\mathrm{H}$

$$\frac{d^2v}{dt^2} + 50\frac{dv}{dt} + 600v = 0$$

Use of the trial solution $v = Ve^{st}$ gives

$$s^2 + 50s + 600 = 0$$

and with $\sigma = 50/2 = 25$ and $\omega_n^2 = 600$, it is observed that

$$\sigma^2 > \omega_n^2$$

so that the response is overdamped. Thus

$$s^2 + 50s + 600 = 0 = (s+20)(s+30)$$

so that

$$v(t) = V_1e^{-20t} + V_2e^{-30t}$$

is the general solution.

The arbitrary constants, V_1 and V_2 are evaluated from the initial conditions. First,

$$v(0^+) = V_1e^0 + V_2e^0 = 0\,\mathrm{V}$$

with the derivative

$$v'(t) = -20V_1e^{-20t} - 30V_2e^{-30t}$$

and with $v'(t = 0^+) = -1600\,\mathrm{V/s}$

$$v'(t = 0^+) = -20V_1e^0 - 30V_2e^0 = -1600\,\mathrm{V/s}$$

There are two equations in the unknown constants

$$\begin{aligned} V_1 \quad + V_2 \ &= \quad 0\,\mathrm{V} \\ -20V_1 \ - 30V_2 \ &= \ -1600\,\mathrm{V/s} \end{aligned}$$

From the first of these

$$V_1 = -V_2$$

and with this in the second

$$-10V_2 = -1600 \text{ V/s}$$

so that

$$V_2 = 160$$

and then

$$V_1 = -160$$

The particular solution for the voltage response will be

$$v(t) = 160[e^{-30t} - e^{-20t}] \text{ V} \Leftarrow$$

DRIVEN SECOND ORDER NETWORKS

9.19: In Fig 9.9, the current is applied to the network instantaneously at $t = 0$ when the voltage across the capacitor is 16 V and no current is flowing in the network. With $R = 4/5\,\Omega$, $L = 1/3\,\text{H}$, $C = 1/8\,\text{F}$ and $i_s = 6\,\text{A}$, determine $v(t)$ for all $t \geq 0$.

Using KCL for all time $t \geq 0$

$$i_C + i_R + i_L = i_s$$

$$C\frac{dv}{dt} + \frac{v}{R} + \frac{1}{L}\int i\,dt = i_s$$

$$\frac{dv}{dt} + \frac{v}{RC} + \frac{1}{LC}\int i\,dt = \frac{i_s}{C}$$

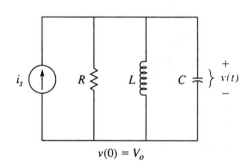

$$v(0) = V_o$$

With the flux, ϕ, defined as

$$\phi \equiv \int v\,dt$$

the differential equation becomes

$$\frac{d^2\phi}{dt^2} + \frac{1}{RC}\frac{d\phi}{dt} + \frac{1}{LC}\phi = \frac{i_s}{C}$$

and with $i_s = 6\,\text{A}$ and

$$RC = \left(\frac{4}{5}\,\Omega\right)\left(\frac{1}{8}\,\text{F}\right) = \frac{1}{10}\,\text{s}$$

$$LC = \left(\frac{1}{3}\,\text{H}\right)\left(\frac{1}{8}\,\text{F}\right) = \frac{1}{24}\,\text{s}^2$$

and

$$\frac{i_s}{C} = \frac{6\,\text{A}}{1/8\,\text{F}} = 48\,\text{Wb/s}^2$$

the differential equation is

$$\frac{d^2\phi}{dt^2} + 10\frac{d\phi}{dt} + 24\phi = 48 \text{ Wb/s}^2$$

This differential equation is to be solved using the initial conditions

$$v(0) = \frac{d\phi}{dt}\bigg|_{t=0} = V_o = 16\,V$$

and because $i_L(0) = 0$

$$\phi(0) = 0$$

For the complementary function, use of the trial solution, $\phi = \Phi e^{st}$ gives

$$s^2 + 10s + 24 = 0$$

and with $\sigma = 10/2 = 5$ and $\omega_n^2 = 24$, it is observed that

$$\sigma^2 > \omega_n^2$$

so that the response is overdamped. Thus

$$s^2 + 10s + 24 = 0 = (s+4)(s+6)$$

and

$$\phi_c(t) = \Phi_1 e^{-4t} + \Phi_2 e^{-6t}$$

is the complementary function.

The particular integral is obtained by assuming that

$$\phi_{\text{pi}} = A$$

to match the forcing function. When this and its derivatives are substituted into the governing differential equation, the result is

$$\frac{d^2 A}{dt^2} + 10\frac{dA}{dt} + 24A = 48$$

so that the particular integral is

$$\phi_{\text{pi}} = A = 2\,\text{Wb}$$

and the complete general solution will be the sum of the complementary function and the particular integral

$$\phi(t) = \Phi_1 e^{-4t} + \Phi_2 e^{-6t} + 2\,\text{Wb}$$

is the complementary function.

The particular integral is obtained by assuming that

$$q_{pi} = A$$

to match the forcing function. This assumed solution and its derivatives may be substituted into the governing differential equation to obtain

$$\frac{d^2 A}{dt^2} + 2\frac{dA}{dt} + 5A = 20$$

so that the particular integral is

$$q_{pi} = A = 4\,\mathrm{C}$$

and the complete general solution will be the sum of the complementary function and the particular integral

$$q(t) = e^{-t}[Q_1 \cos 2t + Q_2 \sin 2t] + 4\,\mathrm{C}$$

Using the complete general solution, the arbitrary constants, Q_1 and Q_2 are evaluated from the initial conditions. First,

$$q(0) = e^0[Q_1 \cos 0 + Q_2 \sin 0] + 4 = \frac{2}{5}\,\mathrm{C}$$

so that

$$Q_1 + 4 = \frac{2}{5}$$

or

$$Q_1 = -\frac{18}{5}$$

Then with the derivative

$$q'(t) = 2e^{-t}\left[Q_2 \cos 2t - Q_1 \sin 2t\right] - e^{-t}\left[Q_1 \cos 2t + Q_2 \sin 2t\right]\ \mathrm{C/s}$$

so that at $q'(t = 0) = 0\,\mathrm{C}$

$$q'(0) = 2e^0\left[Q_2 \cos 0 - Q_1 \sin 0\right] - e^0\left[Q_1 \cos 0 + Q_2 \sin 0\right] = 0\,\mathrm{C/s}$$

$$
\begin{aligned}
2Q_2 - Q_1 &= 0 \\
Q_2 &= \frac{1}{2}Q_1 \\
&= \frac{1}{2}\left(-\frac{18}{5}\right) \\
&= -\frac{9}{5}
\end{aligned}
$$

The particular solution for the charge, q, is

$$q(t) = 4 - \frac{1}{5}e^{-t}\left[18\cos 2t + 9\sin 2t\right] \text{ C}$$

and for the current, $i = dq/dt$

$$i(t) = \frac{1}{5}e^{-t}\left[18\cos 2t + 9\sin 2t\right] - \frac{2}{5}e^{-t}\left[-18\sin 2t + 9\cos 2t\right]$$

or

$$i(t) = 9e^{-t}\sin 2t \text{ A} \Leftarrow$$

9.21: In Fig 9.19, the current is applied to the network instantaneously at $t = 0$ when the voltage across the capacitor is 12 V and no current is flowing in the network. With $R = 5/8\,\Omega$, $L = 1/5\,\text{H}$, $C = 1/5\,\text{F}$ and $i_s = 20\,\text{A}$, determine $v(t)$ for all $t \geq 0$.

It was shown in Problem 9.19, that with the flux, ϕ, defined as

$$\phi \equiv \int v \, dt$$

the differential equation for this network is

$$\frac{d^2\phi}{dt^2} + \frac{1}{RC}\frac{d\phi}{dt} + \frac{1}{LC}\phi = \frac{i_s}{C}$$

and with $i_s = 20\,\text{A}$

$$RC = \left(\frac{5}{8}\,\Omega\right)\left(\frac{1}{5}\,\text{F}\right) = \frac{1}{8}\,\text{s}$$

$$LC = \left(\frac{1}{5}\,\text{H}\right)\left(\frac{1}{5}\,\text{F}\right) = \frac{1}{25}\,\text{s}^2$$

and

$$\frac{i_s}{C} = \frac{20\,\text{A}}{1/5\,\text{F}} = 100\,\text{Wb/s}^2$$

the differential equation is

$$\frac{d^2\phi}{dt^2} + 8\frac{d\phi}{dt} + 25\phi = 100\,\text{Wb/s}^2$$

This differential equation is to be solved using the initial conditions

$$v(0) = \left.\frac{d\phi}{dt}\right|_{t=0} = V_o = 12\,\text{V}$$

and because $i_L(0) = 0$

$$\phi(0) = 0$$

$$LC = \left(\frac{1}{2}\,\mathrm{H}\right)\left(\frac{1}{2}\,\mathrm{F}\right) = \frac{1}{4}\,\mathrm{s}^2$$

and

$$\frac{v_s}{L} = \frac{12\,\mathrm{V}}{1/2\,\mathrm{H}} = 24\,\mathrm{C/s}^2$$

the differential equation is

$$\frac{d^2q}{dt^2} + 4\frac{dq}{dt} + 4q = 24\,\mathrm{C/s}^2$$

This differential equation is to be solved using the initial conditions

$$q(0) = Cv_C(0) = \left(\frac{1}{2}\,\mathrm{F}\right)(8\,\mathrm{V}) = 4\,\mathrm{C}$$

and because $i_L(0) = 0$

$$q'(0) = \frac{di_L}{dt}\bigg|_{t=0} = 0\,\mathrm{C}$$

For the complementary function, use of the trial solution, $q = Qe^{st}$ gives

$$s^2 + 4s + 4 = 0$$

and with $\sigma = 4/2 = 2$ and $\omega_n^2 = 4$, it is observed that

$$\sigma^2 = \omega_n^2$$

so that the response is critically damped. Thus

$$s^2 + 2s + 4 = 0 = (s+2)^2$$

and

$$q_c(t) = Q_1 e^{-2t} + Q_1 t e^{-2t}$$

is the complementary function.

The particular integral is obtained by assuming that

$$q_{\mathrm{pi}} = A$$

to match the forcing function. This assumed solution and its derivatives may be substituted into the governing differential equation to obtain

$$\frac{d^2 A}{dt^2} + 4\frac{dA}{dt} + 4A = 24$$

so that the particular integral is

$$\phi_{\text{pi}} = A = 6 \, \text{C}$$

and the complete general solution will be the sum of the complementary function and the particular integral

$$q(t) = Q_1 e^{-2t} + Q_2 t e^{-2t} + 6 \, \text{C}$$

Using the complete general solution, the arbitrary constants, Q_1 and Q_2 are evaluated from the initial conditions. First,

$$q(0) = Q_1 e^0 + Q_2(0) e^0 + 6 = 4 \, \text{C}$$

so that

$$Q_1 + 6 = 4$$

or

$$Q_1 = -2$$

and with the derivative

$$q'(t) = -2Q_1 e^{-t} + Q_2 e^{-2t} - 2Q_2 t e^{-2t}$$

so that at $q'(t = 0) = 0 \, \text{C}$

$$q'(0) = -2Q_1 e^0 + Q_2 e^0 - 2Q_2(0) e^0 = 0$$

$$
\begin{aligned}
-2Q_1 + Q_2 &= 0 \\
Q_2 &= 2Q_1 \\
&= 2(-2) \\
&= -4
\end{aligned}
$$

The particular solution for the charge, q is

$$q(t) = 6 - 2e^{-2t} - 4te^{-2t} \, \text{C}$$

and for the current, $i = dq/dt$

$$i(t) = 4e^{-2t} + 8te^{-2t} - 4e^{-2t} = 8te^{-2t} \, \text{A} \Leftarrow$$

9.23: In Fig 9.9, the current is applied to the network instantaneously at $t = 0$ when the voltage across the capacitor is $8 \, \text{V}$ and no current is flowing in the network. With $R = 1\,\Omega, L = 1\,\text{H}, C = 1/4\,\text{F}$ and $i_s = 36\,\text{A}$, determine $v(t)$ for all $t \geq 0$.

$$\phi_2 = -64$$

The particular solution for the flux, ϕ is

$$\phi(t) = 36 - 36e^{-2t} - 64te^{-2t} \text{ C}$$

and

$$v(t) = \frac{d\phi}{dt} = 72e^{-2t} - 64e^{-2t} + 128te^{-2t}$$

or

$$v(t) = 8e^{-2t} + 128te^{-2t} \text{ V} \Leftarrow$$

9.24: In Fig 9.10, the voltage is applied to the network instantaneously at $t = 0$ when there is no current flowing in the network. With the voltage across the capacitor at 48 V, $R = 7\,\Omega$, $L = 1\,\text{H}$, $C = 1/12\,\text{F}$ and $v_s = 24\,\text{V}$, determine $i(t)$ for all $t \geq 0$.

It was shown in Problem 19.20, that with the charge, q, defined as

$$q \equiv \int i\,dt$$

the differential equation becomes

$$\frac{d^2q}{dt^2} + \frac{R}{L}\frac{dq}{dt} + \frac{1}{LC}q = \frac{v_s}{L}$$

and with $v_s = 48\,\text{V}$

$$\frac{R}{L} = \frac{7\,\Omega}{1\,\text{H}} = 7\,\text{s}$$

$$LC = (1\,\text{H})\left(\frac{1}{12}\,\text{F}\right) = \frac{1}{12}\,\text{s}^2$$

and

$$\frac{v_s}{L} = \frac{24\,\text{V}}{1\,\text{H}} = 24\,\text{C/s}^2$$

the differential equation is

$$\frac{d^2q}{dt^2} + 7\frac{dq}{dt} + 12q = 24\,\text{C/s}^2$$

This differential equation is to be solved using the initial conditions

$$q(0) = Cv_C(0) = \left(\frac{1}{12}\,\text{F}\right)(48\,\text{V}) = 4\,\text{C}$$

and because $i_L(0) = 0$

$$q'(0) = \left.\frac{di_L}{dt}\right|_{t=0} = 0\,\text{C}$$

For the complementary function, use of the trial solution, $q = Qe^{st}$ gives

$$s^2 + 7s + 12 = 0$$

and with $\sigma = 7/2$ and $\omega_n^2 = 12$, it is observed that

$$\sigma^2 > \omega_n^2$$

so that the response is overdamped. Thus

$$s^2 + 7s + 12 = 0 = (s+3)(s+4)$$

and

$$q_c(t) = Q_1 e^{-3t} + Q_2 e^{-4t}$$

is the complementary function.

The particular integral is obtained by assuming that

$$q_{\text{pi}} = A$$

to match the forcing function. This assumed solution and its derivatives may be substituted into the governing differential equation to obtain

$$\frac{d^2 A}{dt^2} + 7\frac{dA}{dt} + 12A = 24$$

so that the particular integral is

$$\phi_{\text{pi}} = A = 2\,\text{C}$$

and the complete general solution will be the sum of the complementary function and the particular integral

$$q(t) = Q_1 e^{-3t} + Q_2 e^{-4t} + 2\,\text{C}$$

Using the complete general solution, the arbitrary constants, Q_1 and Q_2 are evaluated from the initial conditions. First,

$$q(0) = Q_1 e^0 + Q_2 e^0 + 2 = 4\,\text{C}$$

so that

$$Q_1 + Q_2 = 2$$

10.2: If $N_1 = 4 - j3$, $N_2 = 2\sqrt{2}\underline{/45°}$ and $N_3 = 5e^{90°}$, determine

$$N = \frac{N_1 N_2}{N_3}$$

In rectangular coordinates

$$N = \frac{(4 - j3)(2 + j2)}{0 + j5}$$

and then in polar coordinates

$$
\begin{aligned}
N &= \frac{(5\underline{/-36.87°})(2\sqrt{2}\underline{/45°})}{5\underline{/90°}} \\
&= \frac{14.12\underline{/8.13°}}{5\underline{/90°}} \\
&= 2.828\underline{/-81.87°} \Leftarrow \\
&= 0.40 - j2.80 \Leftarrow \\
&= 2.828e^{-81.87°} \Leftarrow
\end{aligned}
$$

10.3: If $N_1 = 4 - j3, N_2 = 2\sqrt{2}\underline{/45°}$ and $N_3 = 5e^{90°}$, determine

$$N = (N_1 - N_2)^2 N_3$$

Begin with rectangular coordinates

$$
\begin{aligned}
N &= [(4 - j3) - (2 + j2)]^2(0 + j5) \\
&= (2 - j5)^2(0 + j5) \\
&= (5.385\underline{/-68.20°})^2(5\underline{/90°}) \\
&= (29.000\underline{/-136.40°})(5\underline{/90°}) \\
&= 145.00\underline{/-46.40°} \Leftarrow \\
&= 100 - j105 \Leftarrow \\
&= 145e^{-46.40°} \Leftarrow
\end{aligned}
$$

10.4: If $N_1 = 4 - j3, N_2 = 2\sqrt{2}\underline{/45°}$ and $N_3 = 5e^{90°}$, determine

$$N = 2N_1 N_2 N_3$$

Begin with rectangular coordinates

$$N = (2 + j0)(4 - j3)(2 + j2)(0 + j5)$$

Then, in polar coordinates

$$
\begin{aligned}
N &= (2\underline{/0°})(5\underline{/-36.87°})(2\sqrt{2}\underline{/45°})(5\underline{/90°}) \\
&= 100\sqrt{2}\underline{/98.13°} \Leftarrow \\
&= -20 + j140 \Leftarrow \\
&= 100\sqrt{2}e^{98.13°} \Leftarrow
\end{aligned}
$$

10.5: If $N_1 = 4 - j3, N_2 = 2\sqrt{2}\underline{/45°}$ and $N_3 = 5e^{90°}$, determine

$$N = \frac{N_1}{N_2} + \frac{N_1}{N_3}$$

In polar form

$$
\begin{aligned}
N &= \frac{5\underline{/-36.87°}}{2\sqrt{2}\underline{/45°}} + \frac{5\underline{/-36.87°}}{5\underline{/90°}} \\
&= 1.768\underline{/-81.87°} + 1.0000\underline{/-126.87°}
\end{aligned}
$$

Here

$$\begin{aligned} N &= (5\underline{/-36.87°})^3 + (2\sqrt{2}\underline{/45°})^2 + 5\underline{/90°} \\ &= 125\underline{/-110.61°} + 8\underline{/90°} + 5\underline{/90°} \\ &= (-44 - \jmath117) + (0 + \jmath8) + (0 + \jmath5) \\ &= -44 - \jmath104 \Leftarrow \\ &= 112.93\underline{/-112.93°} \Leftarrow \\ &= 112.93e^{-122.93°} \Leftarrow \end{aligned}$$

10.10: If $N_1 = 4 - \jmath3$, $N_2 = 2\sqrt{2}\underline{/45°}$ and $N_3 = 5e^{90°}$, determine

$$N = N_1N_2 + N_2N_3 + N_1N_3$$

$$\begin{aligned} N &= (5\underline{/-36.87°})(2\sqrt{2}\underline{/45°}) + (2\sqrt{2}\underline{/45°})(5\underline{/90°}) + (5\underline{/-36.87°})(5\underline{/90°}) \\ &= 14.142\underline{/8.13°} + 14.142\underline{/135°} + 25\underline{/53.13°} \\ &= (14 + \jmath2) + (-10 + \jmath10) + (15 + \jmath20) \\ &= 19 + \jmath32 \Leftarrow \\ &= 37.22\underline{/59.30°} \Leftarrow \\ &= 37.22e^{59.30°} \Leftarrow \end{aligned}$$

IMPEDANCE AND ADMITTANCE

10.11: If $\omega = 400$ rad/s, find the driving point impedance at terminal-a and -b in the network of Fig 10.1.

Let

$$\begin{aligned} R_1 &= 10\,\Omega \\ L_1 &= 25\,\text{mH} \\ R_2 &= 20\,\Omega \\ L_2 &= 50\,\text{mH} \end{aligned}$$

and

$$C = 125\,\mu\text{F}$$

First, for the parallel combination in terms of the complex frequency, s

$$Z_p(s) = \frac{R_2L_2s}{L_2s + R_2} = \frac{(20)(0.05)s}{20 + 0.05s} = \frac{s}{0.05s + 20}\,\Omega$$

Then

$$Z(s) \;=\; L_1 s + R_1 + \frac{1}{Cs} + Z_p(s)\,\Omega$$

$$=\; \frac{L_1 C s^2 + R_1 C s + 1}{Cs} + Z_p(s)\,\Omega$$

$$=\; \frac{0.025 s^2 + 10 s + 8000}{s} + \frac{s}{0.05 s + 20}\,\Omega$$

With $\omega = 400\,\text{rad/s}$

$$Z(\omega = 400\,\text{rad/s}) \;=\; \frac{-4000 + j4000 + 8000}{j400} + \frac{j400}{j20 + 20}\,\Omega$$

$$=\; \frac{4000\sqrt{2}\underline{/45^\circ}}{400\underline{/90^\circ}} + \frac{400\underline{/90^\circ}}{20\sqrt{2}\underline{/45^\circ}}\,\Omega$$

$$=\; 10\sqrt{2}\underline{/-45^\circ} + 10\sqrt{2}\underline{/45^\circ}\,\Omega$$

$$=\; (10 - j10) + (10 + j10)\,\Omega$$

$$=\; 20\,\Omega \;\Leftarrow$$

10.12: If $\omega = 200$ rad/s, find the driving point impedance at terminal-a and -b in the network of Fig 10.2.

Let

$$R_1 \;=\; 20\,\Omega$$
$$L_1 \;=\; 100\,\text{mH}$$
$$C \;=\; 125\,\mu\text{F}$$
$$R_2 \;=\; 30\,\Omega$$

and

$$L_2 \;=\; 22\,\text{mH}$$

Then in terms of the complex frequency, s

$$Z_{p1}(s) = \frac{R_1 L_1 s}{L_1 s + R_1} = \frac{2s}{0.10s + 20}\,\Omega$$

and

$$Z_{p2}(s) = \frac{R_2/Cs}{1/Cs + R_2} = \frac{R_2}{R_2 Cs + 1} = \frac{30}{(30)(1.25 \times 10^{-4})s + 1} = \frac{30}{3.75 \times 10^{-3}s + 1}$$

For the series combination of the 80 mh inductor and the 30 Ω resistor

$$Z_1(s) = 30 + 0.080s \; \Omega$$

and the for the parallel combination

$$
\begin{aligned}
Z_p(s) &= \frac{(1/Cs)(0.08s + 30)}{1/Cs + 0.08s + 30} \; \Omega \\
&= \frac{(20,000/s)(0.08s + 30)}{20.000/s + 0.08s + 30} \; \Omega \\
&= \frac{1600s + 600,000}{0.08s^2 + 30s + 20,000} \; \Omega
\end{aligned}
$$

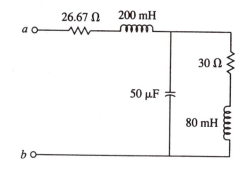

and with the impedance of the series combination of the 26.67 Ω resistor and the 200 mH inductor equal to

$$Z_s(s) = 0.20s + 26.67 \; \Omega$$

the impedance between terminals-a and -b at $\omega = 500 \, \text{rad/s}$ will be

$$
\begin{aligned}
Z(s = \jmath 500 \, \text{rad/s}) &= Z_p(s = \jmath 500 \, \text{rad/s}) + Z_s(s = \jmath 500 \, \text{rad/s}) \\
&= 26.67 \, \Omega + 0.2(\jmath 500) \, \Omega + \frac{1600(\jmath 500) + 600,000}{0.08(\jmath 500)^2 + 30(\jmath 500) + 20,000} \; \Omega \\
&= 26.67 \, \Omega + \jmath 100 \, \Omega + \frac{600,000 + \jmath 800,000}{-20,000 + \jmath 15,000 + 20,000} \; \Omega \\
&= 26.67 \, \Omega + \jmath 100 \, \Omega + \frac{10^6 \underline{/53.13°}}{15,000 \underline{/90°}} \; \Omega \\
&= 26.67 \, \Omega + \jmath 100 \, \Omega + 66.67 \underline{/-36.87°} \; \Omega \\
&= 26.67 \, \Omega + \jmath 100 \, \Omega + 53.33 \, \Omega - \jmath 40 \, \Omega \\
&= 80 \, \Omega + \jmath 60 \, \Omega = 100 \underline{/36.87°} \; \Omega
\end{aligned}
$$

and the admittance will be

$$Y(s = \jmath 500 \, \text{rad/s}) = \frac{1}{Z(s = \jmath 500 \, \text{rad/s})} = \frac{1}{100 \underline{/36.87°} \; \Omega}$$

or

$$Y(s = \jmath 500 \, \text{rad/s}) = 0.01 \underline{/-36.87°} \; \mho = 0.008 - \jmath 0.006 \; \mho \; \Leftarrow$$

10.15: If $\omega = 1000 \text{ rad/s}$, find the driving point impedance at terminal-a and -b in the network of Fig 10.5.

In the phasor domain with $\omega = 1000 \text{ rad/s}$

Element	Reactance, X
$C = 12.5 \,\mu\text{F}$	$80 \,\Omega$
$C = 8.33 \,\mu\text{F}$	$120 \,\Omega$
$L = 120 \,\text{mH}$	$120 \,\Omega$

Then

$$Z_{p1}(\omega = 1000 \text{ rad/s}) = \frac{(50 - j120)(50 + j120)}{(50 - j120) + (50 + j120)} \,\Omega$$

$$= \frac{(130\underline{/-67.38°})(130\underline{/67.38°})}{100} \,\Omega$$

$$= \frac{16,900}{100} \,\Omega = 169 \,\Omega$$

and

$$Z_{p2}(\omega = 1000 \text{ rad/s}) = \frac{(507 \,\Omega)(169 \,\Omega)}{507 \,\Omega + 169 \,\Omega} = \frac{85,683 \,\Omega^2}{676 \,\Omega} = 126.75 \,\Omega$$

Then

$$Z(\omega = 1000 \text{ rad/s}) = 23.25 - j80 + 126.75 \,\Omega$$

or

$$Z(\omega = 1000 \text{ rad/s}) = 150 - j80 \,\Omega = 170\underline{/-28.07°} \,\Leftarrow$$

PHASORS

10.16: If, in the form

$$f(t) = A\cos\omega t + B\sin\omega t$$

$$A = -4 \quad \text{and} \quad B = 4$$

determine F and ϕ in

$$f(t) = F\cos(\omega t + \phi)$$

Here

$$f(t) = -4\cos\omega t + 4\sin\omega t$$

so that

$$F = \sqrt{(-4)^2 + (4)^2} = \sqrt{16 + 16} = \sqrt{32} = 4\sqrt{2}$$

10.21: If $F = 275$ and $\phi = -143.13°$ in the instantaneous form

$$f(t) = F\cos(\omega t + \phi)$$

determine A and B in

$$f(t) = A\cos\omega t + B\sin\omega t$$

Here

$$A = 275\cos(-143.13°) = (275)(-0.80) = -220$$

and

$$B = -275\sin(-143.13°) = -(275)(-0.60) = 165$$

so that

$$f(t) = -220\cos\omega t + 165\sin\omega t \Leftarrow$$

10.22: If $F = 390$ and $\phi = 112.62°$ in the instantaneous form

$$f(t) = F\cos(\omega t + \phi)$$

determine A and B in

$$f(t) = A\cos\omega t + B\sin\omega t$$

Here

$$A = 390\cos 112.62° = (390)(-0.3846) = -150$$

and

$$B = -390\sin 112.62° = -(390)(0.9231) = -360$$

so that

$$f(t) = -150\cos\omega t - 360\sin\omega t \Leftarrow$$

10.23: If $F = 800$ and $\phi = -196.26°$ in the instantaneous form

$$f(t) = F\cos(\omega t + \phi)$$

determine A and B in

$$f(t) = A\cos\omega t + B\sin\omega t$$

Here

$$A = 800\cos(-196.26°) = (800)(-0.96) = -768$$

and

$$B = -800\sin(-196.26°) = -(800)(0.28) = -224$$

so that

$$f(t) = -768 \cos \omega t - 224 \sin \omega t$$

10.24: If $F = 120$ and $\phi = -150°$ in the instantaneous form

$$f(t) = F \cos(\omega t + \phi)$$

determine A and B in

$$f(t) = A \cos \omega t + B \sin \omega t$$

Here

$$A = 120 \cos 150° = (120)(-0.8660) = -103.92$$

and

$$B = -120 \sin 150° = -(120)(0.5000) = -60$$

so that

$$f(t) = -103.92 \cos \omega t - 60 \sin \omega t \Leftarrow$$

10.25: If $F = 1000$ and $\phi = -135°$ in the instantaneous form

$$f(t) = F \cos(\omega t + \phi)$$

determine A and B in

$$f(t) = A \cos \omega t + B \sin \omega t$$

Here

$$A = 1000 \cos -135° = (1000)(-\sqrt{2}/2) = -500\sqrt{2}$$

and

$$B = -1000 \sin -135° = -(1000)(-\sqrt{2}/2) = 500\sqrt{2}$$

so that

$$f(t) = -500\sqrt{2} \cos \omega t + 500\sqrt{2} \sin \omega t$$

10.26: If a voltage is given by

$$v(t) = 208 \cos(377t + 60°)$$

determine the voltage phasor in rectangular, polar and exponential form.

Here

$$\mathbf{V} = 208\underline{/60°} \text{ V} \Leftarrow$$
$$\mathbf{V} = 104 + \jmath 180.13 \text{ V} \Leftarrow$$

and

$$\mathbf{V} = 208e^{60°} \text{ V} \Longleftarrow$$

10.27: If a current is given by

$$i(t) = 4\sqrt{2}\cos(400t - 135°)$$

determine the current phasor in rectangular, polar and exponential form.

Here

$$\mathbf{I} = 4\sqrt{2}\underline{/-135°} \text{ A} \Longleftarrow$$
$$\mathbf{I} = -4 - j4 \text{ A} \Longleftarrow$$

and

$$\mathbf{I} = 4\sqrt{2}e^{-135°} \text{ A} \Longleftarrow$$

10.28: If a current is given by

$$i(t) = 120\sqrt{3}\cos(600t + 15°)$$

determine the current phasor in rectangular, polar and exponential form.

Here

$$\mathbf{I} = 120\sqrt{3}\underline{/15°} \text{ A} \Longleftarrow$$
$$\mathbf{I} = 200.76 + j53.79 \text{ A} \Longleftarrow$$

and

$$\mathbf{I} = 120\sqrt{3}e^{15°} \text{ A} \Longleftarrow$$

10.29: If a voltage is given by

$$v(t) = 1750\cos(200t - 163.74°)$$

determine the voltage phasor in rectangular, polar and exponential form.

Here

$$\mathbf{V} = 1750\underline{/-163.74°} \text{ V} \Longleftarrow$$
$$\mathbf{V} = 1680 - j490 \text{ V} \Longleftarrow$$

and

$$\mathbf{V} = 1750e^{-163.74°} \text{ V} \Leftarrow$$

10.30: If a voltage is given by

$$v(t) = 400\cos(1200t + 90°)$$

determine the voltage phasor in rectangular, polar and exponential form.

Here

$$\mathbf{V} = 400\underline{/90°} \text{ V} \Leftarrow$$
$$\mathbf{V} = 0 + j400 \text{ V} \Leftarrow$$

and

$$\mathbf{V} = 400e^{90°} \text{ V} \Leftarrow$$

10.31: Two elements are connected in parallel as indicated in Fig 10.6. The line current is $i(t) = 10\sqrt{2}\cos(377t + 135)° \text{ A}$ and $i_1(t) = 6\cos(377t + 30)°$. Express $i_2(t)$ instantaneous form.

Using KCL

$$i(t) = i_1(t) + i_2(t)$$

so that, using phasors

$$\mathbf{I} = \mathbf{I_1} + \mathbf{I_2}$$

Then

$$10\sqrt{2}\underline{/135°} = 6\underline{/30°} + \mathbf{I_2}$$

so that

$$
\begin{aligned}
-10 + j10 &= 5.196 + j3.000 + \mathbf{I_2} \\
\mathbf{I_2} &= -10 + j10 - 5.196 - j3.000 \\
&= -15.196 + j7.000 \\
&= 16.731\underline{/155.27°} \text{ A}
\end{aligned}
$$

and

$$i_2(t) = 16.731\cos(377t + 155.27°) \text{ A} \Leftarrow$$

10.32: Three elements are connected in series as indicated in Fig 10.7. The line voltage is $v_s(t) = 120\cos(377t+90°)\,\text{V}$, $v_1(t) = 40\sqrt{2}\cos(377t+45°)\,\text{V}$ and $v_3(t) = 60\cos(377t-53.13°)\,\text{V}$. Express $v_2(t)$ instantaneous form.

Using KVL

$$v_2(t) = v_s(t) - v_1(t) - v_3(t)$$

so that, using phasors

$$\mathbf{V}_2 = \mathbf{V}_s - \mathbf{V}_1 - \mathbf{V}_3$$

Then

$$
\begin{aligned}
\mathbf{V}_2 &= 120\underline{/90°} - 40\sqrt{2}\underline{/45°} - 60\underline{/-53.13°} \\
&= j120 - (40 + j40) - (36 - j48) \\
&= -76 + j128 \\
&= 148.86\underline{/120.70°}\,\text{V}
\end{aligned}
$$

and

$$v_2(t) = 148.86\cos(377t + 120.70°)\,\text{V} \Leftarrow$$

10.33: In Fig 10.8, $\mathbf{I}_1 = 2\sqrt{2}\underline{/45°}$ and $\mathbf{V}_2 = 120\underline{/53.13°}\,\text{V}$. If the line current is $i = 4\cos(\omega t + 90°)\,\text{A}$ and the frequency is $f = 250/2\pi$ Hz, Determine the four components that comprise the parallel combination.

The given data is

$$
\begin{aligned}
\mathbf{I} &= 4\underline{/90°}\,\text{A} \\
\mathbf{I}_1 &= 2\sqrt{2}\underline{/45°}\,\text{A}
\end{aligned}
$$

and

$$\mathbf{V}_1 = \mathbf{V}_2 = 120\underline{/53.13°}\,\text{V}$$

and the angular frequency is

$$\omega = 2\pi f = 2\pi\left(\frac{250}{2\pi}\right) = 250\,\text{rad/s}$$

By KCL

$$
\begin{aligned}
\mathbf{I}_2 &= \mathbf{I} - \mathbf{I}_1 \\
&= 4\underline{/90°}\,\text{A} - 2\sqrt{2}\underline{/45°}\,\text{A} \\
&= (0 + j4) - (2 + j2) \\
&= -2 + j2 = 2\sqrt{2}\underline{/135°}\,\text{A}
\end{aligned}
$$

With $Z_1 = Z_a + Z_b$

$$Z_1 = \frac{\mathbf{V}_1}{\mathbf{I}_1} = \frac{120\underline{/53.13°}\,\text{V}}{2\sqrt{2}\underline{/45°}\,\text{A}}$$

or

$$Z_1 = 30\sqrt{2}\underline{/8.13°} = 42 + j6\,\Omega$$

Here, Z_a is resistive

$$Z_a = 42\,\Omega \Leftarrow$$

and Z_b is inductive ($\omega L = 6\,\Omega$) so that

$$L = \frac{6\,\Omega}{250\ \text{rad/s}} = 0.024\,\text{H} \Leftarrow$$

Then, with $Z_2 = Z_c + Z_d$

$$Z_2 = \frac{\mathbf{V}_2}{\mathbf{I}_2} = \frac{120\underline{/53.13°}\,\text{V}}{2\sqrt{2}\underline{/135°}\,\text{A}}$$

or

$$Z_1 = 30\sqrt{2}\underline{/-81.87°} = 6 - j42\,\Omega$$

Here, Z_c is resistive

$$Z_c = 6\,\Omega \Leftarrow$$

and Z_d is capacitive ($1/\omega C = 42\,\Omega$) so that

$$C = \frac{1}{(42\,\Omega)(250\ \text{rad/s})} = 95.24\,\mu\text{F} \Leftarrow$$

10.34: In Fig 10.9, $i_s(t) = 8\cos(400t + 90°)$ A, $v_{s1}(t) = 200\cos(400t + 36.87°)$ V and $v_{s2}(t) = 200\sqrt{2}\cos(400t + 135°)$ A. Combine the three sources to a single voltage source connected across terminals a-b.

The strategy here is to convert v_{s1} to a current source and then combine the two current sources. With $\omega = 400$ rad/s

$$X_L = \omega L = (400\ \text{rad/s})(0.025\,\text{H}) = 10\,\Omega$$

and

$$X_C = \frac{1}{\omega C} = \frac{1}{(400\ \text{rad/s})(50\,\mu\text{F})} = 50\,\Omega$$

The equivalent current source $\mathbf{I}_{s2}$ is

$$
\begin{aligned}
\mathbf{I}_{s2} &= \frac{\mathbf{V}_{s2}}{j10\,\Omega} \\
&= \frac{200\sqrt{2}\underline{/135°}\text{ V}}{10\underline{/90°}\,\Omega} \\
&= 20\sqrt{2}\underline{/45°} = 20 + j20\,\Omega
\end{aligned}
$$

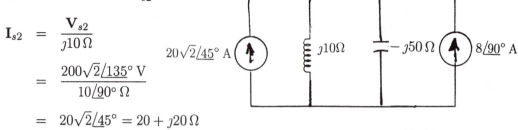

and the combination is shown the right.
Then with

$$
Z_{\text{eq}} = \frac{(j10\,\Omega)(-j50\,\Omega)}{j10\,\Omega - j50\,\Omega} = \frac{500\,\Omega^2}{-j40\,\Omega} = j12.5\,\Omega
$$

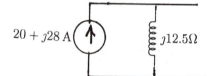

the two current sources can be added to give

$$
\mathbf{I}_{s,\text{eq}} = (20 + j20) + j8 = 20 + j28\text{ A}
$$

The equivalent current source can be
converted to an equivalent voltage source

$$
\begin{aligned}
\mathbf{V}_{s,\text{eq}} &= Z_{\text{eq}}\mathbf{I}_{s,\text{eq}} \\
&= (j12.5\,\Omega)(20 + j28\,\Omega) \\
&= (12.5\underline{/90°})(34.41\underline{/54.46°}) \\
&= 430.12\underline{/144.46°}\text{ V} \\
&= -350 + j250\text{ V}
\end{aligned}
$$

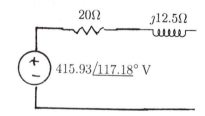

The final step is to combine the two
voltage sources to a single voltage source
across terminals a-b

$$
\begin{aligned}
\mathbf{V} &= \mathbf{V}_{s1} + \mathbf{V}_{s,\text{eq}} \\
&= 200\underline{/36.87°}\text{ V} + 430.12\underline{/144.46°}\text{ V} \\
&= (160 + j120)\text{ V} + (-350 + j250)\text{ V} \\
&= -190 + j370\text{ V} \\
&= 415.93\underline{/117.18°} \Leftarrow
\end{aligned}
$$

with

$$Z_{\text{eq}} = 20 + j12.5\,\Omega$$

LADDER NETWORKS AND PHASOR DIAGRAMS

10.35: Determine all branch currents and branch voltages in the network of Fig 10.10 and then draw a phasor diagram.

The network can be put into the phasor domain by noting that $\omega = 400$ rad/s and that

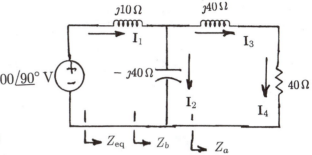

Component	Reactance, X
$C = 62.5\,\mu F$	$40\,\Omega$
$L_1 = 25\,\text{mH}$	$10\,\Omega$
$L_1 = 100\,\text{mH}$	$40\,\Omega$

Then, for the network in the phasor domain

$$Z_a = 40 + j40\,\Omega = 40\sqrt{2}\underline{/45^\circ}$$

$$Z_b = \frac{(-j40\,\Omega)(40 + j40\,\Omega)}{j40\,\Omega + (40 + j40\,\Omega)} = \frac{(40\underline{/-90^\circ})(40\sqrt{2}\underline{/45^\circ})}{40}\,\Omega$$

or

$$Z_b = 40\sqrt{2}\underline{/45^\circ} = 40 - j40\,\Omega$$

and

$$Z_{\text{eq}} = 40 - j40 + j10\,\Omega = 40 - j30\,\Omega = 50\underline{/-36.87^\circ}\,\Omega$$

Then

$$\mathbf{I}_1 = \frac{\mathbf{V}_s}{Z_{\text{eq}}} = \frac{400\underline{/90^\circ}\,\text{V}}{50\underline{/-36.87^\circ}\,\Omega} = 8\underline{/126.87^\circ}\,\text{A} = -4.80 + j6.40\,\text{A} \Leftarrow$$

and

$$\mathbf{V}_1 = (10\underline{/90^\circ}\,\Omega)(8\underline{/126.87^\circ})\,\text{A} = 80\underline{/216.87^\circ}\,\text{V} = -(64 + j48)\,\text{V} \Leftarrow$$

Then

$$\mathbf{V}_2 = \mathbf{V}_s - \mathbf{V}_1 = (0 + j400) - 80\underline{/216.87^\circ} = (0 + j400) - (-64 - j48)$$

or

$$\mathbf{V}_2 = 64 + j448\,\text{V} = 452.55\underline{/81.87^\circ}\,\text{V}$$

$$\mathbf{I}_2 = \frac{\mathbf{V}_2}{40\underline{/-90^\circ}} = \frac{452.55\underline{/81.87^\circ}\,\text{V}}{40\underline{/-90^\circ}\,\Omega} = 11.31\underline{/171.87^\circ}\,\text{A} = -11.20 + j1.60\,\text{A} \Leftarrow$$

and

$$\mathbf{I}_3 = \mathbf{I}_4 = \frac{\mathbf{V}_2}{40\sqrt{2}\underline{/45°}} = \frac{452.55\underline{/81.87°}}{40\sqrt{2}\underline{/45°}} = 8\underline{/36.87°}\,\text{A} = 6.40 + j4.80\,\text{A} \Leftarrow$$

The two voltages, $\mathbf{V}_3$ and $\mathbf{V}_4$ are

$$\mathbf{V}_3 = (j40)(\mathbf{I}_3) = (40\underline{/90°})(8\underline{/36.87°}) = 320\underline{/126.87°}\,\text{V} = -192 + j256\,\text{V} \Leftarrow$$

and

$$\mathbf{V}_4 = (40)(\mathbf{I}_4) = (40)(8\underline{/36.87°}) = 320\underline{/36.87°}\,\text{V} = 256 + j192\,\text{V} \Leftarrow$$

The phasor diagram is shown below.

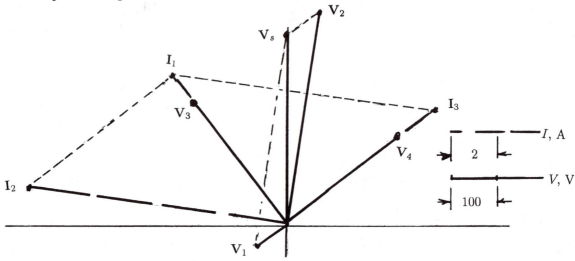

10.36: Determine all branch currents and branch voltages in the network of Fig 10.11 and then draw a phasor diagram.

The network can be put into the phasor domain by noting that $\omega = 250$ rad/s and that

Component	Reactance, X
$C = 33.33\,\mu\text{F}$	$120\,\Omega$
$L = 960\,\text{mH}$	$240\,\Omega$

The network in the phasor domain is shown at the right. Here,

$$Z_a = 160 - j120 = 200\underline{/-36.87°}\,\Omega$$

and by current division with $\mathbf{I}_s = 400\underline{/-36.87°}\,\text{mA}$

$$\mathbf{I}_1 = \left[\frac{160 - j120}{(160 - j120) + j240}\right](400\underline{/-36.87°}\,\text{mA})$$

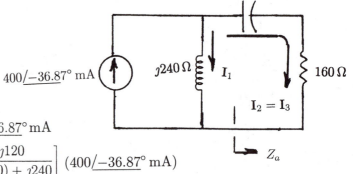

$$= \left[\frac{200\underline{/-36.87°}}{200\underline{/36.87°}} \right] (400\underline{/-36.87°}) \, \text{mA}$$

$$= 400\underline{/-110.61°} \, \text{mA} = -140.80 - j374.40 \, \text{mA} \Leftarrow$$

and by KCL

$$
\begin{aligned}
\mathbf{I}_2 = \mathbf{I}_3 &= \mathbf{I}_s - \mathbf{I}_1 \\
&= 400\underline{/-36.87°} \, \text{mA} - 400\underline{/-110.61°} \, \text{mA} \\
&= (320 - j240) \, \text{mA} - (-140.80 - j374.40) \, \text{mA} \\
&= 460.80 + j134.40 \, \text{mA} \Leftarrow \\
&= 480\underline{/16.26°} \, \text{mA} \Leftarrow
\end{aligned}
$$

Then

$$
\begin{aligned}
\mathbf{V}_1 &= (j240 \, \Omega)\mathbf{I}_1 = (240\underline{/90°} \, \Omega)(400\underline{/-110.61°} \, \text{mA}) \\
&= 96\underline{/-20.61°} \, \text{V} = 89.86 - j33.92 \, \text{V} \Leftarrow \\
\mathbf{V}_2 &= (-j120 \, \Omega)\mathbf{I}_1 = (120\underline{/-90°} \, \Omega)(480\underline{/16.26°} \, \text{mA}) \\
&= 57.60\underline{/-73.74°} \, \text{V} = 16.13 - j55.30 \, \text{V} \Leftarrow
\end{aligned}
$$

and

$$
\begin{aligned}
\mathbf{V}_3 &= (160 \, \Omega)\mathbf{I}_3 = (160\underline{/0°} \, \Omega)(480\underline{/16.26°} \, \text{mA}) \\
&= 76.80\underline{/16.26°} \, \text{V} = 73.73 + j21.50 \, \text{V} \Leftarrow
\end{aligned}
$$

The phasor diagram is shown below.

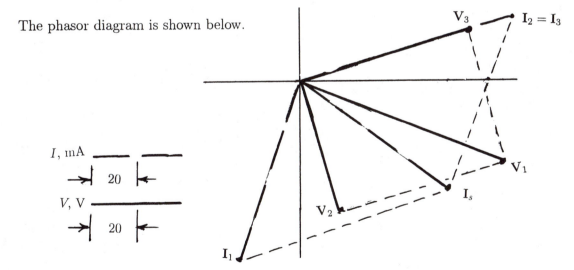

10.37: Determine all branch currents and branch voltages in the network of Fig 10.12 and then draw a phasor diagram.

The network can be put into the phasor domain by noting that $\omega = 1000$ rad/s and that

Component	Reactance, X
$C = 62.5\,\mu\text{F}$	$16\,\Omega$
$L_1 = 58\,\text{mH}$	$58\,\Omega$
$L_2 = 16\,\text{mH}$	$16\,\Omega$

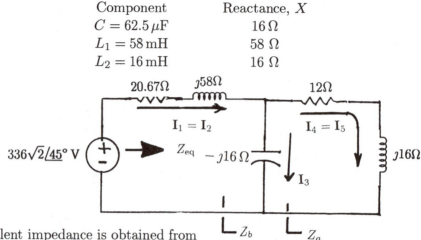

The equivalent impedance is obtained from

$$Z_a = 12 + j16\,\Omega = 20\underline{/53.13°}\,\Omega$$

$$Z_b = \frac{(-j16)(20\underline{/53.13°})}{-j16 + (12 + j16)}$$

$$= \frac{(16\underline{/-90°})(20\underline{/53.13°})}{12}$$

$$= \frac{320\underline{/-36.87°}}{12}\,\Omega$$

$$= 26.67\underline{/-36.87°}\,\Omega = 21.33 - j16\,\Omega$$

$$Z_{\text{eq}} = (21.33 - j16) + (20.67 + j58) = 42 + j42\,\Omega = 42\sqrt{2}\underline{/45°}\,\Omega$$

Then, with $\mathbf{V}_s = 336\sqrt{2}\underline{/135°}$ V

$$\mathbf{I}_1 = \mathbf{I}_2 = \frac{\mathbf{V}_s}{Z_{\text{eq}}} = \frac{336\sqrt{2}\underline{/135°}\text{ V}}{42\sqrt{2}\underline{/45°}\,\Omega} = 8\underline{/90°}\text{ A} = j8\text{ A} \Leftarrow$$

$$\mathbf{V}_1 = (20.67\,\Omega)\mathbf{I}_1 = (20.67\,\Omega)(8\underline{/90°}\text{ A}) = 165.36\underline{/90°}\text{ V} = j165.36\text{ V} \Leftarrow$$

and

$$\mathbf{V}_2 = (j58\,\Omega)\mathbf{I}_1 = (58\underline{/90°})(8\underline{/90°}\text{ A}) = -464 + j0\text{ V} = 464\underline{/180°}\text{ V} \Leftarrow$$

By current division

$$\mathbf{I}_3 = \left[\frac{12 + j16}{(12 + j16) - j16}\right]\mathbf{I}_2 == \frac{20\underline{/53.13°}}{12}(j8\,\mathrm{A}) = (1.67\underline{/53.13°})(8\underline{/90°}\,\mathrm{A})$$

or

$$\mathbf{I}_3 = 13.33\underline{/143.13°}\,\mathrm{A} = -10.67 + j8\,\mathrm{A} \Leftarrow$$

and

$$\mathbf{V}_3 = (-j16)\mathbf{I}_3 = (16\underline{/-90°}\,\Omega)(13.33\underline{/143.13°}\,\mathrm{A})$$

or

$$\mathbf{V}_3 = 213.33\underline{/53.13°}\,\mathrm{V} = 128.00 + j170.67\,\mathrm{V} \Leftarrow$$

Then, by KCL

$$\begin{aligned}
\mathbf{I}_4 = \mathbf{I}_5 &= \mathbf{I}_2 - \mathbf{I}_3 \\
&= (0 + j8) - (-10.67 + j8) \\
&= 10.67\,\mathrm{A} = 10.67\underline{/0°}\,\mathrm{A} \Leftarrow \\
\mathbf{V}_4 &= (12\,\Omega)\mathbf{I}_4 \\
&= (12\underline{/0°})(10.67\underline{/0°}) \\
&= 128\,\mathrm{V} = 128\underline{/0°}\,\mathrm{V} \Leftarrow
\end{aligned}$$

and

$$\begin{aligned}
\mathbf{V}_5 &= (j16\,\Omega)\mathbf{I}_5 \\
&= (16\underline{/90°})(10.67\underline{/0°}) \\
&= j170.67\,\mathrm{V} = 170.67\underline{/90°} \Leftarrow
\end{aligned}$$

The phasor diagram is

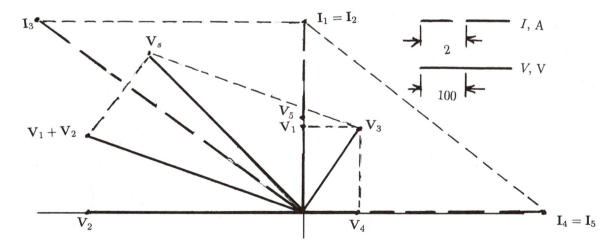

10.38: Determine all branch currents and branch voltages in the network of Fig 10.13 and then draw a phasor diagram.

The network can be put into the phasor domain by noting that $\omega = 250$ rad/s and that

Component	Reactance, X
$C_1 = 20\,\mu\text{F}$	$200\,\Omega$
$C_2 = 40\,\mu\text{F}$	$100\,\Omega$
$L_1 = 400\,\text{mH}$	$100\ \Omega$
$L_2 = 800\,\text{mH}$	$200\ \Omega$

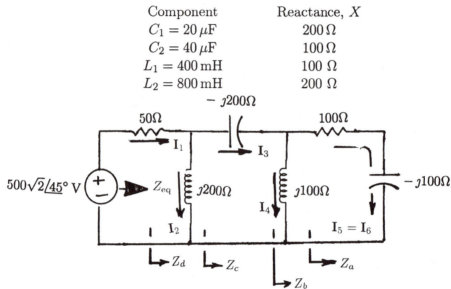

The equivalent impedance is obtained from

$$Z_a = 100 - j100\,\Omega = 100\sqrt{2}\underline{/-45°}$$

$$Z_b = \frac{(j100)(100\sqrt{2}\underline{/-45°})}{j100 + 100 - j100}\,\Omega$$

$$= \frac{10,000\underline{/45°}}{100}\,\Omega$$

$$= 100\underline{/45°}\,\Omega = 100 + j100\,\Omega$$

$$Z_c = -j200 + 100 + j100 = 100 - j100 = 100\sqrt{2}\underline{/-45°}\Omega$$

$$Z_d = \frac{(j200)(100\sqrt{2}\underline{/-45°})}{j200 + 100 - j100}\,\Omega$$

$$= \frac{20,000\sqrt{2}\underline{/45°}}{100\sqrt{2}\underline{/45°}}$$

$$= 200\,\Omega = 200\underline{/0°}\,\Omega$$

and

$$Z_{\text{eq}} = 50\,\Omega + 200\,\Omega = 250\,\Omega$$

Then, with $\mathbf{V}_s = 500\sqrt{2}\underline{/45°}\,\text{V}$

$$\mathbf{I}_1 = \frac{\mathbf{V}_s}{Z_{\text{eq}}} = \frac{500\sqrt{2}\underline{/45°}\,\text{V}}{250\,\Omega} = 2\sqrt{2}\underline{/45°}\,\text{A} = 2 + \jmath2\,\text{A} \Leftarrow$$

$$\mathbf{V}_1 = (50\,\Omega)\mathbf{I}_1 = (50\,\Omega)(2\sqrt{2}\underline{/45°}\,\text{A}) = 100\sqrt{2}\underline{/45°}\,\text{V} = 100 + \jmath100\,\text{V} \Leftarrow$$

By current division

$$\mathbf{I}_2 = \left[\frac{100 - \jmath100}{\jmath200 + 100 - \jmath100)}\right]\mathbf{I}_1 = \left[\frac{100\sqrt{2}\underline{/-45°}}{100\sqrt{2}\underline{/45°}}\right](2\sqrt{2}\underline{/45°})$$

or

$$\mathbf{I}_2 = (1\underline{/-90°})(2\sqrt{2}\underline{/45°}) = 2\sqrt{2}\underline{/-45°}\,\text{A} = 2 - \jmath2\,\text{A} \Leftarrow$$

Then

$$\mathbf{V}_2 = (\jmath200\,\Omega)\mathbf{I}_2 = (200\underline{/90°})(2\sqrt{2}\underline{/-45°})) = 400\sqrt{2}\underline{/45°}) = 400 + \jmath400\,\text{V} \Leftarrow$$

By KCL

$$
\begin{aligned}
\mathbf{I}_3 &= \mathbf{I}_1 - \mathbf{I}_2 \\
&= (2 + \jmath2) - (2 - \jmath2) \\
&= \jmath4\,\text{A} = 4\underline{/90°}\,\text{A} \Leftarrow
\end{aligned}
$$

and

$$
\begin{aligned}
\mathbf{V}_3 &= (-\jmath200\,\Omega)\mathbf{I}_3 \\
&= (200\underline{/-90°})(4\underline{/90°}) \\
&= 800\,\text{V} = 800\underline{/0°}\,\text{V} \Leftarrow
\end{aligned}
$$

Another current division gives

$$\mathbf{I}_4 = \left[\frac{100 - \jmath100}{\jmath100 + 100 - \jmath100)}\right]\mathbf{I}_3 = \left[\frac{100\sqrt{2}\underline{/-45°}}{100}\right](4\underline{/90°}) = 4\sqrt{2}\underline{/45°}\,\text{A} = 4 + \jmath4\,\text{A}$$

and

$$\mathbf{V}_4 = (\jmath100\,\Omega)\mathbf{I}_4 = (100\underline{/90°})4\sqrt{2}\underline{/45°} = 400\sqrt{2}\underline{/135°}\,\text{V} = -400 + \jmath400\,\text{V} \Leftarrow$$

By KCL

$$
\begin{aligned}
\mathbf{I}_5 = \mathbf{I}_6 \; &= \; \mathbf{I}_3 - \mathbf{I}_4 \\
&= \; (0 + j4) - (4 + j4) \\
&= \; -4\,\text{A} = 4\underline{/180°}\,\text{A} \Leftarrow \\
\mathbf{V}_5 \; &= \; (100\,\Omega)\mathbf{I}_5 \\
&= \; (100\underline{/0°})(4\underline{/180°}) \\
&= \; -400\,\text{V} = 400\underline{/180°}\,\text{V} \Leftarrow
\end{aligned}
$$

and

$$
\begin{aligned}
\mathbf{V}_6 \; &= \; (-j100\,\Omega)\mathbf{I}_6 \\
&= \; (100\underline{/-90°})(4\underline{/180°}) \\
&= \; 400\underline{/90°}\,\text{V} = j400\,\text{V} \Leftarrow
\end{aligned}
$$

The phasor diagram is shown below.

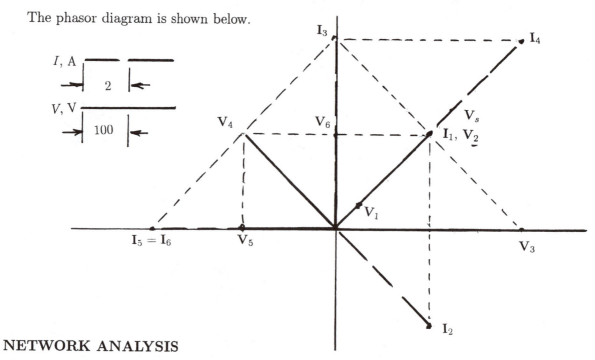

NETWORK ANALYSIS

10.39: Rework Problem 10.35 (Fig 10.10) using a nodal analysis to determine the current through the 25 mH inductor and the voltage across the 62.5 μ F capacitor.

The phasor domain network using the reactances developed in Problem 10.35 and with the voltage source transformed to a current source

$$\mathbf{I}_s = \frac{\mathbf{V}_s}{\jmath 10} = \frac{400\underline{/90°}\text{ V}}{10\underline{/90°}\ \Omega} = 40\text{ A}$$

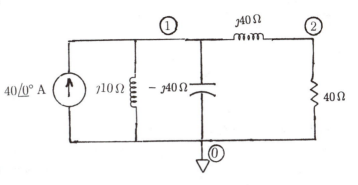

is shown at the right. Note that nodes-1 and -2 have been designated by numerals within circles.

Two node equations can be written using a repeated application of KCL

$$\left(-\jmath\frac{1}{10}+\jmath\frac{1}{40}-\jmath\frac{1}{40}\right)\mathbf{V}_1 \quad -\left(-\jmath\frac{1}{40}\right)\mathbf{V}_2 \quad = \quad 40\text{ A}$$

$$-\left(-\jmath\frac{1}{40}\right)\mathbf{V}_1 \quad +\left(\frac{1}{40}-\jmath\frac{1}{40}\right)\mathbf{V}_2 \quad = \quad 0$$

The node equations may be put into matrix form after multiplying throughout by 40

$$\begin{bmatrix} -\jmath 4 & \jmath \\ \jmath & 1-\jmath \end{bmatrix}\begin{bmatrix} \mathbf{V}_1 \\ \mathbf{V}_2 \end{bmatrix} = \begin{bmatrix} 1600\text{ A} \\ 0 \end{bmatrix}$$

and a matrix inversion can be employed to find the voltage vector

$$\begin{bmatrix} \mathbf{V}_1 \\ \mathbf{V}_2 \end{bmatrix} = \begin{bmatrix} -\jmath 4 & \jmath \\ \jmath & 1-\jmath \end{bmatrix}^{-1}\begin{bmatrix} 1600\text{ A} \\ 0 \end{bmatrix}$$

or

$$\begin{bmatrix} \mathbf{V}_1 \\ \mathbf{V}_2 \end{bmatrix} = \frac{1}{5\underline{/-126.87°}}\begin{bmatrix} \sqrt{2}\underline{/-45°} & 1\underline{/-90°} \\ 1\underline{/-90°} & 4\underline{/-90°} \end{bmatrix}\begin{bmatrix} 1600\text{ A} \\ 0 \end{bmatrix}$$

Thus,

$$\begin{bmatrix} \mathbf{V}_1 \\ \mathbf{V}_2 \end{bmatrix} = \begin{bmatrix} 320\sqrt{2}\underline{/81.87°}\text{ V} \\ 320\underline{/36.87°}\text{ V} \end{bmatrix} = \begin{bmatrix} 64+\jmath 448\text{ V} \\ 256+\jmath 192\text{ V} \end{bmatrix}$$

Then, after a return to the original network containing the voltage source in the phasor domain

$$\begin{aligned}
\mathbf{V}_L &= \mathbf{V}_s - \mathbf{V}_1 \\
&= (0 + j400) - (64 + j448) \\
&= -64 - j48 = 80\underline{/-143.13°}\ \mathrm{V}
\end{aligned}$$

The current through the inductor 25 mH inductor will be

$$\mathbf{I}_L = \frac{80\underline{/-143.13°}\ \mathrm{V}}{10\underline{/90°}\ \Omega} = 8\underline{/126.87°}\ \mathrm{A} = -4.80 + j6.40\ \mathrm{A} \Leftarrow$$

and the voltage across the 62.5 μF capacitor will be

$$\mathbf{V}_C = \mathbf{V}_1 = 320\sqrt{2}\underline{/81.87°}\ \mathrm{V} = 64 + j448\ \mathrm{V} \Leftarrow$$

These confirm the result of Problem 10.35.

10.40: Rework Problem 10.36 (Fig 10.11) using a nodal analysis to determine the current through the 960 mH inductor and the voltage across the 160 Ω resistor.

The phasor domain network using the reactances developed in Problem 10.36 is shown at the right. Two nodes are indicated and by two applications of KCL, the node equations with nodes deignated by numerals within circles.

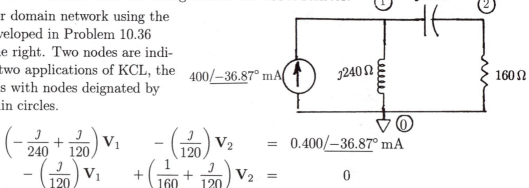

$$\begin{aligned}
\left(-\frac{j}{240} + \frac{j}{120}\right)\mathbf{V}_1 \quad - \left(\frac{j}{120}\right)\mathbf{V}_2 &= 0.400\underline{/-36.87°}\ \mathrm{mA} \\
-\left(\frac{j}{120}\right)\mathbf{V}_1 \quad + \left(\frac{1}{160} + \frac{j}{120}\right)\mathbf{V}_2 &= 0
\end{aligned}$$

The node equations may be put into matrix form after multiplying throughout by 480

$$\begin{bmatrix} j2 & -j4 \\ -j4 & (3+j4) \end{bmatrix}\begin{bmatrix} \mathbf{V}_1 \\ \mathbf{V}_2 \end{bmatrix} = \begin{bmatrix} 192\underline{/-36.87°}\ \mathrm{A} \\ 0 \end{bmatrix}$$

and a matrix inversion can be employed to find the voltage vector

$$\begin{bmatrix} \mathbf{V}_1 \\ \mathbf{V}_2 \end{bmatrix} = \begin{bmatrix} j2 & -j4 \\ -j4 & (3+j4) \end{bmatrix}^{-1}\begin{bmatrix} 192\underline{/-36.87°}\ \mathrm{A} \\ 0 \end{bmatrix}$$

or

$$\begin{bmatrix} \mathbf{V}_1 \\ \mathbf{V}_2 \end{bmatrix} = \frac{1}{10\underline{/36.87^\circ}} \begin{bmatrix} 5\underline{/53.13^\circ} & 4\underline{/90^\circ} \\ 4\underline{/90^\circ} & 2\underline{/90^\circ} \end{bmatrix} \begin{bmatrix} 192\underline{/-36.87^\circ} \text{ A} \\ 0 \end{bmatrix}$$

Thus,

$$\begin{bmatrix} \mathbf{V}_1 \\ \mathbf{V}_2 \end{bmatrix} = \begin{bmatrix} 89.86 - \jmath 33.79 \text{ V} \\ 73.73 + \jmath 21.50 \text{ V} \end{bmatrix}$$

The voltage across the $160\,\Omega$ resistor is

$$\mathbf{V}_R = \mathbf{V}_2 = 73.73 + \jmath 21.50\,\text{V} = 76.80\underline{/16.26^\circ}\,\text{V} \Leftarrow$$

and the current through the $960\,\text{mH}$ inductor will be

$$\mathbf{I}_L = \frac{\mathbf{V}_1}{\jmath 240\,\Omega} = \frac{89.86 - \jmath 33.79\,\text{V}}{\jmath 240\,\Omega} = \frac{96\underline{/-20.61^\circ}}{240\underline{/90^\circ}}$$

or

$$I_L = 400\underline{/-110.61^\circ}\,\text{mA} = -140.80 - \jmath 374.4\,\text{mA} \Leftarrow$$

These confirm the result of Problem 10.36.

10.41: Rework Problem 10.38 (Fig 10.13) using a nodal analysis to determine the current through the $800\,\text{mH}$ inductor and the voltage across the $40\,\mu\text{F}$ capacitor.

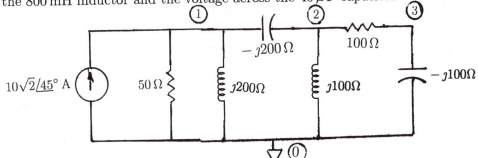

The phasor domain network using the reactances developed in Problem 10.38 is shown with the voltage source transformed to a current source via

$$\mathbf{I}_s = \frac{\mathbf{V}_s}{50\,\Omega} = \frac{500\sqrt{2}\underline{/45^\circ}\,\text{V}}{50\,\Omega} = 10\sqrt{2}\underline{/45^\circ}\,\text{A}$$

By repeated applications of KCL

$$\left(\frac{1}{50} - \frac{\jmath}{200} + \frac{\jmath}{200}\right)\mathbf{V}_1 \quad -\left(\frac{\jmath}{200}\right)\mathbf{V}_2 \qquad\qquad = \ 10\sqrt{2}\underline{/45^\circ}\,\text{A}$$

$$-\left(\frac{\jmath}{200}\right)\mathbf{V}_1 \quad +\left(\frac{\jmath}{200} - \frac{\jmath}{100} + \frac{1}{100}\right)\mathbf{V}_2 \quad -\left(\frac{1}{100}\right)\mathbf{V}_3 \ = \qquad 0$$

$$-\left(\frac{1}{100}\right)\mathbf{V}_2 \quad +\left(\frac{1}{100} + \frac{\jmath}{100}\right)\mathbf{V}_3 \ = \qquad 0$$

If all terms are multiplied by 200, the node equations become

$$
\begin{array}{rrrr}
4\mathbf{V}_1 & -j\mathbf{V}_2 & & = 2000\sqrt{2}\underline{/45^\circ}\ \mathrm{A} \\
-j\mathbf{V}_1 & +(2-j)\mathbf{V}_2 & -2\mathbf{V}_3 & = 0 \\
& -2\mathbf{V}_2 & +(2+j2)\mathbf{V}_3 & = 0
\end{array}
$$

A Cramer's rule solution gives

$$
\mathbf{V}_1 = \frac{\begin{vmatrix} 2000\sqrt{2}\underline{/45^\circ} & 1\underline{/-90^\circ} & 0 \\ 0 & 2.24\underline{/-26.57^\circ} & 2\underline{/180^\circ} \\ 0 & 2\underline{/180^\circ} & 2\sqrt{2}\underline{/45^\circ} \end{vmatrix}}{\begin{vmatrix} 4\underline{/0^\circ} & 1\underline{/-90^\circ} & 0 \\ 1\underline{/-90^\circ} & 2.24\underline{/-26.57^\circ} & 2\underline{/180^\circ} \\ 0 & 2\underline{/180^\circ} & 2\sqrt{2}\underline{/45^\circ} \end{vmatrix}}
$$

or

$$
\mathbf{V}_1 = \frac{8000\underline{/90^\circ}}{10\sqrt{2}\underline{/45^\circ}} = 400\sqrt{2}\underline{/45^\circ}\ \mathrm{V} = 400 + j400\ \mathrm{V}
$$

and

$$
\mathbf{V}_3 = \frac{\begin{vmatrix} 4\underline{/0^\circ} & 1\underline{/-90^\circ} & 2000\sqrt{2}\underline{/45^\circ} \\ 1\underline{/-90^\circ} & 2.24\underline{/-26.57^\circ} & 0 \\ 0 & 2\underline{/180^\circ} & 0 \end{vmatrix}}{10\sqrt{2}\underline{/45^\circ}}
$$

or

$$
\mathbf{V}_3 = \frac{4000\sqrt{2}\underline{/135^\circ}}{10\sqrt{2}\underline{/45^\circ}} = 400\underline{/90^\circ}\ \mathrm{V} = 0 + j400\ \mathrm{V}
$$

The voltage across the $40\,\mu\mathrm{F}$ capacitor is

$$
\mathbf{V}_C = \mathbf{V}_3 = j400 = 400\underline{/90^\circ}\ \mathrm{V} \Leftarrow
$$

and the current through the $800\,\mathrm{mH}$ inductor will be

$$
\mathbf{I}_L = \frac{\mathbf{V}_1}{j200\,\Omega} = \frac{400\sqrt{2}\underline{/45^\circ}\ \mathrm{V}}{200\underline{/90^\circ}\,\Omega}
$$

or

$$
\mathbf{I}_L = 2\sqrt{2}\underline{/-45^\circ}\ \mathrm{A} = 2 - j2\ \mathrm{A} \Leftarrow
$$

These confirm the result of Problem 10.38.

10.42: Use nodal analysis to find the current through the $50\,\mathrm{mH}$ inductor and the voltage across the $40\,\Omega$ resistor in Fig 10.14.

The network can be put into the phasor domain by noting that $\omega = 400$ rad/s and that

The network can be put into the phasor domain by noting that $\omega = 400$ rad/s and that

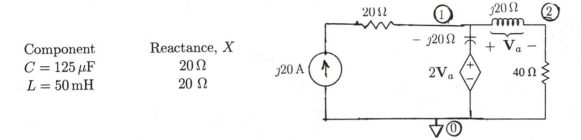

Component	Reactance, X
$C = 125\,\mu\text{F}$	$20\,\Omega$
$L = 50\,\text{mH}$	$20\,\Omega$

with the voltage source transformed to a current source via

$$\mathbf{I}_s = \frac{\mathbf{V}_s}{20\,\Omega} = \frac{400\underline{/90°}\,\text{V}}{20\,\Omega} = 20\underline{/90°}\,\text{A} = 0 + \jmath20\,\text{A}$$

The voltage controlled current source has a strength of $2\mathbf{V}_a = 2(\mathbf{V}_1 - \mathbf{V}_2)$ and it may also be transformed to a current source

$$\mathbf{I}_{CS} = \frac{2(\mathbf{V}_1 - \mathbf{V}_2)}{-\jmath20\,\Omega} = 0.10\jmath(\mathbf{V}_1 - \mathbf{V}_2)$$

Two nodes may be noted and the two node equations may be written using KCL

$$\left(\frac{1}{20} + \frac{\jmath}{20} - \frac{\jmath}{20}\right)\mathbf{V}_1 \quad -\left(-\frac{\jmath}{20}\right)\mathbf{V}_2 \quad = \quad \jmath20 + 0.10\jmath(\mathbf{V}_1 - \mathbf{V}_2)$$

$$-\left(-\frac{\jmath}{20}\right)\mathbf{V}_1 \quad +\left(\frac{1}{40} - \frac{\jmath}{20}\right)\mathbf{V}_2 \quad = \quad 0$$

These may simplified by multiplying all terms by 40 and, after rearrangement, they can be put into matrix form

$$\begin{bmatrix} (2 - \jmath4) & +\jmath6 \\ +\jmath2 & (1 - \jmath2) \end{bmatrix} \begin{bmatrix} \mathbf{V}_1 \\ \mathbf{V}_2 \end{bmatrix} = \begin{bmatrix} \jmath800\,\text{A} \\ 0 \end{bmatrix}$$

and a matrix inversion can be employed to find the voltage vector

$$\begin{bmatrix} \mathbf{V}_1 \\ \mathbf{V}_2 \end{bmatrix} = \begin{bmatrix} (2 - \jmath4) & +\jmath6 \\ +\jmath2 & (1 - \jmath2) \end{bmatrix}^{-1} \begin{bmatrix} \jmath800\,\text{A} \\ 0 \end{bmatrix}$$

$$= \frac{1}{10\underline{/-53.13°}} \begin{bmatrix} 2.24\underline{/-63.43°} & 6\underline{/-90°} \\ 2\underline{/-90°} & 4.47\underline{/-63.43°} \end{bmatrix} \begin{bmatrix} 800\underline{/90°}\,\text{A} \\ 0 \end{bmatrix}$$

or

$$\begin{bmatrix} \mathbf{V}_1 \\ \mathbf{V}_2 \end{bmatrix} = \begin{bmatrix} 178.89\underline{/79.70°}\,\text{V} \\ 160\underline{/53.13°}\,\text{V} \end{bmatrix} = \begin{bmatrix} 32 + \jmath176\,\text{V} \\ 96 + \jmath128\,\text{V} \end{bmatrix}$$

The voltage across the $40\,\Omega$ resistor is

$$\mathbf{V}_R = \mathbf{V}_2 = 160\underline{/53.13°}\,\text{V} = 96 + \jmath128\,\text{V} \Leftarrow$$

and the current through the $50\,\text{mH}$ inductor will be

$$\mathbf{I}_L = \frac{\mathbf{V}_1 - \mathbf{V}_2}{\jmath20\,\Omega} = \frac{(32 + \jmath176\,\text{V}) - (96 + \jmath128\,\text{V})}{\jmath20\,\Omega} = \frac{-64 + \jmath48\,\text{V}}{\jmath20\,\Omega}$$

or

$$\mathbf{I}_L = \frac{80\underline{/143.13°}\,\text{V}}{20\underline{/90°}\,\Omega} = 4\underline{/53.13°}\,\text{V} = 2.40 + \jmath3.20\,\text{V} \Leftarrow$$

10.43: Use nodal analysis to find the current through the $10\,\Omega$ resistor in Fig 10.15.

The network can be put into the phasor domain by noting that $\omega = 2000\,\text{rad/s}$ and that

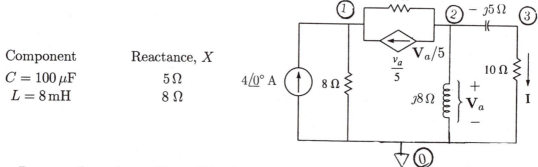

Component	Reactance, X
$C = 100\,\mu\text{F}$	$5\,\Omega$
$L = 8\,\text{mH}$	$8\,\Omega$

Because the voltage, $\mathbf{V}_a = \mathbf{V}_2$, the voltage controlled current source has a strength of $\mathbf{V}_2/5\,\text{V}$ and three node equations may be written using KCL

$$
\begin{aligned}
\left(\frac{1}{8} + \frac{1}{4}\right)\mathbf{V}_1 \quad &- \frac{1}{4}\mathbf{V}_2 &&= 4\,\text{A} + \frac{\mathbf{V}_2}{5} \\
-\frac{1}{4}\mathbf{V}_1 \quad &+ \left(\frac{1}{4} - \frac{\jmath}{8} + \frac{\jmath}{5}\right)\mathbf{V}_2 \quad - \left(\frac{\jmath}{5}\right)\mathbf{V}_3 &&= -\frac{\mathbf{V}_2}{5} \\
&- \left(\frac{\jmath}{5}\right)\mathbf{V}_2 \quad \left(\frac{1}{10} + \frac{\jmath}{5}\right)\mathbf{V}_3 &&= 0
\end{aligned}
$$

These may simplified by multiplying all terms by 40 and, after rearrangement, they become

$$
\begin{aligned}
15\mathbf{V}_1 \quad - 18\mathbf{V}_2 \qquad\qquad &= 160\,\text{A} \\
-10V_1 \quad (18 + \jmath3)\mathbf{V}_2 \quad - \jmath8\mathbf{V}_3 &= 0 \\
-\jmath8\mathbf{V}_2 \quad (4 + \jmath8)\mathbf{V}_3 &= 0
\end{aligned}
$$

Cramer's rule may be used to find $\mathbf{V}_3$

$$\mathbf{V}_3 = \frac{\begin{vmatrix} 15\underline{/0°} & 18\underline{/180°} & 160\underline{/0°} \\ 10\underline{/180°} & 18.24\underline{/9.46°} & 0 \\ 0 & 8\underline{/-90°} & 0 \end{vmatrix}}{\begin{vmatrix} 15\underline{/0°} & 18\underline{/180°} & 0 \\ 10\underline{/180°} & 18.24\underline{/9.46°} & 8\underline{/-90°} \\ 0 & 8\underline{/-90°} & 8.94\underline{/63.44°} \end{vmatrix}} = \frac{j12,800}{960 + j900}$$

or

$$\mathbf{V}_3 = \frac{12,800\underline{/90°}}{1315.90\underline{/43.15°}} = 9.73\underline{/46.85°}\,\text{V} = 6.65 + j7.10\,\text{V}$$

Then, the current through the $10\,\Omega$ resistor will be

$$\mathbf{I}_R = \frac{\mathbf{V}_3}{10\,\Omega} = \frac{9.73\underline{/46.85°}\,\text{V}}{10\underline{/0°}\,\Omega} = 0.973\underline{/46.85°}\,\text{A} = 0.665 + 0.710\,\text{A} \Leftarrow$$

10.44: Rework Problem 10.35 (Fig 10.10) using a mesh analysis to determine the current through the 25 mH inductor and the voltage across the $62.5\,\mu\text{F}$ capacitor.

The phasor domain network using the reactances developed in Problem 10.35 is shown at the right. Note that clockwise mesh currents, $\mathbf{I}_1$ and $\mathbf{I}_2$ have been inserted.

Two mesh equations can be written

$$\begin{array}{llll} (j10 - j40)\mathbf{I}_1 & -(-j40)\mathbf{V}_2 & = & j400\,\text{V} \\ -(-j40)\mathbf{V}_1 & (-j40 + j40 + 40)\mathbf{V}_2 & = & 0 \end{array}$$

The mesh equations may be put into matrix form

$$\begin{bmatrix} -j30 & j40 \\ j40 & 40 \end{bmatrix} \begin{bmatrix} \mathbf{I}_1 \\ \mathbf{I}_2 \end{bmatrix} = \begin{bmatrix} j400\,\text{V} \\ 0 \end{bmatrix}$$

and a matrix inversion can be employed to find the current vector

$$\begin{bmatrix} \mathbf{I}_1 \\ \mathbf{I}_2 \end{bmatrix} = \begin{bmatrix} -j30 & j40 \\ j40 & 40 \end{bmatrix}^{-1} \begin{bmatrix} j400\,\text{V} \\ 0 \end{bmatrix}$$

or

$$\begin{bmatrix} \mathbf{V}_1 \\ \mathbf{V}_2 \end{bmatrix} = \frac{1}{2000\underline{/-36.87°}} \begin{bmatrix} 40\underline{/0°} & 40\underline{/-90°} \\ 40\underline{/-90°} & 30\underline{/-90°} \end{bmatrix} \begin{bmatrix} 400\underline{/90°}\,\text{V} \\ 0 \end{bmatrix}$$

Thus,

$$\begin{bmatrix} \mathbf{I}_1 \\ \mathbf{I}_2 \end{bmatrix} = \frac{1}{2000\underline{/-36.87°}} \begin{bmatrix} 16,000\underline{/90°} \\ 16,000\underline{/0°} \end{bmatrix} = \begin{bmatrix} 8\underline{/126.87°}\,\text{A} \\ 8\underline{/36.87°}\,\text{A} \end{bmatrix} = \begin{bmatrix} -4.80 + j6.40\,\text{A} \\ 6.40 + j4.80\,\text{A} \end{bmatrix}$$

The current through the 25 mH inductor is

$$\mathbf{I}_L = \mathbf{I}_1 = -4.80 + j6.40\,\text{A} \Leftarrow$$

and with

$$\begin{aligned} \mathbf{I}_1 - \mathbf{I}_2 &= (-4.80 + j6.40) - (6.40 + j4.80) \\ &= -11.20 + j1.60\,\text{A} = 11.31\underline{/171.87°} \Leftarrow \end{aligned}$$

the voltage across the 62.5 μF capacitor is

$$\mathbf{V}_C = (-j40)(11.31\underline{/171.87°}) = (40\underline{/-90°})(11.31\underline{/171.87°})$$

or

$$\mathbf{V}_C = 452.54\underline{/81.87°}\,\text{V} = 64 + j448\,\text{V} \Leftarrow$$

These confirm the result of Problems 10.35 and 10.39.

10.45: Rework Problem 10.36 (Fig 10.11) using a mesh analysis to determine the current through the 960 mH inductor and the voltage across the 160 Ω resistor.

The phasor domain network using the reactances developed in Problem 10.36 is repeated here and two clockwise mesh currents can be noted. Using KVL gives

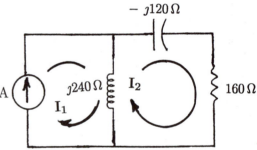

$$\mathbf{I}_1 = \mathbf{I}_s = 400\underline{/-36.87°}\,\text{mA}$$

and the equation for mesh-2 is

$$-j240\mathbf{I}_1 + (j240 - j120 + 160)\mathbf{I}_2 = 0$$

so that

$$-j240\mathbf{I}_1 + (160 + j120)\mathbf{I}_2 = 0$$

and

$$\begin{aligned} \mathbf{I}_2 &= \frac{j240\mathbf{I}_1}{160 + j120} \\ &= \frac{(240\underline{/90°}\,\Omega)(400\underline{/-36.87°}\,\text{mA})}{200\underline{/36.87°}\,\Omega} \\ &= 480\underline{/16.26°}\,\text{mA} = 460.8 + j134.4\,\text{mA} \Leftarrow \end{aligned}$$

The current through the 960 mH inductor is

$$
\begin{aligned}
\mathbf{I}_L &= \mathbf{I}_1 - \mathbf{I}_2 \\
&= 400\underline{/-36.87°} - (460.8 + j134.4) \\
&= (320 - j240) - (460.8 + j134.4) \\
&= -140.8 - j374.4\,\text{mA} = 400\underline{/-110.61°}\,\text{mA} \Leftarrow
\end{aligned}
$$

and the voltage across the $160\,\Omega$ resistor is

$$
\mathbf{V}_R = (160\,\Omega)\mathbf{I}_2 = (160\underline{/0°})(480\underline{/16.26°}\,\text{mA})
$$

or

$$
\mathbf{V}_R = 76.80\underline{/16.26°}\,\text{V} = 73.73 + j21.50\,\text{V} \Leftarrow
$$

These confirm the result of Problems 10.36 and 10.40.

10.46: Rework Problem 10.42 (Fig 10.14) using a mesh analysis to find the current through the 50 mH inductor and the voltage across the $40\,\Omega$ resistor.

The phasor domain network using the reactances developed in Problem 10.42 is shown at the right. The voltage source has not been transformed but the voltage controlled voltage source has been transformed to a current controlled voltage source having strength

$$
2\mathbf{V}_a = 2(j20)\mathbf{I}_2 = j40\mathbf{I}_2\,\text{V}
$$

Two meshes containing clockwise mesh currents are evident and, using KVL, the two mesh currents are written as

$$
\begin{aligned}
(20 - j20)\mathbf{I}_1 \quad - (-j20)\mathbf{I}_2 &= j400\,\text{V} - j40\mathbf{I}_2 \\
-(-j20) \quad + (-j20 + j20 + 40)\mathbf{I}_2 &= j40\mathbf{I}_2
\end{aligned}
$$

and these may be simplified to

$$
\begin{aligned}
(20 - j20)\mathbf{I}_1 \quad + j60\mathbf{I}_2 &= j400\,\text{V} \\
j20 \quad + (40 - j40)\mathbf{I}_2 &= 0
\end{aligned}
$$

Cramer's rule may be used to find $\mathbf{I}_2$

$$
\mathbf{I}_2 = \frac{\begin{vmatrix} 20\sqrt{2}\underline{/-45°} & 400\underline{/90°} \\ 20\underline{/90°} & 0 \end{vmatrix}}{\begin{vmatrix} 20\sqrt{2}\underline{/-45°} & 60\underline{/90°} \\ 20\underline{/90°} & 40\sqrt{2}\underline{/-45°} \end{vmatrix}} = \frac{8000}{1200 - j1600}
$$

or

$$\mathbf{I}_2 = \frac{8000\underline{/0^\circ}}{2000\underline{/-53.13^\circ}} = 4\underline{/53.13^\circ}\,\text{A} = 2.40 + \jmath3.20\,\text{A} \Leftarrow$$

and

$$\mathbf{V}_2 = 40\mathbf{I}_2 = (40\,\Omega)(4\underline{/53.13^\circ}\,\text{A}) = 160\underline{/53.13^\circ} = 96 + \jmath128\,\text{V} \Leftarrow$$

These confirm the results of Problem 10.42.

10.47: Rework Problem 10.43 (Fig 10.15) using mesh analysis to find the current through the $10\,\Omega$ resistor in Fig 10.15.

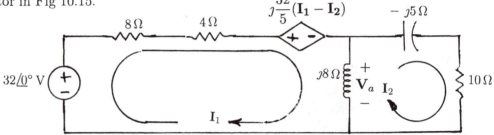

The phasor domain network developed in Problem 10.43 can be used here with two modifications. The current source has been transformed to a a voltage source via

$$\mathbf{V}_s = 8\mathbf{I}_s = (8\,\Omega)(4\underline{/0^\circ}\,\text{A}) = 32\underline{/0^\circ}\,\text{V}$$

and the voltage controlled current source has been transformed to a current controlled voltage source of strength

$$\mathbf{V}_{\mathrm{CS}} = (4\,\Omega)\left(\frac{\mathbf{V}_a}{5}\right) = \frac{4}{5}[\jmath8(\mathbf{I}_1 - \mathbf{I}_2)] = \jmath\frac{32}{5}(\mathbf{I}_1 - \mathbf{I}_2)$$

The two mesh equations may be written using KVL

$$(8 + 4 + \jmath8)\mathbf{I}_1 \qquad - \jmath8\mathbf{I}_2 \qquad = \quad 32\,\text{V} - \jmath\frac{32}{5}(\mathbf{I}_1 - \mathbf{I}_2)$$
$$- \jmath8\mathbf{I}_1 \qquad + (\jmath8 - \jmath5 + 10)\mathbf{I}_2 \quad = \qquad 0$$

These may simplified by multiplying all terms by 5 and, after rearrangement, they become

$$(60 + \jmath72)\mathbf{I}_1 \qquad - \jmath72\mathbf{I}_2 \qquad = \quad 160\,\text{V}$$
$$- \jmath40\mathbf{I}_1 \qquad + (50 + \jmath15)\mathbf{I}_2 \quad = \qquad 0$$

The second of these may be used to obtain

$$\mathbf{I}_1 = \frac{50 + \jmath15}{\jmath40} = \frac{52.20\underline{/16.70^\circ}}{40\underline{/90^\circ}}$$

or

$$\mathbf{I}_1 = 1.305\underline{/-73.30°}\,\text{A} = 0.375 - \jmath 1.250\,\text{A}$$

and with this in the first,

$$(60 + \jmath 72)\mathbf{I}_1 - \jmath 72\mathbf{I}_2 = 160\,\text{V}$$
$$(93.72\underline{/50.19°})(1.305\underline{/-73.30°})\mathbf{I}_2 - \jmath 72\mathbf{I}_2 = 160\,\text{V}$$
$$(122.31\underline{/-23.10°} - \jmath 72)\mathbf{I}_2 = 160\,\text{V}$$
$$(112.5 - \jmath 120.0)\mathbf{I}_2 = 160\,\text{V}$$

or

$$\mathbf{I}_2 = \frac{160\,\text{V}}{164.49\underline{/-46.85°}} = 0.973\underline{/46.85°}\,\text{A} = 0.665 + \jmath 0.710\,\text{A} \Leftarrow$$

This confirms the result of Problem 10.43.

10.48: Use mesh analysis to determine the current through the 20 mH inductor in Fig 10.16.

With $\omega = 10,000$ rad/s, the reactances are

Component	Reactance, X
$C = 1\,\mu\text{F}$	$100\,\Omega$
$L = 20\,\text{mH}$	$200\,\Omega$

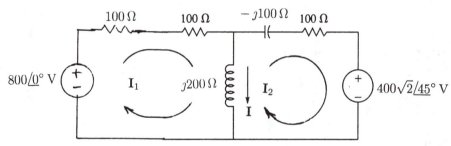

The network is shown above in the phasor domain with the current source transformed to a voltage source via

$$\mathbf{V}_{s1} = (100\,\Omega)(8\underline{/0°}\,\text{A}) = 800\underline{/0°}\,\text{V} = 800 + \jmath 0\,\text{V}$$

The two mesh equations may be written using KVL

$$(100 + 100 + \jmath 200)\mathbf{I}_1 - \jmath 200\mathbf{I}_2 = 800 + \jmath 0\,\text{V}$$
$$- \jmath 200\mathbf{I}_1 + (\jmath 200 - \jmath 100 + 100)\mathbf{I}_2 = -400\sqrt{2}\underline{/45°}\,\text{V}$$

or in matrix form after simplification

$$\begin{bmatrix} 200\sqrt{2}\underline{/45°} & 200\underline{/-90°} \\ 200\underline{/-90°} & 100\sqrt{2}\underline{/45°} \end{bmatrix} \begin{bmatrix} I_1 \\ I_2 \end{bmatrix} = \begin{bmatrix} 800\underline{/0°}\,\text{V} \\ -400\sqrt{2}\underline{/45°}\,\text{V} \end{bmatrix}$$

and a matrix inversion can be employed to find the current vector

$$\begin{bmatrix} I_1 \\ I_2 \end{bmatrix} = \begin{bmatrix} 200\sqrt{2}\underline{/45°} & 200\underline{/-90°} \\ 200\underline{/-90°} & 100\sqrt{2}\underline{/45°} \end{bmatrix}^{-1} \begin{bmatrix} 800\underline{/0°}\,\text{V} \\ -400\sqrt{2}\underline{/45°}\,\text{V} \end{bmatrix}$$

so that

$$\begin{bmatrix} I_1 \\ I_2 \end{bmatrix} = \frac{1}{40,000\sqrt{2}\underline{/45°}} \begin{bmatrix} 100\sqrt{2}\underline{/45°} & 200\underline{/90°} \\ 200\underline{/90°} & 200\sqrt{2}\underline{/45°} \end{bmatrix} \begin{bmatrix} 800\underline{/0°}\,\text{V} \\ -400\sqrt{2}\underline{/45°}\,\text{V} \end{bmatrix}$$

$$= \frac{1}{40,000\sqrt{2}\underline{/45°}} \begin{bmatrix} 160,000\underline{/0°} \\ 0 \end{bmatrix}$$

or

$$\begin{bmatrix} I_1 \\ I_2 \end{bmatrix} = \begin{bmatrix} 2\sqrt{2}\underline{/-45°}\,\text{A} \\ 0\,\text{A} \end{bmatrix} = \begin{bmatrix} 2-j2\,\text{A} \\ 0\,\text{A} \end{bmatrix}$$

Then, the current through the 20 mH inductor will be

$$\begin{aligned} \mathbf{I}_L &= \mathbf{I}_1 - \mathbf{I}_2 \\ &= (2-j2) - 0 \\ &= 2-j2\,\text{A} = 2\sqrt{2}\underline{/-45°} \Leftarrow \end{aligned}$$

10.49: Rework Problem 10.48 (Fig 10.16) using superposition to determine the current through the 20 mH inductor.

The current through the 20 mH inductor is due to the superposition of two currents, $\mathbf{I}_C$, due to the current source acting alone with the voltage source replaced with a short circuit and, $\mathbf{I}_v$, due to the voltage source acting alone with the current source replaced with an open circuit

$$\mathbf{I}_L = \mathbf{I}_C + \mathbf{I}_V$$

For $\mathbf{I}_C$

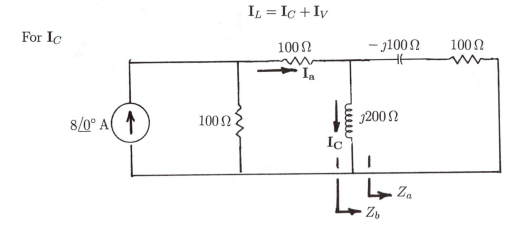

$$Z_a = 100 - j100 \, \Omega = 100\sqrt{2}\underline{/-45°} \, \Omega$$

$$Z_b = \frac{(200\underline{/90°})(100\sqrt{2}\underline{/-45°})}{j200 + 100 - j100} \, \Omega$$

$$= \frac{20,000\sqrt{2}\underline{/45°}}{100\sqrt{2}\underline{/45°}} \, \Omega$$

$$= 200\underline{/0°} \, \Omega = 200 \, \Omega$$

Then, by two current dividers

$$\mathbf{I}_a = \left(\frac{100}{100 + 300}\right)\mathbf{I}_s = \frac{1}{4}8\underline{/0°} = 2\underline{/0°} \, \text{A} = 2 + j0 \, \text{A}$$

and

$$\mathbf{I}_C = \left(\frac{100 - j100}{j200 + 100 - j100}\right)\mathbf{I}_a = \frac{100\sqrt{2}\underline{/-45°}}{100\sqrt{2}\underline{/45°}}(2\underline{/0°}) = (1\underline{/-90°})(2\underline{/0°})$$

or

$$\mathbf{I}_C = 2\underline{/-90°} \, \text{A} = -j2 \, \text{A}$$

For $\mathbf{I}_V$

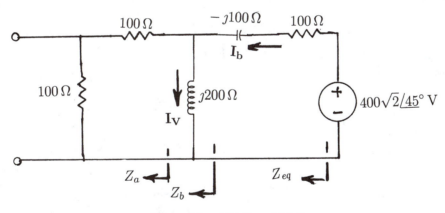

$$Z_a = 100 + 100 \, \Omega = 200 \, \Omega$$

and

$$Z_b = \frac{(200)(j200)}{200 + j200} = \frac{40,000\underline{/90°}}{200\sqrt{2}\underline{/45°}} = 100\sqrt{2}\underline{/45°} \, \Omega$$

$$= 100 + j100 \, \Omega$$

so that

$$Z_{eq} = (100 + j100) + (100 - j100) = 200\,\Omega$$

Then

$$\mathbf{I}_b = \frac{\mathbf{V}_s}{Z_{eq}} = \frac{400\sqrt{2}\underline{/45°}\ \text{V}}{200\,\Omega} = 2\sqrt{2}\underline{/45°} = 2 + j2\ \text{A}$$

and, by a current divider

$$\mathbf{I}_V = \left(\frac{200}{200 + j200}\right)\mathbf{I}_b = \left(\frac{200\underline{/0°}}{200\sqrt{2}\underline{/45°}}\right)(2\sqrt{2}\underline{/45°}) = 2\underline{/0°}\ \text{A} = 2 + j0\ \text{A}$$

Then, the current through the 20 mH inductor is

$$
\begin{aligned}
\mathbf{I}_L &= \mathbf{I}_C + \mathbf{I}_V \\
&= (0 - j2) + (2 + j0) \\
&= 2 - j2\ \text{A} = 2\sqrt{2}\underline{/-45°}\ \text{A} \Leftarrow
\end{aligned}
$$

This confirms the result of Problem 10.48.

10.50: Use superposition to determine the voltage across the 58 mH inductor in Fig 10.17.

With $\omega = 1000$ rad/s, the reactances are

Component	Reactance, X
$C = 62.5\,\mu\text{F}$	$16\,\Omega$
$L = 58\,\text{mH}$	$58\,\Omega$
$L = 16\,\text{mH}$	$16\,\Omega$

The network is shown in the phasor domain and it observed that the current through the 58 mH inductor is due to the superposition of two currents, $\mathbf{I}_{s1}$, due to the left hand voltage source acting alone with the right hand voltage source replaced with a short circuit and, $\mathbf{I}_{s2}$, due to the right hand voltage source acting alone with the left hand voltage source replaced with a short circuit

$$\mathbf{I}_L = \mathbf{I}_{s1} + \mathbf{I}_{s2}$$

For $\mathbf{I}_{V1}$

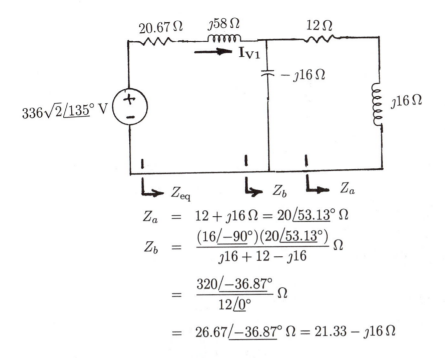

$$Z_a = 12 + j16\,\Omega = 20\underline{/53.13°}\,\Omega$$

$$Z_b = \frac{(16\underline{/-90°})(20\underline{/53.13°})}{j16 + 12 - j16}\,\Omega$$

$$= \frac{320\underline{/-36.87°}}{12\underline{/0°}}\,\Omega$$

$$= 26.67\underline{/-36.87°}\,\Omega = 21.33 - j16\,\Omega$$

and

$$Z_{eq} = (20.67 + j58) + (21.33 - j16)\,\Omega$$

$$= 42 + j42\,\Omega = 42\sqrt{2}\underline{/45°}\,\Omega$$

Then,

$$\mathbf{I}_{V1} = \frac{\mathbf{V}_{s1}}{Z_{eq}} = \frac{336\sqrt{2}\underline{/135°}\,\text{V}}{42\sqrt{2}\underline{/45°}\,\Omega} = 8\underline{/90°}\,\text{A} = 0 + j8\,\text{A}$$

For $\mathbf{I}_{V2}$

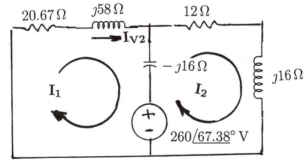

Note two meshes and two clockwise mesh currents. By KVL

$$\begin{array}{rcl}
(20.67 + j58 - j16)\mathbf{I}_1 \quad -(-j16)\mathbf{I}_2 & = & -260\underline{/67.38°}\,\text{V} \\
-(-j16)\mathbf{I}_2 \quad +(-j16 + 12 + j16)\mathbf{I}_2 & = & 260\underline{/67.38°}\,\text{V}
\end{array}$$

or

$$46.81\underline{/63.80°}\mathbf{I}_1 \quad +16\underline{/90°}\mathbf{I}_2 \;=\; 260\underline{/-112.62°}\,\text{V}$$
$$16\underline{/90°}\mathbf{I}_1 \quad +12\underline{/0°}\mathbf{I}_2 \;=\; 260\underline{/67.38°}\,\text{V}$$

and a Cramer's rule solution for $\mathbf{I}_1$ gives

$$\mathbf{I}_1 = \mathbf{I}_{V2} = \frac{\begin{vmatrix} 260\underline{/-112.62°} & 16\underline{/90°} \\ 260\underline{/67.38°} & 12\underline{/0°} \end{vmatrix}}{\begin{vmatrix} 46.81\underline{/63.80°} & 16\underline{/90°} \\ 16\underline{/90°} & 12\underline{/0°} \end{vmatrix}} = \frac{2640 - j4480}{504 + j504}$$

or

$$\mathbf{I}_{V2} = \frac{5200\underline{/-59.49°}}{504\sqrt{2}\underline{/45°}} = 7.30\underline{/-104.49°}\,\text{A} = -1.83 - j7.06\,\text{A}$$

The current through the 58 mH inductor is

$$\begin{aligned} \mathbf{I}_L &= \mathbf{I}_{V1} + \mathbf{I}_{V2} \\ &= (0 + j8) + (-1.83 - j7.06) \\ &= -1.83 + j0.94\,\text{A} = 2.06\underline{/152.81°}\,\text{A} \Leftarrow \end{aligned}$$

and the voltage across the 58 mH inductor will be

$$V_L = (58\underline{/90°})(2.06\underline{/152.81°}) = 118.89\underline{/-117.16°} = -54.32 - j105.87\,\text{V} \Leftarrow$$

10.51: Use superposition to determine the voltage across the capacitor in Fig 10.18.

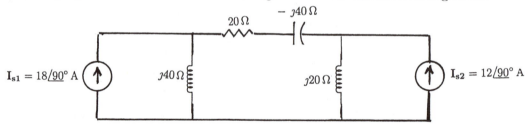

The network is shown in the phasor domain and it observed that the voltage across the capacitor is due to the superposition of two voltages, $\mathbf{V}_{s1}$, due to the left hand current source acting alone with the right hand current source replaced with an open circuit and $\mathbf{V}_{s2}$, due to the right hand current source acting alone with the left hand current source replaced by an open circuit

$$\mathbf{V}_C = \mathbf{V}_{s1} + \mathbf{V}_{s2}$$

For $\mathbf{V}_{s1}$

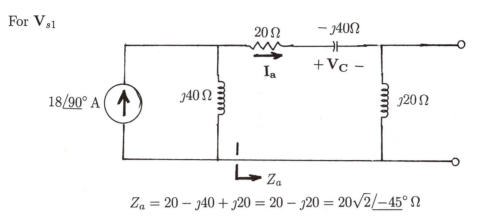

$$Z_a = 20 - j40 + j20 = 20 - j20 = 20\sqrt{2}\underline{/-45°}\ \Omega$$

By a current divider with $\mathbf{I}_{s1} = 18\underline{/90°}\ \text{A}$

$$\mathbf{I}_a = \left(\frac{j40}{j40 + 20 - j20}\right)\mathbf{I}_{s1} = \left(\frac{40\underline{/90°}}{20\sqrt{2}\underline{/45°}}\right)(18\underline{/90°}) = 18\sqrt{2}\underline{/135°}\ \text{A} = -18 + j18\ \text{A}$$

and

$$\mathbf{V}_{s1} = (-j40)\mathbf{I}_a = (40\underline{/-90°})(18\sqrt{2}\underline{/135°}\ \text{A}) = 720\sqrt{2}\underline{/45°}\ \text{V} = 720 + j720\ \text{V}$$

For $\mathbf{V}_{s2}$

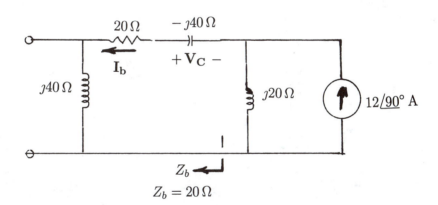

$$Z_b = 20\ \Omega$$

so that by a current divider

$$\mathbf{I}_b = \left(\frac{j20}{20 + j20}\right)\mathbf{I}_{s2} = \left(\frac{20\underline{/90°}}{20\sqrt{2}\underline{/45°}}\right)(12\underline{/90°}) = 6\sqrt{2}\underline{/135°}\ \text{A} = -6 + j6\ \text{A}$$

and

$$\mathbf{V}_{s2} = -(-j40)\mathbf{I}_a = (40\underline{/90°})(6\sqrt{2}\underline{/135°}\ \text{A}) = 240\sqrt{2}\underline{/-135°}\ \text{V} = -240 - j240\ \text{V}$$

The voltage across the capacitor is

$$
\begin{aligned}
\mathbf{V}_C &= \mathbf{V}_{s1} + \mathbf{V}_{s2} \\
&= (720 + \jmath720) + (-240 - \jmath240) \\
&= 480 + \jmath480\,\text{V} = 480\sqrt{2}\underline{/45°}\,\text{V} \Leftarrow
\end{aligned}
$$

10.52: Use superposition to determine the current through the 20 mH inductor in Fig 10.19.

With $\omega = 2000$ rad/s, the reactances are

Component	Reactance, X
$C = 14.29\,\mu\text{F}$	$35\,\Omega$
$L = 60\,\text{mH}$	$120\,\Omega$
$L = 20\,\text{mH}$	$40\,\Omega$

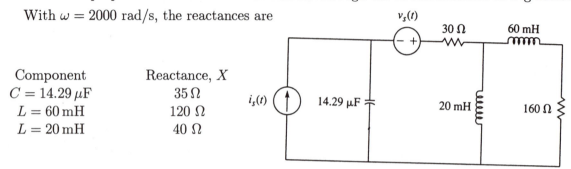

The network is shown in the phasor domain and it observed that the current through the 20 mH inductor is due to the superposition of two currents, $\mathbf{I}_C$, due to the current source acting alone with the voltage source replaced with a short circuit and $\mathbf{I}_V$, due to the voltage source acting alone with the current source replaced by an open circuit

$$
\mathbf{I}_L = \mathbf{I}_C + \mathbf{I}_V
$$

For $\mathbf{I}_C$

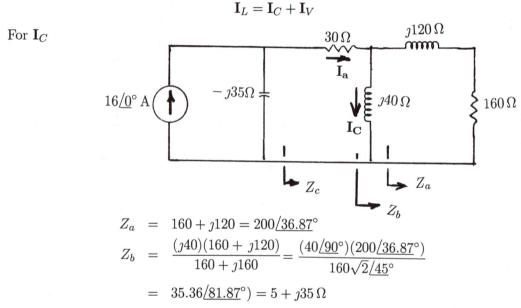

$$
\begin{aligned}
Z_a &= 160 + \jmath120 = 200\underline{/36.87°} \\
Z_b &= \frac{(\jmath40)(160 + \jmath120)}{160 + \jmath160} = \frac{(40\underline{/90°})(200\underline{/36.87°})}{160\sqrt{2}\underline{/45°}} \\
&= 35.36\underline{/81.87°}) = 5 + \jmath35\,\Omega
\end{aligned}
$$

and

$$Z_c = 30 + 5 + j35 = 35 + j35\,\Omega = 35\sqrt{2}\underline{/45°}\,\Omega$$

With $\mathbf{I}_s = 16\underline{/0°}$ A, two current dividers give

$$\mathbf{I}_a = \left(\frac{-j35}{-j35 + 35 + j35}\right)\mathbf{I}_s = \left(\frac{35\underline{/-90°}}{35\underline{/0°}}\right)(16\underline{/0°}\text{ A})$$

$$= 16\underline{/-90°}\text{ A} = 0 - j16\text{ A}$$

and

$$\mathbf{I}_C = \left(\frac{160 + j120}{160 + j120 + j40}\right)\mathbf{I}_a = \left(\frac{200\underline{/36.87°}}{160\sqrt{2}\underline{/45°}}\right)(16\underline{/-90°}\text{ A})$$

$$= 14.14\underline{/-98.13°}\text{ A} = -2 - j14\text{ A}$$

For $\mathbf{I}_V$

$$Z_a = 160 + j120\,\Omega$$

and

$$Z_b = 5 + j35\,\Omega$$

and with $\mathbf{V}_s = 210\underline{/90°}$ V, the current $\mathbf{I}_b$ will be

$$\mathbf{I}_b = \frac{\mathbf{V}_s}{Z_b + 30 - j35} = \frac{210\underline{/90°}\text{ V}}{5 + j35 + 30 - j35} = \frac{210\underline{/90°}\text{ V}}{35\underline{/0°}}$$

or

$$\mathbf{I}_b = 6\underline{/90°}\text{ A} = 0 = j6\text{ A}$$

Then, by a current divider

$$\mathbf{I}_v = \left(\frac{160 + j120}{j40 + 160 + j120}\right)\mathbf{I}_b = \left(\frac{200\underline{/36.87°}}{160\sqrt{2}\underline{/45°}}\right)(6\underline{/90°}) = 0.75 + j5.25\text{ A}$$

The current through the 20 mH inductor is

$$\begin{aligned}
\mathbf{I}_L &= \mathbf{I}_c + \mathbf{I}_v \\
&= (-2 - j14) + (0.75 + j5.25) \\
&= -1.25 - j8.75\text{ A} = 8.34\underline{/-98.13°}\text{ A} \Leftarrow
\end{aligned}$$

10.53: Rework Problem 10.35 (Fig 10.10) using Thevenin's theorem to determine the current through the $40\,\Omega$ resistor.

The $40\,\Omega$ resistor will be treated as the load and with reactances developed in Problem 10.35, the phasor domain network with this resistor removed is shown at the right. The open circuit voltage can be obtained from a voltage divider

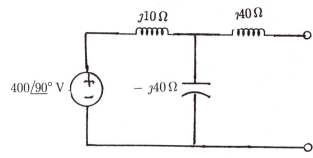

$$\mathbf{V}_{oc} = V_T = \left(\frac{-\jmath 40}{\jmath 10 - \jmath 40}\right)\mathbf{V}_s = \frac{40\underline{/-90^\circ}}{30\underline{/-90^\circ}}(400\underline{/90^\circ}) = \frac{1600}{3}\underline{/90^\circ}\,\text{V}$$

The Thevenin equivalent impedance is the impedance looking back into the open-circuited terminals with the voltage source removed and replaced with a short circuit

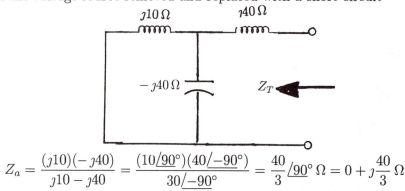

With

$$Z_a = \frac{(\jmath 10)(-\jmath 40)}{\jmath 10 - \jmath 40} = \frac{(10\underline{/90^\circ})(40\underline{/-90^\circ})}{30\underline{/-90^\circ}} = \frac{40}{3}\underline{/90^\circ}\,\Omega = 0 + \jmath\frac{40}{3}\,\Omega$$

the Thevenin equivalent impedance is

$$Z_T = \left(0 + \jmath\frac{40}{3}\right) + \jmath 40 = \jmath\frac{160}{3}\,\Omega$$

The Thevenin equivalent circuit with the load resistance of $40\,\Omega$ reconnected is shown at the right. Here

$$\mathbf{I}_R = \frac{\mathbf{V}_T}{Z_T + 40}$$

or

$$\mathbf{I}_R = \frac{1600/3\underline{/90^\circ}\,\text{V}}{\jmath 160/3 + 40} = \frac{1600\underline{/90^\circ}}{120 + \jmath 160} = \frac{1600\underline{/90^\circ}}{200\underline{/53.13^\circ}}$$

Hence

$$\mathbf{I}_R = 8\underline{/36.87°}\,\text{A} = 6.40 + j4.80\,\Omega\,\text{A} \Leftarrow$$

This confirms the result of Problem 10.35.

10.54: Rework Problem 10.51 (Fig 10.18) using Thevenin's theorem to determine the voltage across the capacitor.

The capacitor will be treated treated as the load and with the reactances developed in Problem 10.51, the phasor domain network with capacitor removed is shown at the right. Here

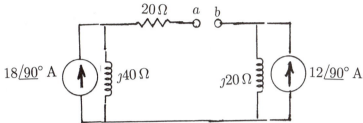

$$\mathbf{V}_a = (j40)(18\underline{/90°}) = -720\,\text{V}$$

and

$$\mathbf{V}_b = (j20)(12\underline{/90°}) = -240\,\text{V}$$

so that

$$\mathbf{V}_T = \mathbf{V}_a - \mathbf{V}_b$$
$$= -720 - (-240)$$
$$= -480\,\text{V}$$

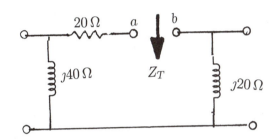

Looking back into terminals a-b gives

$$Z_T = (20 + j40) + j20$$
$$= 20 + j60 = 63.25\underline{/71.57°}\,\Omega$$

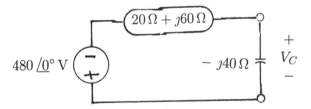

The Thevenin equivalent network with the capacitor reconnected is shown at at the right. Note the polarity of $\mathbf{V}_T$ and by a voltage divider

$$\mathbf{V}_C = -\left(\frac{-j40}{20 + j60 - j40}\right)\mathbf{V}_T = -\left(\frac{40\underline{/-90°}}{20\sqrt{2}\underline{/45°}}\right)(-480\underline{/0°})$$

or

$$\mathbf{V}_C = 480\sqrt{2}\underline{/45°}\,\text{V} = 480 + j480\,\text{V} \Leftarrow$$

This confirms the result of Problem 10.51.

10.55: Rework Problem 10.43 (Fig 10.15) using Thevenin's theorem to determine the current through the $10\,\Omega$ resistor.

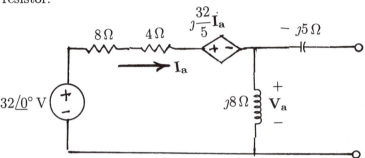

The $10\,\Omega$ resistor will be treated as the load and with reactances developed in Problem 10.43, the phasor domain network with this resistor removed is shown here. The current source has been transformed to a voltage source and the voltage controlled current source has been transformed to a voltage controlled voltage source with strength

$$\mathbf{V}_{VS} = 4\left(\frac{\mathbf{V}_a}{5}\right) = \frac{4}{5}\mathbf{V}_a = \frac{4}{5}(j8\mathbf{I}_a) = j\frac{32}{5}\mathbf{I}_a$$

KVL can be employed to find $\mathbf{V}_T = \mathbf{V}_{\text{oc}}$

$$-32\underline{/0^\circ} + 8\mathbf{I}_a + j\frac{32}{5}\mathbf{I}_a + 4\mathbf{I}_a + j8\mathbf{I}_a = 0$$

$$\left(8 + j\frac{32}{5} + 4 + j8\right)\mathbf{I}_a = 32\underline{/0^\circ}$$

$$(60 + j72)\mathbf{I}_a = 160\underline{/0^\circ}$$

so that

$$\mathbf{I}_a = \frac{160\underline{/0^\circ}}{60 + j72} = \frac{160\underline{/0^\circ}\text{ V}}{93.72\underline{/50.19^\circ}\,\Omega} = 1.71\underline{/-50.19^\circ}\text{ A}$$

and

$$\mathbf{V}_T = \mathbf{V}_{\text{oc}} = (j8)\mathbf{I}_a = (8\underline{/90^\circ})(1.71\underline{/-50.19^\circ}) = 13.66\underline{/39.81^\circ}\text{ V} = 10.49 + j8.74\text{ V}$$

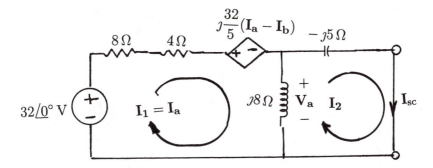

A mesh analysis may now be used to find $\mathbf{I}_{sc}$. Notice that because $\mathbf{V}_a = \jmath 8(\mathbf{I}_1 - \mathbf{I}_2)$, the strength of the controlled voltage source is

$$\mathbf{V}_{VS} = \jmath\frac{32}{5}(\mathbf{I}_1 - \mathbf{I}_2)$$

and that the two clockwise mesh currents, after a repeated application of KVL, have mesh equations given by

$$
\begin{aligned}
(8 + 4 + \jmath 8)\mathbf{I}_1 \quad - \jmath 8 \mathbf{I}_2 \quad &= \quad 32\,\mathrm{V} - \jmath\frac{32}{5}(\mathbf{I}_1 - \mathbf{I}_2) \\
-\jmath 8 \mathbf{I}_1 \quad + (\jmath 8 - \jmath 5)\mathbf{I}_2 \quad &= \quad 0
\end{aligned}
$$

These may simplified by multiplying all terms by 5 and, after rearrangement, they become

$$
\begin{aligned}
(60 + \jmath 72)\mathbf{I}_1 \quad -\jmath 72 \mathbf{I}_2 \quad &= \quad 160\,\mathrm{V} \\
-\jmath 40 \mathbf{I}_1 \quad +\jmath 15 \mathbf{I}_2 \quad &= \quad 0
\end{aligned}
$$

The second of these may be used to obtain

$$\mathbf{I}_1 = \frac{\jmath 15}{\jmath 40}\mathbf{I}_2 = \frac{3}{8}\mathbf{I}_2$$

and with this in the first equation,

$$(60 + \jmath 72)\left(\frac{3}{8}\mathbf{I}_2\right) - \jmath 72 \mathbf{I}_2 = 160\underline{/0^\circ}\,\mathrm{V}$$

$$(22.5 - \jmath 45)\mathbf{I}_2 = 160\underline{/0^\circ}\,\mathrm{V}$$

and

$$\mathbf{I}_{sc} = \mathbf{I}_2 = \frac{160\,\mathrm{V}}{50.31\underline{/-63.43^\circ}} = 3.18\underline{/63.43^\circ}\,\mathrm{A} = 1.42 + \jmath 2.84\,\mathrm{A}$$

Thus,

$$Z_T = \frac{\mathbf{V}_T}{\mathbf{I}_{sc}} = \frac{13.66\underline{/39.81^\circ}\,\mathrm{V}}{3.18\underline{/63.43^\circ}\,\mathrm{A}} = 4.29\underline{/-23.63^\circ}\,\Omega = 3.93 - \jmath 1.72\,\Omega$$

The Thevenin equivalent network with the $10\,\Omega$ resistor reconnected is shown at at the right. The current through the $10\,\Omega$ resistor will be

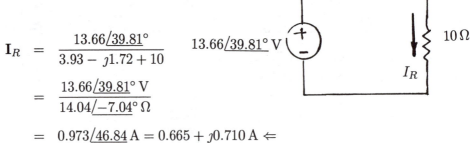

$$\mathbf{I}_R = \frac{13.66\underline{/39.81°}}{3.93 - \jmath1.72 + 10}$$

$$= \frac{13.66\underline{/39.81°}\,\text{V}}{14.04\underline{/-7.04°}\,\Omega}$$

$$= 0.973\underline{/46.84}\,\text{A} = 0.665 + \jmath0.710\,\text{A} \Leftarrow$$

This confirms the result of Problem 10.43.

10.56: Rework Problem 10.37 (Fig 10.12) using Thevenin's theorem to determine the current through the 16 mH inductor.

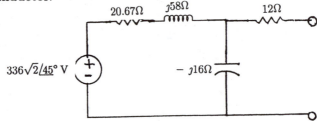

The 16 mH inductor will be treated as the load and, with reactances developed in Problem 10.37, the phasor domain network with this inductor removed is shown. The open circuit voltage can be obtained from a voltage divider

$$\mathbf{V}_{\text{oc}} = \mathbf{V}_T = \left(\frac{-\jmath16}{20.67 + \jmath58 - \jmath16}\right)\mathbf{V}_s = \left(\frac{16\underline{/-90°}}{46.81\underline{/63.80°}}\right)(336\sqrt{2}\underline{/135°})$$

or

$$\mathbf{V}_T = 162.42\underline{/-18.80°}\,\text{V}$$

The Thevenin equivalent impedance is the impedance looking back into the open-circuited terminals with the voltage source removed and replaced with a short circuit

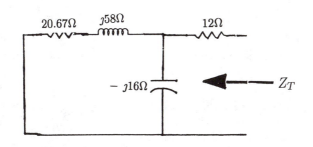

With

$$Z_a = 20.67 + j58 \, \Omega$$

and

$$Z_b = \frac{(20.67 + j58)(-j16)}{20.67 + j58 - j16} = \frac{(61.57\underline{/70.39°})(16\underline{/-90°})}{46.81\underline{/63.80°}}$$

or

$$Z_b = 21.05\underline{/-83.41°} \, \Omega = 2.42 - j20.91 \, \Omega$$

the Thevenin equivalent impedance will be

$$Z_T = 2.42 - j20.91 + 12 = 14.42 - j20.91 \, \Omega$$

The Thevenin equivalent circuit with the load resistance of $j16 \, \Omega$ reconnected is shown at the right. Here

$$\mathbf{I} = \frac{\mathbf{V}_T}{Z_T + j16}$$

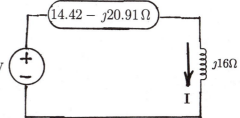

or

$$\mathbf{I} = \frac{162.48\underline{/-18.80°} \, \text{V}}{14.42 - j20.91 + j16} = \frac{162.48\underline{/-18.80°} \, \text{V}}{15.23\underline{/-18.80°} \, \Omega}$$

or

$$\mathbf{I} = 10.67\underline{/0°} = 10.67 + j0 \, \text{A} \Leftarrow$$

This confirms the result of Problem 10.37.
10.57: Rework Problem 10.38 (Fig 10.13) using Thevenin's theorem to determine the current through the $40 \, \mu$F capacitor.

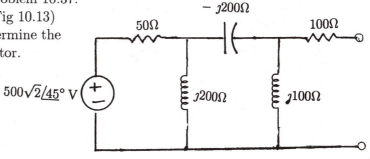

The $40\,\mu F$ capacitor will be treated as the load and, with reactances developed in Problem 10.38, the phasor domain network with this capacitor removed is shown. The open circuit voltage can be obtained from a mesh analysis using the two clockwise mesh currents

$$(50 + j200)\mathbf{I}_1 \qquad -j200\mathbf{I}_2 \qquad = \quad 500\sqrt{2}\underline{/45^\circ}\ \text{V}$$
$$-j200\mathbf{I}_1 \quad + (j200 - j200 + j100)\mathbf{I}_2 = \qquad 0$$

and after simplification, the two mesh equations are

$$(50 + j200)\mathbf{I}_1 \quad -j200\mathbf{I}_2 \quad = \quad 500\sqrt{2}\underline{/45^\circ}\ \text{V}$$
$$-j200\mathbf{I}_1 \quad + j100\mathbf{I}_2 \quad = \qquad 0$$

The second mesh equation yields an expression for $\mathbf{I}_1$ in terms of $\mathbf{I}_2$

$$\mathbf{I}_1 = \frac{-j100}{-j200}\mathbf{I}_2 = \frac{1}{2}\mathbf{I}_2$$

With this result in the first mesh equation

$$(50 + j200)\left(\frac{1}{2}\right)\mathbf{I}_2 - j200\mathbf{I}_2 = 500\sqrt{2}\underline{/45^\circ}\ \text{V}$$

$$(25 - j100)\mathbf{I}_2 = 500\sqrt{2}\underline{/45^\circ}\ \text{V}$$

$$\mathbf{I}_2 = \frac{500\sqrt{2}\underline{/45^\circ}\ \text{V}}{103.08\underline{/-75.96^\circ}\ \Omega}$$

$$= 6.86\underline{/120.96^\circ}\ \text{A}$$

The open circuit voltage is the Thevenin equivalent voltage

$$\mathbf{V}_T = \mathbf{V}_{oc} = (j100)\mathbf{I}_2 = (100\underline{/90^\circ}\ \Omega)(6.86\underline{/120.96^\circ}\ \text{A}) = 686\underline{/-149.04^\circ}\ \text{V}$$

The Thevenin equivalent impedance is the impedance looking back into the open-circuited terminals with the voltage soiurce removed and replaced with a short circuit. With

$$Z_a = \frac{(50)(j200)}{50 + j200} = \frac{(50\underline{/0^\circ})(200\underline{/90^\circ})}{206.15\underline{/75.96^\circ}}$$

$$= 48.51\underline{/14.04^\circ}\ \Omega = 47.06 + j11.76\ \Omega$$

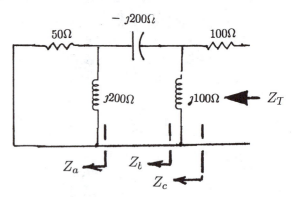

$$Z_b = 47.06 + j11.76 - j200 = 47.06 - j188.24\,\Omega$$

$$Z_c = \frac{(47.06 - j188.24)(j100)}{47.06 - j188.24 + j100} = \frac{(194.03\underline{/-75.96°})(100\underline{/90°})}{100\underline{/-61.93°}}$$

or

$$Z_c = 194.03\underline{/75.96°}\,\Omega = 47.06 + j188.24\,\Omega$$

and

$$Z_T = 47.06 + j188.24 + 100 = 147.06 + j188.24\,\Omega$$

The Thevenin equivalent circuit with the load capacitance of $-j100\,\Omega$ reconnected is illustrated.

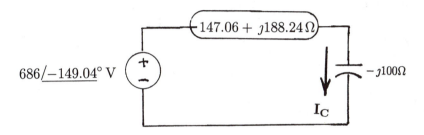

Here

$$\mathbf{I}_C = \frac{\mathbf{V}_T}{Z_T - j100}$$

or

$$\mathbf{I}_C = \frac{686\underline{/-149.04°}\ \mathrm{V}}{171.50\underline{/30.96°}\ \Omega} = 4\underline{/180°}\ \mathrm{A} = -4 + j0\ \mathrm{A} \Leftarrow$$

This confirms the result of Problem 10.38.

CHAPTER ELEVEN
SINUSOIDAL STEADY STATE POWER CALCULATIONS

AVERAGE AND EFFECTIVE VALUES

11.1: If the period, $T = 10\,\text{s}$, determine the average and effective values for the voltage wave shown in Fig 11.1.

Here

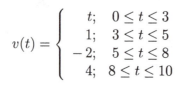

$$v(t) = \begin{cases} t; & 0 \le t \le 3 \\ 1; & 3 \le t \le 5 \\ -2; & 5 \le t \le 8 \\ 4; & 8 \le t \le 10 \end{cases}$$

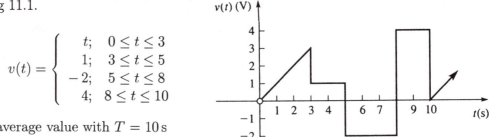

For the average value with $T = 10\,\text{s}$

$$\begin{aligned}
V_{\text{avg}} &= \frac{1}{10\,\text{s}}\left[\int_0^3 t\,dt + \int_3^5 1\,dt + \int_5^8 (-2)\,dt + \int_8^{10} 4\,dt\right] \\
&= \frac{1}{10\,\text{s}}\left[\frac{1}{2}t^2\Big|_0^3 + t\Big|_3^5 - 2t\Big|_5^8 + 4t\Big|_8^{10}\right] \\
&= \frac{1}{10\,\text{s}}\left[\frac{1}{2}(9) + (5-3) - 2(8-5) + 4(10-8)\right] \\
&= \frac{1}{10\,\text{s}}\left[\frac{9}{2} + 2 - 6 + 8\right] \\
&= \left(\frac{1}{10\,\text{s}}\right)\left(\frac{17}{2}\text{V-s}\right) \\
&= 0.850\,\text{V} \Leftarrow
\end{aligned}$$

For the effective value with $T = 10\,\text{s}$

$$\begin{aligned}
V_{\text{eff}}^2 &= \frac{1}{10\,\text{s}}\left[\int_0^3 t^2\,dt + \int_3^5 1\,dt + \int_5^8 4\,dt + \int_8^{10} 16\,dt\right] \\
&= \frac{1}{10\,\text{s}}\left[\frac{1}{3}t^3\Big|_0^3 + t\Big|_3^5 + 4t\Big|_5^8 + 16t\Big|_8^{10}\right] \\
&= \frac{1}{10\,\text{s}}\left[\frac{1}{3}(27) + (5-3) + 4(8-5) + 16(10-8)\right]
\end{aligned}$$

493

$$= \frac{1}{10\,\text{s}}[9 + 2 + 12 + 32]$$

$$= \left(\frac{1}{10\,\text{s}}\right)(55\text{V}^2\text{-s})$$

and

$$V_{\text{eff}} = \sqrt{\frac{55}{10}\text{V}^2} = 2.345\,\text{V} \Leftarrow$$

11.2: If the period, $T = 10\,\text{s}$, determine the average and effective values for the current wave shown in Fig 11.2.

Here

$$i(t) = \begin{cases} 2; & 0 \leq t \leq 2 \\ 4; & 2 \leq t \leq 4 \\ -3; & 4 \leq t \leq 6 \\ 1; & 6 \leq t \leq 8 \\ 0; & 8 \leq t \leq 10 \end{cases}$$

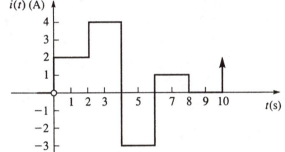

For the average value with $T = 10\,\text{s}$

$$I_{\text{avg}} = \frac{1}{10\,\text{s}}\left[\int_0^2 2\,dt + \int_2^4 4\,dt + \int_4^6 (-3)\,dt + \int_6^8 1\,dt + \int_8^{10} 0\,dt\right]$$

$$= \frac{1}{10\,\text{s}}\left[2t\Big|_0^2 + 4t\Big|_2^4 - 3t\Big|_4^6 + t\Big|_6^8\right]$$

$$= \frac{1}{10\,\text{s}}[2(2-0) + 4(4-2) - 3(6-4) + (8-6)]$$

$$= \frac{1}{10\,\text{s}}[4 + 8 - 6 + 2]$$

$$= \left(\frac{1}{10\,\text{s}}\right)(8\text{A-s})$$

$$= 0.800\,\text{A} \Leftarrow$$

For the effective value with $T = 10\,\text{s}$

$$I_{\text{eff}}^2 = \frac{1}{10\,\text{s}}\left[\int_0^2 (2)^2\,dt + \int_2^4 (4)^2\,dt + \int_4^6 (-3)^2\,dt + \int_6^8 (1)^2\,dt + \int_8^{10} (0)^2\,dt\right]$$

$$= \frac{1}{10\,\text{s}}\left[4t\Big|_0^2 + 16t\Big|_2^4 + 9t\Big|_4^6 + t\Big|_6^8\right]$$

$$= \frac{1}{10\,\mathrm{s}}\left[4(2-0)+16(4-2)+9(6-4)+(8-6)\right]$$

$$= \frac{1}{10\,\mathrm{s}}\left[8+32+18+2\right]$$

$$= \left(\frac{1}{10\,\mathrm{s}}\right)(60\mathrm{A}^2\text{-s})$$

and

$$I_{\text{eff}} = \sqrt{6\,\mathrm{A}^2} = 2.450\,\mathrm{A} \Leftarrow$$

11.3: If the period, $T = 10\,\mathrm{s}$, determine the average and effective values for the voltage wave shown in Fig 11.3.

Here

$$v(t) = \begin{cases} 4e^{-t/4}; & 0 \le t \le 3 \\ 1.8895; & 3 \le t \le 6 \\ -2; & 6 \le t \le 10 \end{cases}$$

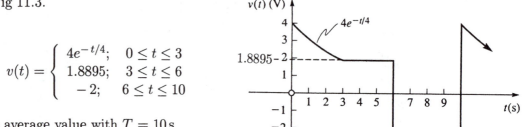

For the average value with $T = 10\,\mathrm{s}$

$$V_{\text{avg}} = \frac{1}{10\,\mathrm{s}}\left[\int_0^3 4e^{-t/4}\,dt + \int_3^6 1.8895\,dt + \int_6^{10}(-2)\,dt\right]$$

$$= \frac{1}{10\,\mathrm{s}}\left[-16e^{-t/4}\Big|_0^3 + 1.8895t\Big|_3^6 - 2t\Big|_6^{10}\right]$$

$$= \frac{1}{10\,\mathrm{s}}\left[-16(0.4724-1)+1.8895(6-3)-2(10-6)\right]$$

$$= \frac{1}{10\,\mathrm{s}}\left[8.4421+5.6685-8\right]$$

$$= \left(\frac{1}{10\,\mathrm{s}}\right)(6.1106\,\mathrm{V}\text{-s})$$

$$= 0.611\,\mathrm{V} \Leftarrow$$

For the effective value with $T = 10\,\mathrm{s}$

$$V_{\text{eff}}^2 = \frac{1}{10\,\mathrm{s}}\left[\int_0^3 \left(4e^{-t/4}\right)^2 dt + \int_2^4 (1.8895)^2\,dt + \int_6^{10}(-2)^2\,dt\right]$$

$$= \frac{1}{10\,\mathrm{s}}\left[-32e^{-t/2}\Big|_0^3 + 3.5702t\Big|_3^6 + 4t\Big|_6^{10}\right]$$

$$= \frac{1}{10\,\text{s}}[-32(0.2231-1)+3.5702(6-3)+4(10-6)]$$

$$= \frac{1}{10\,\text{s}}[24.8608+10.7106+16]$$

$$= \left(\frac{1}{10\,\text{s}}\right)(51.5714\text{V}^2\text{-s})$$

and

$$V_{\text{eff}} = \sqrt{5.1571\,\text{V}^2} = 2.271\,\text{V} \Leftarrow$$

11.4: If the period, $T=10\,\text{s}$, determine the average and effective values for the current wave shown in Fig 11.1.

Here

$$i(t) = \begin{cases} 2t; & 0 \le t \le 2 \\ -2; & 2 \le t \le 8 \\ 1; & 8 \le t \le 10 \end{cases}$$

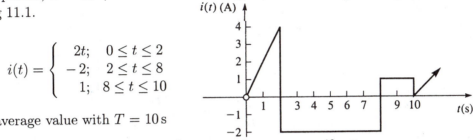

For the average value with $T=10\,\text{s}$

$$I_{\text{avg}} = \frac{1}{10\,\text{s}}\left[\int_0^2 2t\,dt + \int_2^8(-2)\,dt + \int_8^{10} dt\right]$$

$$= \frac{1}{10\,\text{s}}\left[t^2\Big|_0^2 - 2t\Big|_2^8 + t\Big|_8^{10}\right]$$

$$= \frac{1}{10\,\text{s}}[(4-0)-2(8-2)+(10-8)]$$

$$= \frac{1}{10\,\text{s}}[4-12+2]$$

$$= \left(\frac{1}{10\,\text{s}}\right)(-6\,\text{A-s})$$

$$= -0.600\,\text{A} \Leftarrow$$

For the effective value with $T=10\,\text{s}$

$$I_{\text{eff}} = \frac{1}{10\,\text{s}}\left[\int_0^2 4t^2\,dt + \int_2^8 4\,dt + \int_8^{10} dt\right]$$

$$= \frac{1}{10\,\text{s}}\left[\frac{4}{3}t^3\Big|_0^2 + 4t\Big|_2^8 + t\Big|_8^{10}\right]$$

$$= \frac{1}{10\,\mathrm{s}}\left[\frac{32}{3} + 4(8-2) + (10-8)\right]$$

$$= \frac{1}{10\,\mathrm{s}}\left[\frac{32}{3} + 24 + 2\right]$$

$$= \left(\frac{1}{10\,\mathrm{s}}\right)\left(\frac{110}{3}\mathrm{A}^2\text{-s}\right)$$

$$I_{\mathrm{eff}} = \sqrt{\frac{110}{30}\,\mathrm{A}^2} = 1.915\,\mathrm{A} \Leftarrow$$

11.5: If the period, $T = 40\,\mathrm{ms}$, determine the average and effective values for the voltage wave shown in Fig 11.5.

Here, with all times in s

$$v(t) = \begin{cases} 4\sin 50\pi t; & 0 \le t \le 0.01 \\ 4; & 0.01 \le t \le .02 \\ 2; & 0.02 \le t \le 0.03 \\ -2; & .03 \le t \le 0.04 \end{cases}$$

For the average value with $T = 0.04\,\mathrm{s}$

$$V_{\mathrm{avg}} = \frac{1}{0.04\,\mathrm{s}}\left[\int_0^{-.0.01} 4\sin 50\pi t\, dt + \int_{0.01}^{0.02} 4\, dt + \int_{0.02}^{0.03} 2\, dt + \int_{0.03}^{0.04} (-2)\, dt\right]$$

$$= 25\mathrm{s}^{-1}\left[-\frac{4}{50\pi}\cos 50\pi t\Big|_0^{0.01} + 4t\Big|_{0.01}^{0.02} + 2t\Big|_{0.02}^{0.03} - 2t\Big|_{0.03}^{0.04}\right]$$

$$= 25\mathrm{s}^{-1}\left[-\frac{4}{50\pi}(0-1) + 4(0.02-0.01) + 2(0.03-0.02) - 2(0.04-0.03)\right]$$

$$= 25\mathrm{s}^{-1}(0.0255 + 0.0400 + 0.0200 - 0.0200)$$

$$= 25\mathrm{s}^{-1}(0.0655\mathrm{V}\text{-s}) = 1.637\,\mathrm{V}$$

For the effective value with $T = 0.04\,\mathrm{s}$

$$V_{\mathrm{eff}}^2 = \frac{1}{0.04\,\mathrm{s}}\left[\int_0^{0.01} 16\sin^2 50\pi t\, dt + \int_{0.01}^{0.02} 16\, dt + \int_{0.02}^{0.03} 4\, dt + \int_{0.03}^{0.04} 4\, dt\right]$$

With the trigonometric identity

$$\sin^2 at = \frac{1}{2}[1 - \cos 2at]$$

this becomes

$$V_{\text{eff}}^2 = 25\text{s}^{-1}\left[\frac{16}{2}\int_0^{0.01}dt - \frac{16}{2}\int_0^{0.01}\cos 100\pi t\,dt + 16t\Big|_{0.01}^{0.02} + 4t\Big|_{0.02}^{0.03} + 4t\Big|_{0.03}^{0.04}\right]$$

$$= 25\text{s}^{-1}\left[8t\Big|_0^{0.01} - \frac{8}{100\pi}\sin 100\pi t\Big|_0^{0.01} + (16 + 4 + 4)(0.01)\right]$$

$$= 25\text{s}^{-1}\left[8(0.01) - \frac{8}{100\pi}(0 - 0) + 24(0.01)\right]$$

$$= 25\,\text{s}^{-1}[32(0.01)\text{V}^2\text{-s}]$$

and

$$V_{\text{eff}} = \sqrt{8\,\text{V}^2} = 2.828\,\text{V} \Leftarrow$$

11.6: If the period, $T = 6\,\text{s}$, determine the average and effective values for the voltage wave shown in Fig 11.6.

Here

$$v(t) = \begin{cases} t; & 0 \leq t \leq 2 \\ 1; & 2 \leq t \leq 3 \\ -2; & 3 \leq t \leq 5 \\ -3; & 5 \leq t \leq 6 \end{cases}$$

For the average value with $T = 6\,\text{s}$

$$V_{\text{avg}} = \frac{1}{6\,\text{s}}\left[\int_0^2 t\,dt + \int_2^3 1\,dt + \int_3^5 (-2)\,dt + \int_5^6 (-3)\,dt\right]$$

$$= \frac{1}{6\,\text{s}}\left[\frac{1}{2}t^2\Big|_0^2 + t\Big|_2^3 - 2t\Big|_3^5 - 3t\Big|_5^6\right]$$

$$= \frac{1}{6\,\text{s}}\left[\frac{1}{2}(2^2 - 0) + (3 - 2) - 2(5 - 3) - 3(6 - 5)\right]$$

$$= \frac{1}{6\,\text{s}}(2 + 1 - 4 - 3)$$

$$= \left(\frac{1}{6\,\text{s}}\right)(-4\text{V-s})$$

$$= -\frac{2}{3}\,\text{V} \Leftarrow$$

For the effective value with $T = 6\,\mathrm{s}$

$$V_{\mathrm{eff}}^2 = \frac{1}{6\,\mathrm{s}}\left[\int_0^2 t^2\, dt + \int_2^3 1\, dt + \int_3^5 4\, dt + \int_5^5 9\, dt\right]$$

$$= \frac{1}{6\,\mathrm{s}}\left[\frac{1}{3}t^3\Big|_0^2 + t\Big|_2^3 + 4t\Big|_3^5 + 9t\Big|_5^6\right]$$

$$= \frac{1}{6\,\mathrm{s}}\left[\frac{1}{3}(8) + (3-2) + 4(5-3) + 9(6-5)\right]$$

$$= \frac{1}{6\,\mathrm{s}}\left[\frac{8}{3} + 1 + 8 + 9\right]$$

$$= \left(\frac{1}{6\,\mathrm{s}}\right)\left(\frac{62}{3}\mathrm{V}^2\text{-s}\right)$$

and

$$V_{\mathrm{eff}} = \sqrt{\frac{31}{9}\mathrm{V}^2} = 1.856\,\mathrm{V} \Leftarrow$$

AVERAGE POWER IN THE SINUSOIDAL STEADY STATE

11.7: Determine the average power and the power factor for an RLC series network operating at $400\,\mathrm{Hz}$ with $R = 100\,\Omega, L = 39.79\,\mathrm{mH}$ and $C = 15.92\mu\mathrm{F}$ if the network is subjected to a sinusoidal voltage with a maximum amplitude of $180\,\mathrm{V}$.

With $f = 400\,\mathrm{Hz}$, $\omega = 2\pi f = 2\pi(400\,\mathrm{Hz}) = 800\pi\,\mathrm{rad/s}$ and

Component	Reactance, X
$C = 15.92\,\mu\mathrm{F}$	$25\,\Omega$
$L = 39.79\,\mathrm{mH}$	$100\,\Omega$

Here,

$$Z_{\mathrm{total}} = 100 + \jmath100 - \jmath25 = 100 + \jmath75 = 125\underline{/36.87^\circ}\,\Omega$$

and

$$\mathbf{I} = \frac{\mathbf{V}_s}{Z_{\mathrm{total}}} = \frac{180\underline{/0^\circ}\,\mathrm{V}}{125\underline{/36.87^\circ}\,\Omega} = 1.44\underline{/-36.87^\circ}\,\mathrm{A}$$

Then with $V_m = 180\,\mathrm{V}$, $I_m = 1.44\,\mathrm{A}$ and the impedance angle, $\theta = 36.78^\circ$,

$$PF = \cos\theta = \cos 36.87^\circ = 0.800 \Leftarrow$$

and

$$P_{\text{avg}} = \frac{V_m I_m}{2}\cos\theta = \frac{(180)(1.44)}{2}\cos 36.87° = \frac{(180)(1.44)}{2}(0.80) = 103.68\,\text{W} \Leftarrow$$

11.8: Determine the average power and the power factor for an RLC parallel network operating at $250\,\text{Hz}$ with $R = 50\,\Omega, L = 50.93\,\text{mH}$ and $C = 21.22\mu\text{F}$ if the network is subjected to a sinusoidal current with a maximum amplitude of $16\,\text{A}$.

With $f = 250\,\text{Hz}$, $\omega = 2\pi f = 2\pi(250\,\text{Hz}) = 500\pi\,\text{rad/s}$ and

Component	Reactance, X
$C = 21.22\,\mu\text{F}$	$30\,\Omega$
$L = 50.93\,\text{mH}$	$80\,\Omega$

Here,

$$Y_{\text{total}} = \frac{1}{50} - \jmath\frac{1}{80} + \jmath\frac{1}{30} = 0.0200 + \jmath 0.0208 = 0.0289\underline{/46.12°}\,\mho$$

and

$$\mathbf{V} = \frac{\mathbf{I}_s}{Y_{\text{total}}} = \frac{16\underline{/0°}\,\text{A}}{0.0289\underline{/46.12°}\,\mho} = 554.49\underline{/-46.12°}\,\text{V}$$

Then with $I_m = 16\,\text{A}$, $V_m = 554.49\,\text{V}$ and the admittance angle, $\theta = -46.12°$,

$$PF = \cos\theta = \cos(-46.12°) = 0.6932 \Leftarrow$$

and

$$P_{\text{avg}} = \frac{V_m I_m}{2}\cos\theta = \frac{(554.49)(16)}{2}(0.6932) = 3075\,\text{W} \Leftarrow$$

11.9 Two impedances, $Z_1 = 40\sqrt{2}\underline{/-45°}\,\Omega$ and $Z_2 = 80 - \jmath 60\,\Omega$ are connected in parallel and the combination is then connected in series with a $24\,\Omega$ resistor. If the combination is connected across a source having an rms value of $120\,\text{V}$, how much power is drawn and what is the power factor?

Here

$$Z_1 = 40\sqrt{2}\underline{/-45°}\,\Omega = 40 - \jmath 40\,\Omega$$
$$Z_2 = 100\underline{/-36.87°}\,\Omega = 80 - \jmath 60\,\Omega$$

so that

$$Z_p = \frac{(40\sqrt{2}\underline{/-45°})(100\underline{/-36.87°})}{40 - \jmath 40 + 80 - \jmath 60} = \frac{4000\sqrt{2}\underline{/-81.87°}}{156.21\underline{/-39.81°}} = 36.21\underline{/-42.06°}\,\Omega = 26.89 - \jmath 24.26\,\Omega$$

and

$$Z_{\text{total}} = 26.89 - \jmath 24.26 + 24 = 50.89 - \jmath 24.26\,\Omega = 56.37\underline{/-25.49°}\,\Omega$$

Then, with the maximum value of the voltage equal to $V_m = 120\sqrt{2}\,\text{V}$, the maximum value of the current will be

$$\mathbf{I} = \frac{V_m}{|Z|} = \frac{120\sqrt{2}\underline{/0°}\,\text{V}}{56.37\underline{/-25.49°}\,\Omega} = 3.011\underline{/25.49°}\,\text{A}$$

With $I_m = 3.011\,\text{A}$, $V_m = 120\sqrt{2}\,\text{V}$ and the impedance angle, $\theta = -25.49°$,

$$PF = \cos\theta = \cos(-25.49°) = 0.9027 \Leftarrow$$

and

$$P_{\text{avg}} = \frac{V_m I_m}{2}\cos\theta = \frac{(120\sqrt{2})(3.011)}{2}(0.9027) = 230.6\,\text{W} \Leftarrow$$

11.10: Two impedances, $Z_1 = 25\underline{/53.13°}\,\Omega$ and $Z_2 = 39\underline{/67.38°}\,\Omega$ are connected in parallel and the combination is then connected in series with an impedance of $17\underline{/28.07°}\,\Omega$. If the combination is connected across a source having an rms value of $20\,\text{A}$, how much power is drawn and what is the power factor?

Here

$$Z_1 = 25\underline{/53.13°}\,\Omega = 15 + \jmath 20\,\Omega$$
$$Z_2 = 39\underline{/67.38°}\,\Omega = 15 + \jmath 36\,\Omega$$

and

$$Z_3 = 17\underline{/28.07°}\,\Omega = 15 + \jmath 8\,\Omega$$

so that the impedance of the parallel combination will be

$$Z_p = \frac{(25\underline{/53.13°})(39\underline{/67.38°})}{(15 + \jmath 20) + (15 + \jmath 36)} = \frac{975\underline{/120.51°}}{63.53\underline{/61.82°}} = 15.35\underline{/58.69°}\,\Omega = 7.98 + \jmath 13.11\,\Omega$$

and the total impedance is

$$Z_{\text{total}} = (7.98 + \jmath 13.11) + (15 + \jmath 8) = 22.98 + \jmath 21.11\,\Omega = 31.20\underline{/42.57°}\,\Omega$$

The impedance angle is $42.57°$ so that

$$PF = \cos\theta = \cos 42.57° = 0.7365 \Leftarrow$$

and using the rms value of the current and with

$$Z = R + \jmath X = 22.98 + \jmath 21.11\,\Omega$$

$$P_{\text{avg}} = (I_{\text{eff}})^2 R = (20\,\text{A})^2 (22.98\,\Omega) = 9192\,\text{W} \Leftarrow$$

11.11: An RLC series network is subjected to a voltage of

$$v(t) = 200 \cos 400t \text{ V}$$

An impedance meter has measured the impedance angle as $36.87°$. If, $R = 40\,\Omega$, $L = 387.5\,\text{mH}$ and the power dissipated is $640\,\text{W}$, determine the value of C.

Here

$$X_L = \omega L = (400\,\text{rad/s})(0.3875\,\text{H}) = 155\,\Omega$$

and with $Z = |Z|\underline{/36.87°}$ and $V_m = 200\,\text{V}$

$$P_R = (I_m)^2 R = \left(\frac{V_m}{|Z|}\right)^2 R = \left(\frac{200\,\text{V}}{|Z|}\right)^2 R = 640\,\text{W}$$

Thus

$$|Z|^2 = \frac{(200\,\text{V})^2(40\,\Omega)}{640\,\text{W}} = 2500\,\Omega^2$$

and

$$|Z| = 50\,\Omega$$

Then

$$Z = 50\underline{/36.87°} = 40 + j30$$

and in looking at just the reactive part of the impedance, it is seen that

$$X = 30\,\Omega = X_L - X_C = 155\,\Omega - X_C$$

or

$$X_C = 125\,\Omega$$

With $\omega = 400\,\text{rad/s}$

$$C = \frac{1}{\omega X} = \frac{1}{(400\,\text{rad/s})(125\,\Omega)} = 20\,\mu\text{F} \Leftarrow$$

11.12: A series RC circuit absorbs $22.15\,\text{W}$ at a power factor of 0.9231 when it is connected to a voltage source of

$$v(t) = 120 \cos 400t \text{ V}$$

Determine the values of R and C.

If the power factor is 0.9231, the impedance angle for this capacitive circuit is

$$\theta = -\arccos 0.9231 = -22.62°$$

so that with $\mathbf{S} = 22.15\underline{/-22.62°}$ W and

$$\mathbf{I} = \frac{\mathbf{S}}{\mathbf{V}} = \frac{22.15\underline{/-22.62°}\text{ W}}{120\underline{/0°}\text{ V}} = 0.1846\underline{/-22.62°}\text{ A}$$

the magnitude of the impedance is, Z is

$$Z = \frac{|V|}{|I|} = \frac{120\text{ V}}{0.1846\text{ A}} = 650\,\Omega$$

Because

$$Z = R - jX = 650\underline{/-22.62°}\,\Omega = 600 - j250\,\Omega$$

$$R = 600\,\Omega \Leftarrow$$

and with $\omega = 400$ rad/s

$$C = \frac{1}{\omega X} = \frac{1}{(400\text{ rad/s})(250\,\Omega)} = 10\,\mu\text{F} \Leftarrow$$

11.13: Determine the total power dissipated by all of the resistors in the phasor domain network of Fig 11.7.

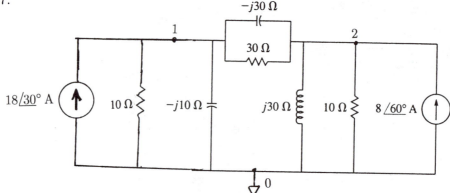

The phasor domain network with the voltage source transformed to a current source and two node voltages indicated is shown in the figure. Here, by a repeated application of KCL, the two node equations are

$$\left(\frac{1}{10} + \frac{j}{10} + \frac{j}{30} + \frac{1}{30}\right)\mathbf{V}_1 \qquad -\left(\frac{1}{30} + \frac{j}{30}\right)\mathbf{V}_2 = 18\underline{/30°}\text{ A}$$

$$-\left(\frac{1}{30} + \frac{j}{30}\right)\mathbf{V}_1 \qquad +\left(\frac{1}{30} + \frac{j}{30} + \frac{1}{10} - \frac{j}{30}\right)\mathbf{V}_1 = 8\underline{/60°}\text{ A}$$

These may be simplified and, after multiplication of all terms by 30, they may be put into matrix form

$$\begin{bmatrix} (4+j4) & -(1+j) \\ -(1+j) & 4 \end{bmatrix} \begin{bmatrix} V_1 \\ V_2 \end{bmatrix} = \begin{bmatrix} 467.65 + j270\,\text{A} \\ 120 + j207.851\,\text{A} \end{bmatrix}$$

and a matrix inversion can be employed to find the voltage vector

$$\begin{bmatrix} V_1 \\ V_2 \end{bmatrix} = \begin{bmatrix} (4+j4) & -(1+j) \\ -(1+j) & 4 \end{bmatrix}^{-1} \begin{bmatrix} 467.65 + j270\,\text{A} \\ 120 + j207.851\,\text{A} \end{bmatrix}$$

$$\begin{bmatrix} V_1 \\ V_2 \end{bmatrix} \begin{bmatrix} (0.1416 - j0.1239) & (0.0664 + j0.0044) \\ (0.0664 + j0.0044) & (0.2655 + j0.0177) \end{bmatrix} \begin{bmatrix} 467.65 + j270\,\text{A} \\ 120 + j207.851\,\text{A} \end{bmatrix}$$

or

$$\begin{bmatrix} V_1 \\ V_2 \end{bmatrix} = \begin{bmatrix} 106.71 - j5.38\,\text{V} \\ 58.02 + j77.29\,\text{V} \end{bmatrix} = \begin{bmatrix} 106.85\underline{/-2.89°}\,\text{V} \\ 96.64\underline{/53.13°}\,\text{V} \end{bmatrix}$$

Here, with the voltage magnitudes taken as rms values

$$|V_1| = 106.85\,\text{V}$$

$$|V_2| = 96.64\,\text{V}$$

and

$$\begin{aligned} V_{12} &= V_1 - V_2 \\ &= (106.71 - j5.38) - (58.02 + j77.29) \\ &= 48.69 - j82.67\,\text{V} = 95.94\underline{/-59.50°}\,\text{V} \end{aligned}$$

so that

$$|V_{12}| = 95.94\,\text{V}$$

For the $10\,\Omega$ resistor associated with the voltage source

$$P_{10\,\Omega} = \frac{(|V_1|)^2}{10\,\Omega} = \frac{(106.85\,\text{V})^2}{10\,\Omega} = 1141.7\,\text{W} \Leftarrow$$

for the $10\,\Omega$ resistor associated with the current source

$$P_{10\,\Omega} = \frac{(|V_2|)^2}{10\,\Omega} = \frac{(96.64\,\text{V})^2}{10\,\Omega} = 933.9\,\text{W} \Leftarrow$$

and for the $30\,\Omega$ resistor

$$P_{30\,\Omega} = \frac{(|V_{12}|)^2}{30\,\Omega} = \frac{(95.94\,\text{V})^2}{30\,\Omega} = 306.8\,\text{W} \Leftarrow$$

11.14: Determine the complex power delivered to the phasor domain network in Fig 11.8.

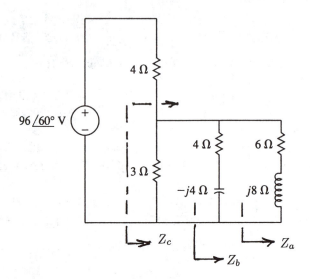

Here

$$Z_a = 6 + j8\,\Omega = 10\underline{/53.13°}\,\Omega$$

$$Z_b = \frac{(4-j4)(10\underline{/53.13°})}{(6+j8)+(4-j4)} = \frac{(4\sqrt{2}\underline{/-45°}\,\Omega)(10\underline{/53.13°})}{10.77\underline{/21.80}}$$

$$= 5.25\underline{/-13.67°}\,\Omega = 5.10 - j1.24\,\Omega$$

$$Z_c = \frac{(3\underline{/0°})(5.25\underline{/-13.67°})}{5.10 - j3.24 + 3}$$

$$= \frac{15.75\underline{/-13.67°})}{8.20\underline{/-8.71°}} = 1.92\underline{/-4.96°}\,\Omega = 1.92 - j0.17\,\Omega$$

and

$$Z_{\text{eq}} = 4 + 1.92 - j0.17 = 5.92 - j0.17\,\Omega = 5.92\underline{/-1.61}\,\Omega$$

Then

$$\mathbf{I} = \frac{\mathbf{V}}{Z} = \frac{96\underline{/60°}\,\text{V}}{5.92\underline{/-1.61}\,\Omega} = 16.22\underline{/61.61°}\,\text{A}$$

and the complex power is

$$\mathbf{S} = \mathbf{VI}^* = (96\underline{/60°}\,\text{V})(16.22\underline{/-61.61°}\,\text{A}) = 1556.9\underline{/-1.61°}\,\text{VA} \Leftarrow$$

11.15: Determine the complex power delivered to the network in Fig 11.9.

With $\omega = 400$ rad/s

Component	Reactance, X
$C = 62.5\,\mu\text{F}$	$40\,\Omega$
$L = 25\,\text{mH}$	$10\,\Omega$
$L = 35\,\text{mH}$	$14\,\Omega$
$L = 100\,\text{mH}$	$40\,\Omega$

The phasor domain network is shown at the right. In it,

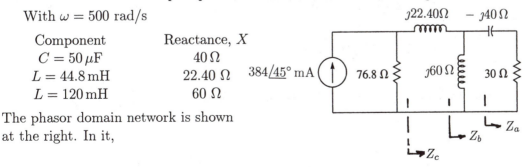

$$Z_a = 40 + j40\,\Omega = 40\sqrt{2}\underline{/45°}\,\Omega$$

$$Z_b = \frac{(-j40)(40\sqrt{2}\underline{/45°})}{-j40 + 40 + j40} = \frac{(40\underline{/-90°})(40\sqrt{2}\underline{/45°})}{40} = 40\sqrt{2}\underline{/-45°}\,\Omega = 40 - j40\,\Omega$$

$$Z_c = j10 + 40 - j40 = 40 - j30\,\Omega = 50\underline{/-36.87°}\,\Omega$$

and

$$Z_d = \frac{(j14)(50\underline{/-36.87°})}{j14 + 40 - j30} = \frac{(14\underline{/90°})(50\underline{/-36.87°})}{43.08\underline{/-21.80°}} = 16.24\underline{/74.93°}14\,\Omega = 4.22 + j15.69\,\Omega$$

and

$$Z_{\text{eq}} = 4.22 + j15.69 + 17.5 = 21.72 + j15.69\,\Omega = 26.80\underline{/35.84°}\,\Omega$$

The current leaving the source will be

$$\mathbf{I} = \frac{\mathbf{V}}{Z_{\text{total}}} = \frac{400\underline{/90°}\text{ V}}{26.80\underline{/35.84°}\,\Omega} = 14.93\underline{/54.16°}\text{ A}$$

and the complex power is

$$\mathbf{S} = \mathbf{VI}^* = (400\underline{/90°}\text{ V})(14.93\underline{/-54.16°}\text{ A}) = 5971.4\underline{/35.84°}\text{ VA} \Leftarrow$$

11.16: Determine the complex power delivered to the network in Fig 11.10.

With $\omega = 500$ rad/s

Component	Reactance, X
$C = 50\,\mu\text{F}$	$40\,\Omega$
$L = 44.8\,\text{mH}$	$22.40\,\Omega$
$L = 120\,\text{mH}$	$60\,\Omega$

The phasor domain network is shown at the right. In it,

$$Z_a = 30 - j40\,\Omega = 50\underline{/-53.13°}\,\Omega$$

$$Z_b = \frac{(j60)(50\underline{/-53.13°})}{j60 + 30 - j40} = \frac{(60\underline{/90°})(50\underline{/-53.13°})}{36.06\underline{/33.69°}}$$

or

$$Z_b = 83.21\underline{/3.18°}\,\Omega = 83.08 + j4.62\,\Omega$$

$$Z_c = j22.4 + 83.08 + j4.62\,\Omega = 83.08 + j27.02\,\Omega = 87.36\underline{/18.01°}\,\Omega$$

and

$$Z_{eq} = \frac{(76.8)(87.36\underline{/18.01°})}{76.8 + 83.08 + j27.02} = \frac{(76.8\underline{/0°})(87.36\underline{/18.01°}}{162.14\underline{/9.59°}} = 41.37\underline{/8.42°}\,\Omega$$

The voltage across the current source will be

$$\mathbf{V} = Z_{eq}\mathbf{I} = (41.37\underline{/8.42°}\,\Omega)(0.384\underline{/45°}\,\text{A}) = 22.47\underline{/53.42°}\,\text{V}$$

and the complex power is

$$\mathbf{S} = \mathbf{VI}^* = (22.47\underline{/53.42°}\,\text{V})(0.384\underline{/-45°}) = 12.21\underline{/8.42°}\,\text{VA} \Leftarrow$$

11.17: Determine the complex power delivered to the network in Fig 11.11.

With $\omega = 1000$ rad/s

Component	Reactance, X
$C = 62.5\,\mu\text{F}$	$16\,\Omega$
$L = 58\,\text{mH}$	$58\,\Omega$
$L = 16\,\text{mH}$	$16\,\Omega$

The phasor domain network is shown
at the right. In it,

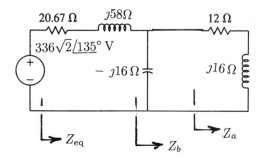

$$Z_a = 12 + j16\,\Omega = 20\underline{/53.13°}\,\Omega$$

$$Z_b = \frac{(-j16)(50\underline{/53.13°})}{-j16 + 12 - j16} = \frac{(16\underline{/-90°})(50\underline{/53.13°})}{12\underline{/0°}}$$

or

$$Z_b = 26.67\underline{/-36.87°}\,\Omega = 21.33 - j16\,\Omega$$

and

$$Z_{eq} = (20.67 + j58) + (21.33 - j16) = 42 + j42\,\Omega = 42\sqrt{2}\underline{/45°}\,\Omega$$

The current leaving the source will be

$$\mathbf{I} = \frac{\mathbf{V}}{Z_{eq}} = \frac{336\sqrt{2}\underline{/135°}\,\text{V}}{42\sqrt{2}\underline{/45°}\,\Omega} = 8\underline{/90°}\,\text{A} = 0 + j8\,\text{A}$$

and this makes the complex power

$$\mathbf{S} = \mathbf{VI}^* = (336\sqrt{2}\underline{/135°}\,\text{V})(8\underline{/-90°}\,\text{A}) = 2688\sqrt{2}\underline{/45°}\,\text{VA} \Leftarrow$$

11.18: For the phasor domain network shown in Fig 11.12, Determine the real power, the apparent power, the magnitude of the apparent power and the power factor.

In the phasor domain network shown at the right, the impedance presented to the voltage source is equal to

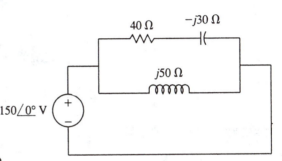

$$\begin{aligned} Z_{eq} &= \frac{(40 - j30)(j50)}{40 - j30 + j50} \\[2mm] &= \frac{(50\underline{/-36.87°})(50\underline{/90°})}{44.72\underline{/26.57°}} \\[2mm] &= 55.90\underline{/26.57°}\,\Omega = 50 + j25\,\Omega \end{aligned}$$

The current drawn by the network will be

$$\mathbf{I} = \frac{\mathbf{V}}{Z_{total}} = \frac{150\underline{/0°}}{55.90\underline{/26.57°}\,\Omega} = 2.68\underline{/-26.57°}\,\text{A}$$

The apparent power is the complex power

$$\mathbf{S} = \mathbf{VI}^* = (150\underline{/0°}\,\text{V})(2.68\underline{/26.57°}\,\text{A}) = 402.49\underline{/26.57°}\,\text{VA}) = 360 + j180\,\text{VA} \Leftarrow$$

The real power is

$$360\,\text{W} \Leftarrow$$

The magnitude of the apparent power is

$$402.49\,\text{VA} \Leftarrow$$

and the power factor is

$$PF = \cos 26.57° = 0.8944 \Leftarrow$$

POWER FACTOR CORRECTION

11.19: In the power distribution system shown in Fig 11.13, the frequency is 60 Hz. Determine:

(a) the real power,
(b) the reactive power,
(c) the magnitude of the apparent power,
(d) the power factor,
(e) the correction necessary to make the power factor 0.950 and
(f) the component necessary to achieve correction in part (e)

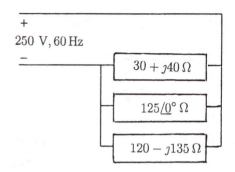

Here

$$Z_1 = 30 + j40\,\Omega = 50\underline{/53.13°}\,\Omega$$

$$Z_2 = 125\underline{/0°}\,\Omega = 125 + j0\,\Omega$$

and

$$Z_3 = 180.62\underline{/-48.37°}\,\Omega$$

The current magnitudes are

$$I_1 = \frac{V}{|Z_1|} = \frac{250\,\text{V}}{50\,\Omega} = 5\,\text{A}$$

$$I_2 = \frac{V}{|Z_2|} = \frac{250\,\text{V}}{125\,\Omega} = 2\,\text{A}$$

and

$$I_3 = \frac{V}{|Z_3|} = \frac{250\,\text{V}}{180.62\,\Omega} = 1.384\,\text{A}$$

(a)

$$
\begin{aligned}
P &= P_1 + P_2 + P_3 \\
&= (I_1)^2 R_1 + (I_2)^2 R_2 + (I_3)^2 R_3 \\
&= (5\,\text{A})^2(30\,\Omega) + (2\,\text{A})^2(125\,\Omega) + (1.384\,\text{A})^2(120\,\Omega) \\
&= 750\,\text{W} + 500\,\text{W} + 229.85\,\text{W} = 1479.85\,\text{W} \Leftarrow
\end{aligned}
$$

(b)

$$
\begin{aligned}
Q &= Q_1 + Q_2 + Q_3 \\
&= (I_1)^2 X_1 + (I_2)^2 X_2 + (I_3)^2 X_3 \\
&= (5\,\text{A})^2(40\,\Omega) + (2\,\text{A})^2(0\,\Omega) - (1.384\,\text{A})^2(135\,\Omega) \\
&= 1000\,\text{VAR} + 0\,\text{VAR} - 258.59\,\text{VAR} = 741.41\,\text{VAR} \Leftarrow
\end{aligned}
$$

510

(c)
$$\mathbf{S} = P + jQ = 1479.85 + j741.41 = 1655.19\underline{/26.61°}$$

The magnitude of the apparent power is

$$|S| = 1655.19 \,\text{VA} \Leftarrow$$

(d) The power factor is
$$PF = \cos 26.61° = 0.8941 \Leftarrow$$

(e) The subscript, d, may be used to indicate desired values. Then, in order to make the power factor equal to 0.950,

$$\theta_d = \arccos 0.95 = 18.19°$$

$$\tan \theta_d = \tan 18.19° = 0.3287$$

$$Q_d = P \tan \theta_d = (1479.85 \,\text{W})(0.3287) = 486.41 \,\text{VAR}$$

and the correction required will be

$$Q_{\text{corr}} = Q_d - Q = 486.41 - 741.41 = -255 \,\text{VAR} \Leftarrow$$

The minus sign indicates that the correction must be capacitive.

(f) With the correction earmarked as capacitive and $\omega = 2\pi(60 \,\text{Hz}) = 377 \,\text{rad/s}$

$$X_{\text{corr}} = \frac{V^2}{|Q_{\text{corr}}|} = \frac{(250 \,\text{V})^2}{|255 \,\text{VAR}|} = 245.10 \,\Omega$$

the value of a single capacitor will be

$$C = \frac{1}{\omega X_{\text{corr}}} = \frac{1}{(377 \,\text{rad/s})(245.10 \,\Omega)} = 10.82 \,\mu\text{F} \Leftarrow$$

11.20: In the power distribution system shown in Fig 11.14, the frequency is 60 Hz. Determine:

    (a) the real power,
    (b) the reactive power,
    (c) the magnitude of the apparent power,
    (d) the power factor,
    (e) the correction necessary to make
        the power factor 0.935 and
    (f) the component necessary to achieve
        correction in part (e)

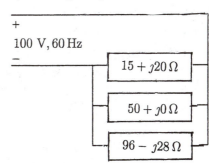

A convenient tabular form for making complex power calculations may be used.

Item	Load-1	Load-2	Load-3	Σ
Z, Ω	$15 + j20$	$50 + j0$	$96 - j28$	
$|Z|, \Omega$	25	50	100	
$I = V/|Z|, \text{A}$	4	2	1	
R, Ω	15	50	96	
X, Ω	20	0	-28	
$P = I^2 R, \text{W}$	240	200	96	536
$Q = I^2 X, \text{VAR}$	320	0	-28	292

Here, from the foregoing tabulation

$$\text{(a)} \qquad P = 536 \, \text{W} \Leftarrow$$

$$\text{(b)} \qquad Q = 292 \, \text{VAR} \Leftarrow$$

and

$$\text{(c)} \qquad \mathbf{S} = 536 + j192 \, \text{VA} = 610.38 \underline{/28.58°} \, \text{VA}$$

which makes the magnitude of the apparent power

$$|S| = 610.38 \, \text{VA} \Leftarrow$$

(d) The power factor is

$$PF = \cos 28.58° = 0.8781 \Leftarrow$$

(e) The subscript, d, may be used to indicate desired values. Then, in order to make the power factor equal to 0.935,

$$\theta_d = \arccos 0.935 = 20.77°$$

$$\tan \theta_d = \tan 20.77° = 0.3793$$

$$Q_d = P \tan \theta_d = (536 \, \text{W})(0.3793) = 203.3 \, \text{VAR}$$

and the correction required will be

$$Q_{\text{corr}} = Q_d - Q = 203.3 - 292 = -88.7 \, \text{VAR} \Leftarrow$$

The minus sign indicates that the correction must be capacitive.

(f) With the correction earmarked as capacitive and $\omega = 2\pi(60\,\mathrm{Hz}) = 377\,\mathrm{rad/s}$

$$X_{\mathrm{corr}} = \frac{V^2}{|Q_{\mathrm{corr}}|} = \frac{(100\,\mathrm{V})^2}{|88.7\,\mathrm{VAR}|} = 112.7\,\Omega$$

the value of a single capacitor will be

$$C = \frac{1}{\omega X_{\mathrm{corr}}} = \frac{1}{(377\,\mathrm{rad/s})(112.7\,\Omega)} = 23.53\,\mu\mathrm{F} \Leftarrow$$

11.21: In the power distribution system shown in Fig 11.15, the frequency is 60 Hz. Determine:

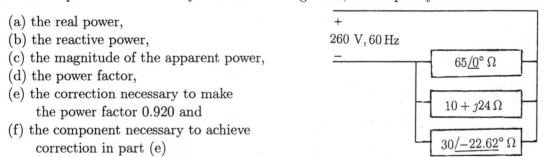

(a) the real power,
(b) the reactive power,
(c) the magnitude of the apparent power,
(d) the power factor,
(e) the correction necessary to make
    the power factor 0.920 and
(f) the component necessary to achieve
    correction in part (e)

A convenient tabular form for making complex power calculations may be used.

Item	Load-1	Load-2	Load-3	Σ
Z, Ω	$65\underline{/0^\circ}$	$10 + j24$	$130\underline{/-22.62^\circ}$	
$|Z|, \Omega$	65	26	130	
$I = V/|Z|, \mathrm{A}$	4	10	2	
R, Ω	65	10	120	
X, Ω	0	24	-50	
$P = I^2 R, \mathrm{W}$	1040	1000	480	2520
$Q = I^2 X, \mathrm{VAR}$	0	2400	-200	2200

Here, from the foregoing tabulation

$$\text{(a)} \qquad P = 2520\,\mathrm{W} \Leftarrow$$

$$\text{(b)} \qquad Q = 2200\,\text{VAR} \Leftarrow$$

and

$$\text{(c)} \qquad \mathbf{S} = 2520 + j2200\,\text{VA} = 3345.2\underline{/41.12°}\,\text{VA}$$

which makes the magnitude of the apparent power

$$|S| = 3345.2\,\text{VA} \Leftarrow$$

(d) The power factor is

$$PF = \cos 41.12° = 0.7533 \Leftarrow$$

(e) The subscript, d, may be used to indicate desired values. Then, in order to make the power factor equal to 0.935,

$$\theta_d = \arccos 0.920 = 23.07°$$

$$\tan \theta_d = \tan 23.07° = 0.4260$$

$$Q_d = P \tan \theta_d = (2520\,\text{W})(0.4260) = 1073.5\,\text{VAR}$$

and the correction required will be

$$Q_\text{corr} = Q_d - Q = 1073.5 - 2200 = -1126.5\,\text{VAR} \Leftarrow$$

The minus sign indicates that the correction must be capacitive.

(f) With the correction earmarked as capacitive and $\omega = 2\pi(60\,\text{Hz}) = 377\,\text{rad/s}$

$$X_\text{corr} = \frac{V^2}{|Q_\text{corr}|} = \frac{(260\,\text{V})^2}{|1126.5\,\text{VAR}|} = 60.01\,\Omega$$

the value of a single capacitor will be

$$C = \frac{1}{\omega X_\text{corr}} = \frac{1}{(377\,\text{rad/s})(60.01\,\Omega)} = 44.20\,\mu\text{F} \Leftarrow$$

11.22: In the power distribution system shown in Fig 11.16, the frequency is 100 Hz. Determine:

(a) the real power,
(b) the reactive power,
(c) the magnitude of the apparent power,
(d) the power factor,
(e) the correction necessary to make
   the power factor 0.945 and
(f) the component necessary to achieve
   correction in part (e)

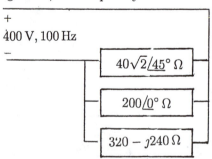

A convenient tabular form for making complex power calculations may be used.

Item	Load-1	Load-2	Load-3	Σ
Z, Ω	$40\sqrt{2}\underline{/45°}$	$200\underline{/0°}$	$320 - j240$	
$|Z|, \Omega$	$40\sqrt{2}$	200	400	
$I = V/|Z|, A$	$5\sqrt{2}$	2	1	
R, Ω	40	200	320	
X, Ω	40	0	-240	
$P = I^2 R, W$	2000	800	320	3120
$Q = I^2 X, VAR$	2000	0	-240	1760

Here, from the foregoing tabulation

$$\text{(a)} \qquad P = 3120\,\text{W} \Leftarrow$$

$$\text{(b)} \qquad Q = 1760\,\text{VAR} \Leftarrow$$

and

$$\text{(c)} \qquad \mathbf{S} = 3120 + j1760 \text{ VA} = 3582.2\underline{/29.43°}\text{ VA}$$

which makes the magnitude of the apparent power

$$|S| = 3582.2\,\text{VA} \Leftarrow$$

(d) The power factor is

$$PF = \cos 29.43° = 0.8710 \Leftarrow$$

(e) The subscript, d, may be used to indicate desired values. Then, in order to make the power factor equal to 0.945,

$$\theta_d = \arccos 0.945 = 19.09°$$

$$\tan \theta_d = \tan 19.09° = 0.3461$$

$$Q_d = P \tan \theta_d = (3120\,\text{W})(0.3461) = 1079.9\,\text{VAR}$$

and the correction required will be

$$Q_{\text{corr}} = Q_d - Q = 1079.9 - 1760 = -680.1\,\text{VAR} \Leftarrow$$

The minus sign indicates that the correction must be capacitive.

(f) With the correction earmarked as capacitive and $\omega = 2\pi(100\,\text{Hz}) = 628.3\,\text{rad/s}$

$$X_{\text{corr}} = \frac{V^2}{|Q_{\text{corr}}|} = \frac{(400\,\text{V})^2}{|680.1\,\text{VAR}|} = 235.4\,\Omega$$

the value of a single capacitor will be

$$C = \frac{1}{\omega X_{\text{corr}}} = \frac{1}{(628.3\,\text{rad/s})(235.4\,\Omega)} = 6.77\,\mu\text{F} \Leftarrow$$

MAXIMUM POWER TRANSFER

11.23: In the network of Fig 11.17, determine what elements should be placed across the terminals a-b to make the power factor unity.

With $\omega = 250$ rad/s

Component	Reactance, X
$C_1 = 20\,\mu\text{F}$	$200\,\Omega$
$C_1 = 40\,\mu\text{F}$	$100\,\Omega$
$L_1 = 400\,\text{mH}$	$100\,\Omega$
$L_2 = 800\,\text{mH}$	$200\,\Omega$

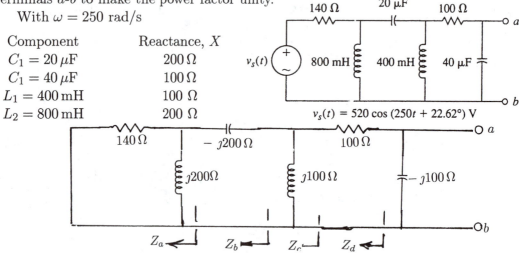

$v_s(t) = 520 \cos(250t + 22.62°)$ V

The phasor domain network with the voltage source replaced by a short circuit is shown above. In it,

$$Z_a = \frac{(140)(j200)}{140 + j200} = \frac{28,000\underline{/90°}}{244.13\underline{/55.01°}} = 114.69\underline{/34.99°}\,\Omega = 93.96 + j65.77\,\Omega$$

$$Z_b = 93.96 + j65.77 - j200 = 93.96 - j134.22\,\Omega = 163.85\underline{/-55.01°}\,\Omega$$

$$Z_c = \frac{(j100)(163.85\underline{/-55.01°})}{j100 + 93.96 - j134.23} = \frac{16,385\underline{/34.99°}}{100\underline{/-20.02°}} = 163.85\underline{/55.01°}\,\Omega = 93.96 + j134.23\,\Omega$$

$$Z_d = 93.96 + j134.23 + 100 = 193.96 + j134.23\,\Omega = 235.88\underline{/34.68°}$$

and the impedance looking into the terminals (the Thevenin equivalent impedance) is

$$Z_T = \frac{(-j100)(235.88\underline{/34.68°})}{193.96 + j134.23 - j100} := \frac{23,588\underline{/-55.32°}}{196.95\underline{/10.01°}}$$

or

$$Z_T = 119.76\underline{/-65.32°}\ \Omega = 50.00 - j108.22\ \Omega$$

In order for this network to operate at a power factor of unity, the load, designated as Z_o, should be set at the complex conjugate of the foregoing equivalent impedance

$$Z_o = 50 + j108.22\ \Omega \Leftarrow$$

11.24: In the phasor domain network of Fig 11.18, determine the load to be placed across terminals a-b to make the power drawn by the load a maximum and then determine the value of this maximum power.

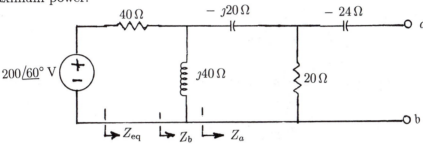

The problem requires the development of a Thevenin equivalent circuit. The phasor domain network shown above permits the evaluation of the Thevenin equivalent voltage. Here

$$Z_a = 20 - j20\ \Omega$$

$$Z_b = \frac{(20 - j20)(j40)}{20 - j20 + j40} = \frac{20\sqrt{2}\underline{/-45°}(40\underline{/90°})}{20\sqrt{2}\underline{/45°}} = 40\underline{/0°}\ \Omega = 40 + j0\ \Omega$$

$$Z_{eq} = 40 + 40 = 80\ \Omega$$

Then

$$\mathbf{I}_{40\,\Omega} = \frac{\mathbf{V}_s}{Z_{eq}} = \frac{200\underline{/60°}}{80\ \Omega} = \frac{5}{2}\underline{/60°}\ \text{A}$$

and

$$\mathbf{V}_{40\,\Omega} = (40\ \Omega)(\mathbf{I}_{40\,\Omega}) = (40\ \Omega)\left(\frac{5}{2}\underline{/60°}\ \text{A}\right) = 100\underline{/60°}\ \text{V}$$

By KVL,

$$
\begin{aligned}
\mathbf{V}_{j40\,\Omega} &= \mathbf{V}_s - \mathbf{V}_{40\,\Omega} \\
&= 200\underline{/60°} - 100\underline{/60°} \\
&= (100 + j173.21) - (50 + j86.60) \\
&= 50 + j86.60\ \text{V} = 100\underline{/60°}\ \text{V}
\end{aligned}
$$

and a voltage division yields

$$\mathbf{V}_{oc} = \mathbf{V}_T = \left(\frac{20}{20 - j20} \right) (100\underline{/60°}) = \frac{(20\underline{/0°})(100\underline{/60°})}{20\sqrt{2}\underline{/45°}} = 50\sqrt{2}\underline{/105°} \text{ V}$$

For Z_T, remove the voltage source and replace it with a short circuit. Then work looking into terminals a-b

$$Z_a = \frac{(40)(j40)}{40 + j40} = \frac{(40\underline{/0°})(40\underline{/90°}}{40\sqrt{2}\underline{/45°})} = 20\sqrt{2}\underline{/45°} \; \Omega = 20 + j20 \; \Omega$$

$$Z_b = 20 + j20 - j20 = 20 \; \Omega$$

$$Z_c = \frac{(20)(20)}{20 + 20} = \frac{400 \; \Omega^2}{40 \; \Omega} = 10 \; \Omega$$

and

$$Z_T = 10 - j24 \; \Omega$$

In order to make the load draw maximum power, it should be selected as the complex conjugate of the Thevenin equivalent impedance. With the load impedance designated as Z_o

$$Z_o = 10 + j24 \; \Omega \Leftarrow$$

The Thevenin equivalent network with the load connected across terminals a-b is shown at the right. The magnitude of the load current is

$$
\begin{aligned}
I &= \frac{V_s}{Z_T + Z_o} = \frac{50\sqrt{2}}{(10 - j24) + (10 + j24)} \\
&= \frac{50\sqrt{2} \, \text{V}}{20 \, \Omega} = \frac{5}{2}\sqrt{2} \, \text{A}
\end{aligned}
$$

and the power delivered to the load will be

$$P_o = (I)^2 R_o = \left(\frac{5}{2}\sqrt{2}\,\mathrm{A}\right)^2 (10\,\Omega) = 125\,\mathrm{W} \Leftarrow$$

11.25: In the phasor domain network of Fig 11.19, determine the load to be placed across terminals a-b to make the power drawn by the load a maximum and then determine the value of this maximum power.

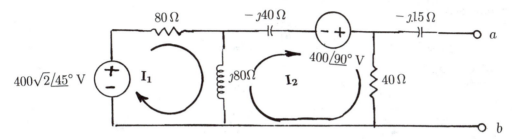

The problem requires the development of a Thevenin equivalent circuit. The phasor domain network shown above permits the evaluation of the Thevenin equivalent voltage. Here, with two meshes with clockwise mesh currents, the mesh equations may be written with two applications of KVL.

$$(80 + j80)\mathbf{I}_1 - j80\mathbf{I}_2 = 400 + j400\,\mathrm{V}$$
$$-j80\mathbf{I}_1 + (j80 - j40 + 40)\mathbf{I}_2 = 0 + j400\,\mathrm{V}$$

After division of these throughout by 40, these may be put into matrix form

$$\begin{bmatrix} (2 + j2) & -j2 \\ -j2 & (1 + j1) \end{bmatrix} \begin{bmatrix} \mathbf{I}_1 \\ \mathbf{I}_2 \end{bmatrix} = \begin{bmatrix} 10 + j10\,\mathrm{V} \\ 0 + j10\,\mathrm{V} \end{bmatrix}$$

and a matrix inversion can be employed to find the current vector

$$\begin{bmatrix} \mathbf{I}_1 \\ \mathbf{I}_2 \end{bmatrix} = \begin{bmatrix} (2 + j2) & -j2 \\ -j2 & (1 + j1) \end{bmatrix}^{-1} \begin{bmatrix} 10 + j10\,\mathrm{V} \\ 0 + j10\,\mathrm{V} \end{bmatrix}$$

or

$$\begin{bmatrix} \mathbf{I}_1 \\ \mathbf{I}_2 \end{bmatrix} = \begin{bmatrix} (0.025 + j0) & (0.25 + j0.25) \\ (0.25 + j0.25) & (0.50 + j0) \end{bmatrix} \begin{bmatrix} 10 + j10\,\mathrm{V} \\ 0 + j10\,\mathrm{V} \end{bmatrix} = \begin{bmatrix} 0 + j5\,\mathrm{A} \\ 0 + j10\,\mathrm{A} \end{bmatrix}$$

Then

$$\mathbf{V}_{\mathrm{oc}} = \mathbf{V}_T = 40\mathbf{I}_2 = (40\,\Omega)(0 + j10\,\mathrm{A}) = j400\,\mathrm{V} = 400\underline{/90^\circ}\,\mathrm{V}$$

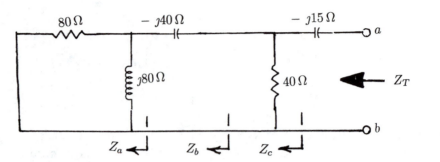

For Z_T, remove the voltage sources and replace them with short circuits. Then work looking into terminals a-b

$$Z_a = \frac{(80)(j80)}{80 + j80} = \frac{(80\underline{/0°})(80\underline{/90°})}{80\sqrt{2}\underline{/45°}} = 40\sqrt{2}\underline{/45°}\ \Omega = 40 + j40\ \Omega$$

$$Z_b = 40 + j40 - j40 = 40\ \Omega$$

$$Z_c = \frac{(40)(40)}{40 + 40} = \frac{1600\ \Omega^2}{80\ \Omega} = 20\ \Omega$$

and

$$Z_T = 20 - j15\ \Omega$$

In order to make the load draw maximum power, it should be selected as the complex conjugate of the Thevenin equivalent impedance. With the load impedance designated as Z_o

$$Z_o = 20 + j15\ \Omega \Leftarrow$$

The Thevenin equivalent network with the load connected across terminals a-b is shown at the right. the magnitude of the load current is

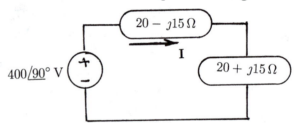

$$I = \frac{V_s}{Z_T + Z_o} = \frac{400}{(20 - j15) + (20 + j15)} = \frac{400\ \mathrm{V}}{40\ \Omega} = 10\ \mathrm{A}$$

and the power delivered to the load will be

$$P_o = (I)^2 R_o = (10\ \mathrm{A})^2 (20\ \Omega) = 2000\ \mathrm{W} \Leftarrow$$

11.26: In the phasor domain network of Fig 11.20, determine the load to be placed across terminals a-b to make the power drawn by the load a maximum and then determine the value of this maximum power.

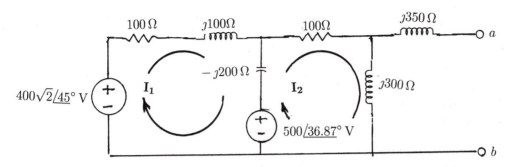

The problem requires the development of a Thevenin equivalent circuit. The phasor domain network shown above permits the evaluation of the Thevenin equivalent voltage. With the current source transformed to a voltage sourece, two meshes with clockwise mesh currents are observed. The mesh equations may be written with two applications of KVL.

$$(100 + j100 - j200)\mathbf{I}_1 - (-j200)\mathbf{I}_2 = (400 + j400) - (400 + j300)$$
$$-(-j200)\mathbf{I}_1 + (-j200 + 100 + j300)\mathbf{I}_2 = 400 + j300$$

After division of these throughout by 100, these may be put into matrix form

$$\begin{bmatrix} (1 - j1) & +j2 \\ +j2 & (1 + j1) \end{bmatrix} \begin{bmatrix} \mathbf{I}_1 \\ \mathbf{I}_2 \end{bmatrix} = \begin{bmatrix} j1\,\text{V} \\ 4 + j3\,\text{V} \end{bmatrix}$$

and a matrix inversion can be employed to find the current vector

$$\begin{bmatrix} \mathbf{I}_1 \\ \mathbf{I}_2 \end{bmatrix} = \begin{bmatrix} (1 - j1) & +j2 \\ +j2 & (1 + j1) \end{bmatrix}^{-1} \begin{bmatrix} j1\,\text{V} \\ 4 + j3\,\text{V} \end{bmatrix}$$

$$\begin{bmatrix} \mathbf{I}_1 \\ \mathbf{I}_2 \end{bmatrix} = \begin{bmatrix} (0.167 + j0.167) & (0.000 - j0.333) \\ (0.000 - j0.333) & (0.167 - j0.167) \end{bmatrix} \begin{bmatrix} j1\,\text{V} \\ 4 + j3\,\text{V} \end{bmatrix}$$

or

$$\begin{bmatrix} \mathbf{I}_1 \\ \mathbf{I}_2 \end{bmatrix} = \begin{bmatrix} 0.833 - j1.167\,\text{A} \\ 1.500 - j0.167\,\text{A} \end{bmatrix} = \begin{bmatrix} 1.434\underline{/-54.46°}\,\text{A} \\ 1.509\underline{/-6.34°}\,\text{A} \end{bmatrix}$$

Then, the Thevenin equivalent voltage will be

$$\mathbf{V}_{oc} = \mathbf{V}_T = j300\mathbf{I}_2 = (j300\,\Omega)(1.509\underline{/-6.34°}\,\text{A}) = 452.77\underline{/-83.66°}\,\text{V}$$

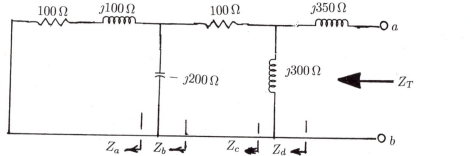

For Z_T, remove the voltage sources and replace them with short circuits. Then work looking into terminals a-b

$$Z_a = 100 + j100\,\Omega = 100\sqrt{2}\underline{/45°}\,\Omega$$

$$Z_b = \frac{(100 + j100)(-j200)}{100 + j100 - j200} = \frac{(100\sqrt{2}\underline{/45°})(200\underline{/-90°})}{100\sqrt{2}\underline{/-45°}} = 200\underline{/0°}\,\Omega = 200\,\Omega$$

$$Z_c = 200 + 100 = 300\,\Omega$$

$$Z_d = \frac{(300)(-j300)}{300 - j300} = \frac{90,000\underline{/-90°}}{300\sqrt{2}\underline{/-45°}} = 150\sqrt{2}\underline{/-45°}\,\Omega = 150 - j150\,\Omega$$

and

$$Z_T = 150 - j150 + j350 = 150 + j200\,\Omega$$

In order to make the load draw maximum power, it should be selected as the complex conjugate of the Thevenin equivalent impedance. With the load impedance designated as Z_o

$$Z_o = 150 - j200\,\Omega \Leftarrow$$

The Thevenin equivalent network with the load connected across terminals a-b is shown at the right. The magnitude of the load current is

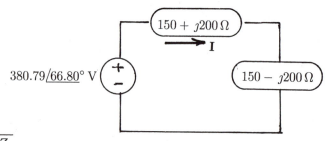

$$
\begin{aligned}
I &= \frac{V_T}{Z_T + Z_o} \\[2mm]
&= \frac{452.77}{(150 + j200) + (150 - j200)} \\[2mm]
&= \frac{452.79\text{ V}}{300\,\Omega} = 1.509\text{ A}
\end{aligned}
$$

and the power delivered to the load will be

$$P_o = (I)^2 R_o = (1.509\text{ A})^2 (150\,\Omega) = 341.6\text{ W} \Leftarrow$$

CHAPTER TWELVE

THREE PHASE POWER SYSTEMS

BALANCED SYSTEMS

12.1: In Fig 12.1, $\mathbf{V}_{AN} = \mathbf{V}_{NB} = 120\underline{/0°}$ V. If $Z_1 = Z_2 = Z_3 = 40\,\Omega$, determine the total power delivered to the system of loads.

Here, $Z = 40\underline{/0°} = 40 + \jmath 0\,\Omega$ and with $I = V/|Z|$

$$I_1 = \frac{120\,\text{V}}{40\,\Omega} = 3\,\text{A}$$

$$I_2 = \frac{120\,\text{V}}{40\,\Omega} = 3\,\text{A}$$

and

$$I_3 = \frac{240\,\text{V}}{40\,\Omega} = 6\,\text{A}$$

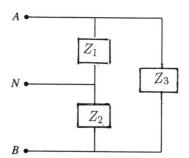

Then

$$
\begin{aligned}
P &= R[I_1^2 + I_2^2 + I_3^2] \\
&= (40\,\Omega)[(3\,\text{A})^2 + (3\,\text{A})^2 + (6\,\text{A})^2] \\
&= (40\,\Omega)(9\,\text{A}^2 + 9\,\text{A}^2 + 36\,\text{A}^2) \\
&= (40\,\Omega)(54\,\text{A}^2) = 2160\,\text{W} \Leftarrow
\end{aligned}
$$

12.2: In Fig 12.1, $\mathbf{V}_{AN} = \mathbf{V}_{NB} = 120\underline{/0^\circ}$ V. If $Z_1 = Z_2 = Z_3 = 30\underline{/16.26^\circ}\,\Omega$, determine the total power delivered to the system of loads.

Here, $Z = 30\underline{/16.26^\circ}\,\Omega = 28.80 + \jmath 8.40\,\Omega$ and with $I = V/|Z|$

$$
I_1 = \frac{120\,\text{V}}{30\,\Omega} = 4\,\text{A}
$$

$$
I_2 = \frac{120\,\text{V}}{30\,\Omega} = 4\,\text{A}
$$

and

$$
I_3 = \frac{240\,\text{V}}{30\,\Omega} = 8\,\text{A}
$$

Then

$$\begin{aligned} P &= R[I_1^2 + I_2^2 + I_3^2] \\ &= (28.80\,\Omega)[(4\,\text{A})^2 + (4\,\text{A})^2 + (8\,\text{A})^2] \\ &= (28.80\,\Omega)(16\,\text{A}^2 + 16\,\text{A}^2 + 64\,\text{A}^2) \\ &= (28.80\,\Omega)(96\,\text{A}^2) = 2764.8\,\text{W} \Leftarrow \end{aligned}$$

12.3: For the balanced three phase generator connected as shown in Fig 12.2, the phase voltages have an rms amplitude of 440 V and are connected in the positive sequence. Determine the currents, $\mathbf{I}_a, \mathbf{I}_b$ and $\mathbf{I}_c$ if the load is balanced with all $Z = 44\underline{/0°}\,\Omega$.

The generator is connected in wye and because the load is balanced, no current will flow in a neutral wire which does not need to be shown. The load arrangement is shown in the figure and for the positive sequence

$$\begin{aligned} \mathbf{V}_{an} &= V\underline{/0°}\text{V} \\ \mathbf{V}_{bn} &= V\underline{/-120°}\text{V} \end{aligned}$$

and

$$\mathbf{V}_{cn} = V\underline{/120°}\text{V}$$

Hence

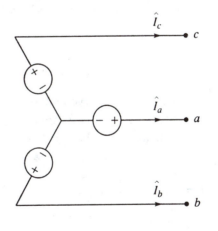

$$\begin{aligned} \mathbf{I}_1 &= \frac{\mathbf{V}_{cn} - \mathbf{V}_{an}}{44} \\[2mm] &= \frac{440\underline{/120°} - 440\underline{/0°}}{44} \\[2mm] &= \frac{-220 + j381.05 - 440}{44} \\[2mm] &= \frac{-660 + j381.05}{44} \\[2mm] &= \frac{762.08\underline{/150°}\ \text{V}}{44\underline{/0°}\ \Omega} \\[2mm] &= 17.32\underline{/150°}\ \text{A} = -15 + j8.66\ \text{A} \end{aligned}$$

and, in similar fashion

$$\mathbf{I}_2 = \frac{\mathbf{V}_{an} - \mathbf{V}_{bn}}{44} = \frac{440\underline{/0°} - 440\underline{/-120°}}{44} = \frac{440 - (-220 - j381.05)}{44}$$

or

$$\mathbf{I}_2 = \frac{660 + j381.05\,\text{V}}{44\,\Omega} = 15 + j8.66\,\text{A} = 17.32\underline{/30°}\,\text{A}$$

and

$$\mathbf{I}_3 = \frac{\mathbf{V}_{cn} - \mathbf{V}_{bn}}{44\,\Omega} = \frac{440\underline{/120°} - 440\underline{/-120°}}{44} = \frac{(-220 + j381.05) - (-220 - j381.05)}{44}$$

or

$$\mathbf{I}_3 = \frac{j762.10\,\text{V}}{44\,\Omega} = 0 + j17.32 = 17.32\underline{/90°}\,\text{A}$$

Then, by KCL

$$\mathbf{I}_a = \mathbf{I}_2 - \mathbf{I}_1 = (15 + j8.66) - (-15 + j8.66) = 30 + j0\,\text{A} = 30\underline{/0°}\,\text{A}$$

$$\mathbf{I}_b = -(\mathbf{I}_2 + \mathbf{I}_3) = -[(15 + j8.66) + j17.32] = -15 - j25.98\,\text{A} = 30\underline{/-120°}\,\text{A}$$

and

$$\mathbf{I}_c = \mathbf{I}_1 + \mathbf{I}_3 = -15 + j8.66 + j17.32 = -15 + j25.98\,\text{A} = 30\underline{/120°}\,\text{A}$$

In summary

$$\mathbf{I}_a = 30 + j0\,\text{A} = 30\underline{/0°} \Leftarrow$$
$$\mathbf{I}_b = -15 - j25.98\,\text{A} = 30\underline{/-120°}\,\text{A} \Leftarrow$$

and

$$\mathbf{I}_c = -15 + j25.98\,\text{A} = 30\underline{/120°}\,\text{A} \Leftarrow$$

12.4: The line voltages in the wye connected generator that supplies the loads shown in Fig 12.3 have a magnititude of 208 V and are in the positive sequence at 60 Hz. Determine all phase and line voltages in instantaneous form.

Here,

$$\omega = 2\pi f = 2\pi(60\,\text{Hz} = 377\,\text{rad/s}$$

and for the wye connection with

$$V_\ell = 208\,\text{V}$$

the phase voltage is

$$V_p = \frac{V_\ell}{\sqrt{3}} = \frac{208\,\text{V}}{\sqrt{3}} = 120.09\,\text{V}$$

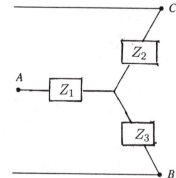

526

For the positive sequence, the phase voltages are

$$\mathbf{V}_{an} = 120.09\underline{/0^\circ} \longrightarrow v_{an}(t) = 120.09\cos(377t + 0^\circ)\,\text{V} \Leftarrow$$

$$\mathbf{V}_{bn} = 120.09\underline{/-120^\circ} \longrightarrow v_{bn}(t) = 120.09\cos(377t - 120^\circ)\,\text{V} \Leftarrow$$

and

$$\mathbf{V}_{cn} = 120.09\underline{/120^\circ} \longrightarrow v_{cn}(t) = 120.09\cos(377t + 120^\circ)\,\text{V} \Leftarrow$$

and the line voltages are

$$\mathbf{V}_{ab} = 208\underline{/30^\circ} \longrightarrow v_{ab}(t) = 208\cos(377t + 30^\circ)\,\text{V} \Leftarrow$$

$$\mathbf{V}_{bc} = 208\underline{/-90^\circ} \longrightarrow v_{bc}(t) = 208\cos(377t - 90^\circ)\,\text{V} \Leftarrow$$

and

$$\mathbf{V}_{ca} = 208\underline{/150^\circ} \longrightarrow v_{ca}(t) = 208\cos(377t + 150^\circ)\,\text{V} \Leftarrow$$

12.5: The line voltages in the wye connected generator that supplies the loads shown in Problem 12.4 (Fig 12.3) have a magnitude of 208 V and are in the *negative sequence* at 60 Hz. Determine all phase and line voltages in instantaneous form.

The only difference between this problem which involves the negative sequence and Problem 12.4 which involves the positive sequence is in the angles of the phase and line voltages. Here, for the negative sequence, the phase voltages are

$$\mathbf{V}_{an} = 120.09\underline{/0^\circ} \longrightarrow v_{an}(t) = 120.09\cos(377t + 0^\circ)\,\text{V} \Leftarrow$$

$$\mathbf{V}_{bn} = 120.09\underline{/120^\circ} \longrightarrow v_{bn}(t) = 120.09\cos(377t + 120^\circ)\,\text{V} \Leftarrow$$

and

$$\mathbf{V}_{cn} = 120.09\underline{/-120^\circ} \longrightarrow v_{cn}(t) = 120.09\cos(377t - 120^\circ)\,\text{V} \Leftarrow$$

and the line voltages are

$$\mathbf{V}_{ab} = 208\underline{/-30^\circ} \longrightarrow v_{ab}(t) = 208\cos(377t - 30^\circ)\,\text{V} \Leftarrow$$

$$\mathbf{V}_{bc} = 208\underline{/90^\circ} \longrightarrow v_{bc}(t) = 208\cos(377t + 90^\circ)\,\text{V} \Leftarrow$$

and

$$\mathbf{V}_{ca} = 208\underline{/-150^\circ} \longrightarrow v_{ca}(t) = 208\cos(377t - 150^\circ)\,\text{V} \Leftarrow$$

12.6: The line voltages in the delta connected generator that supplies the loads shown in Fig 12.4 have a magnitude of 208 V and are in the positive sequence at 60 Hz. Determine all phase and line voltages in instantaneous form.

Here,

$$\omega = 2\pi f = 2\pi(60\,\text{Hz} = 377\ \text{rad/s}$$

and for the delta connection with

$$V_\ell = 208\ \text{V}$$

the line and phase voltages are equal

$$V_p = V_\ell = 208\ \text{V}$$

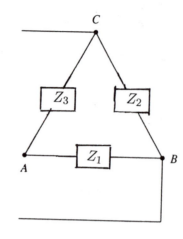

For the positive sequence,

$$\mathbf{V}_{ab} = 208\underline{/0^\circ} \longrightarrow v_{ab}(t) = 208\cos(377t + 0^\circ)\ \text{V} \Longleftarrow$$

$$\mathbf{V}_{bc} = 208\underline{/-120^\circ} \longrightarrow v_{bc}(t) = 208\cos(377t - 120^\circ)\ \text{V} \Longleftarrow$$

and

$$\mathbf{V}_{ca} = 208\underline{/120^\circ} \longrightarrow v_{ca}(t) = 208\cos(377t + 120^\circ)\ \text{V} \Longleftarrow$$

12.7: The line voltages in the delta connected generator that supplies the loads shown in Problem 12.6 (Fig 12.4) have a magnititude of 208 V and are in the *negative sequence* at 60 Hz. Determine all phase and line voltages in instantaneous form.

The only difference between this problem which involves the negative sequence and Problem 12.6 which involves the positive sequence is in the angles of the phase and line voltages. Here, for the negative sequence,

$$\mathbf{V}_{ab} = 208\underline{/0^\circ} \longrightarrow v_{ab}(t) = 208\cos(377t + 0^\circ)\ \text{V} \Longleftarrow$$

$$\mathbf{V}_{bc} = 208\underline{/120^\circ} \longrightarrow v_{bc}(t) = 208\cos(377t + 120^\circ)\ \text{V} \Longleftarrow$$

and

$$\mathbf{V}_{ca} = 208\underline{/-120^\circ} \longrightarrow v_{ca}(t) = 208\cos(377t - 120^\circ)\ \text{V} \Longleftarrow$$

12.8: The line-to-line voltages in Fig 12.5 have a magnititude of 440 V and are in the positive sequence at 60 Hz. The loads are balanced with $Z = Z_1 = Z_2 = Z_3 = 25\underline{/36.87^\circ}\ \Omega$. Determine all phase and line and phase voltages and load currents in instantaneous form.

Here,

$$\omega = 2\pi f = 2\pi(60\,\text{Hz}) = 377\ \text{rad/s}$$

and for the wye connection with

$V_\ell = 440\,\text{V}$

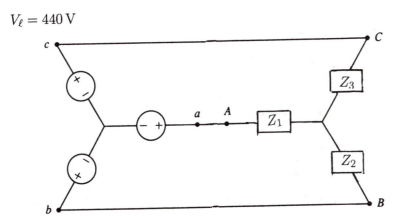

the phase voltage is

$$V_p = \frac{V_\ell}{\sqrt{3}} = \frac{440\,\text{V}}{\sqrt{3}} = 254\,\text{V}$$

For the positive sequence, the phase voltages are

$$\mathbf{V}_{AN} = \mathbf{V}_{an} = 254\underline{/0^\circ} \longrightarrow v_{an}(t) = 254\cos(377t + 0^\circ)\,\text{V} \Longleftarrow$$

$$\mathbf{V}_{BN} = \mathbf{V}_{bn} = 254\underline{/-120^\circ} \longrightarrow v_{bn}(t) = 254\cos(377t - 120^\circ)\,\text{V} \Longleftarrow$$

and

$$\mathbf{V}_{CN} = \mathbf{V}_{cn} = 254\underline{/120^\circ} \longrightarrow v_{cn}(t) = 254\cos(377t + 120^\circ)\,\text{V} \Longleftarrow$$

and the line voltages are

$$\mathbf{V}_{AB} = \mathbf{V}_{ab} = 440\underline{/30^\circ} \longrightarrow v_{ab}(t) = 440\cos(377t + 30^\circ)\,\text{V} \Longleftarrow$$

$$\mathbf{V}_{BC} = \mathbf{V}_{bc} = 440\underline{/-90^\circ} \longrightarrow v_{bc}(t) = 440\cos(377t - 90^\circ)\,\text{V} \Longleftarrow$$

and

$$\mathbf{V}_{CA} = \mathbf{V}_{ca} = 440\underline{/150^\circ} \longrightarrow v_{ca}(t) = 440\cos(377t + 150^\circ)\,\text{V} \Longleftarrow$$

The currents are

$$\mathbf{I}_{AN} = \frac{\mathbf{V}_{AN}}{Z} = \frac{254\underline{/0^\circ}\,\text{V}}{25\underline{/36.87^\circ}\,\Omega} = 10.16\underline{/-36.87^\circ}\,\text{A}$$

and

$$i_{AN}(t) = 10.16\cos(377t - 36.87^\circ)\,\text{A} \Longleftarrow$$

$$\mathbf{I}_{BN} = \frac{\mathbf{V}_{BN}}{Z} = \frac{254\underline{/-120°}\,\text{V}}{25\underline{/36.87°}\,\Omega} = 10.16\underline{/-156.87°}\,\text{A}$$

and

$$i_{BN}(t) = 10.16\cos(377t - 156.87°)\,\text{A} \Leftarrow$$

and

$$\mathbf{I}_{CN} = \frac{\mathbf{V}_{CN}}{Z} = \frac{254\underline{/120°}\,\text{V}}{25\underline{/36.87°}\,\Omega} = 10.16\underline{/83.13°}\,\text{A}$$

and

$$i_{CN}(t) = 10.16\cos(377t + 83.13°)\,\text{A} \Leftarrow$$

12.9 In Problem 12.8, determine the power drawn by the load.

In Problem 12.8,

$$V_p = 254\,\text{V}$$

$$I_p = 10.16\,\text{A}$$

and

$$\theta = 36.87°$$

Thus

$$
\begin{aligned}
P &= 3V_pI_p\cos\theta \\
&= 3(254\,\text{V})(10.16\,\text{A})\cos 36.87° \\
&= (7741.9)(0.8000) = 6193.5\,\text{W} \Leftarrow
\end{aligned}
$$

12.10: In Fig 12.5, the rms amplitude of the line-to-line voltages in the three phase source is 240 V. The source is connected in the positive sequence and operates at 60 Hz. If the balanced load consists of $Z = Z_1 = Z_2 = Z_3 = 30\underline{/36.87°}\Omega$, determine the phase voltages and currents and the power delivered to the load.

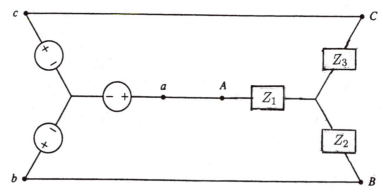

Here,

$$V_\ell = 240\,\text{V} \qquad \text{and} \qquad V_p = \frac{V_\ell}{\sqrt{3}} = 138.6\,\text{V}$$

and the phase voltages are

$$\mathbf{V}_{AN} = \mathbf{V}_{an} = 138.6\underline{/0°}\,\text{V} \Leftarrow$$
$$\mathbf{V}_{BN} = \mathbf{V}_{bn} = 138.6\underline{/-120°}\,\text{V} \Leftarrow$$

and

$$\mathbf{V}_{CN} = \mathbf{V}_{cn} = 138.6\underline{/120°}\,\text{V} \Leftarrow$$

The phase currents are

$$\mathbf{I}_{AN} = \frac{\mathbf{V}_{AN}}{Z} = \frac{138.6\underline{/0°}\,\text{V}}{30\underline{/36.87°}\,\Omega} = 4.62\underline{/-36.87°}\,\text{A} \Leftarrow$$

$$\mathbf{I}_{BN} = \frac{\mathbf{V}_{BN}}{Z} = \frac{138.6\underline{/-120°}\,\text{V}}{30\underline{/36.87°}\,\Omega} = 4.62\underline{/-156.87°}\,\text{A} \Leftarrow$$

and

$$\mathbf{I}_{CN} = \frac{\mathbf{V}_{CN}}{Z} = \frac{138.6\underline{/120°}\,\text{V}}{30\underline{/36.87°}\,\Omega} = 4.62\underline{/83.13°}\,\text{A} \Leftarrow$$

With

$$V_p = 138.6\text{V}$$

$$I_p = 4.62\,\text{A}$$

and

$$\theta = 36.87°$$

Thus

$$\begin{aligned}
P &= 3V_p I_p \cos\theta \\
&= 3(138.6\,\text{V})(4.62\,\text{A})\cos 36.87° \\
&= (1921.0)(0.8000) = 1536.8\,\text{W} \Leftarrow
\end{aligned}$$

12.11: In Fig 12.5, the rms amplitude of the line-to-line voltages in the three phase source is 240 V. The source is connected in the positive sequence and operates at 60 Hz. If the balanced load consists of $Z = Z_1 = Z_2 = Z_3 = 40\underline{/53.13°}\Omega$, determine the phase voltages and currents and the power delivered to the load.

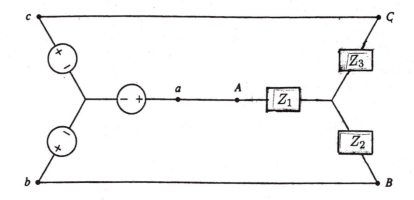

Here,

$$V_\ell = 240\,\text{V} \qquad \text{and} \qquad V_p = \frac{V_\ell}{\sqrt{3}} = 138.6\,\text{V}$$

and the phase voltages are

$$\mathbf{V}_{AN} = \mathbf{V}_{an} = 138.6\underline{/0°}\,\text{V} \Leftarrow$$
$$\mathbf{V}_{BN} = \mathbf{V}_{bn} = 138.6\underline{/-120°}\,\text{V} \Leftarrow$$

and

$$\mathbf{V}_{CN} = \mathbf{V}_{cn} = 138.6\underline{/120°}\,\text{V} \Leftarrow$$

The phase currents are

$$\mathbf{I}_{AN} = \frac{\mathbf{V}_{AN}}{Z} = \frac{138.6\underline{/0°}\,\text{V}}{40\underline{/53.13°}\,\Omega} = 3.47\underline{/-53.13°}\,\text{A} \Leftarrow$$

$$\mathbf{I}_{BN} = \frac{\mathbf{V}_{BN}}{Z} = \frac{138.6\underline{/-120°}\,\text{V}}{40\underline{/53.13°}\,\Omega} = 3.47\underline{/-173.13°}\,\text{A} \Leftarrow$$

and

$$\mathbf{I}_{CN} = \frac{\mathbf{V}_{CN}}{Z} = \frac{138.6\underline{/120°}\,\text{V}}{40\underline{/53.13°}\,\Omega} = 3.47\underline{/66.87°}\,\text{A} \Leftarrow$$

With

$$V_p = 138.6\,\text{V}$$

$$I_p = 3.47\,\text{A}$$

and

$$\theta = 53.13°$$

Thus

$$
\begin{aligned}
P &= 3V_p I_p \cos\theta \\
&= 3(138.6\,\text{V})(3.47\,\text{A})\cos 53.13° \\
&= (1442.6)(0.6000) = 865.7\,\text{W} \Leftarrow
\end{aligned}
$$

12.12: In Fig 12.6, the rms amplitude of the line-to-line voltages in the three phase source is 240 V. The source is connected in the positive sequence and operates at 60 Hz. If the balanced load consists of $Z = Z_1 = Z_2 = Z_3 = 30\underline{/36.87°}\,\Omega$, determine the line and phase voltages, the line currents and the power delivered to the load.

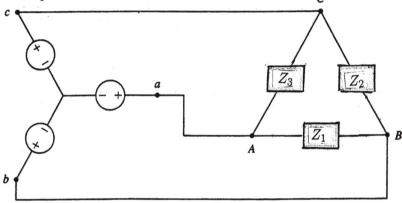

Here,

$$
V_\ell = 240\,\text{V} \qquad \text{and} \qquad V_p = \frac{V_\ell}{\sqrt{3}} = 138.6\,\text{V}
$$

The phase voltages are

$$
\begin{aligned}
\mathbf{V}_{an} &= 138.6\underline{/0°}\,\text{V} \Leftarrow \\
\mathbf{V}_{bn} &= 138.6\underline{/-120°}\,\text{V} \Leftarrow
\end{aligned}
$$

and

$$
\mathbf{V}_{vn} = 138.6\underline{/120°}\,\text{V} \Leftarrow
$$

The line voltages are

$$
\begin{aligned}
\mathbf{V}_{AB} = \mathbf{V}_{ab} &= 240\underline{/30°}\,\text{V} \Leftarrow \\
\mathbf{V}_{BC} = \mathbf{V}_{bc} &= 240\underline{/-90°}\,\text{V} \Leftarrow
\end{aligned}
$$

and

$$\mathbf{V}_{CA} = \mathbf{V}_{ca} = 240\underline{/150^\circ}\,\text{V} \Leftarrow$$

The phase currents are

$$\mathbf{I}_{AB} = \frac{\mathbf{V}_{AB}}{Z} = \frac{240\underline{/30^\circ}\,\text{V}}{30\underline{/36.87^\circ}\,\Omega} = 8\underline{/-6.87^\circ}\,\text{A} \Leftarrow$$

$$\mathbf{I}_{BC} = \frac{\mathbf{V}_{BC}}{Z} = \frac{240\underline{/-90^\circ}\,\text{V}}{30\underline{/36.87^\circ}\,\Omega} = 8\underline{/-126.87^\circ}\,\text{A} \Leftarrow$$

and

$$\mathbf{I}_{CA} = \frac{\mathbf{V}_{CA}}{Z} = \frac{240\underline{/150^\circ}\,\text{V}}{30\underline{/36.87^\circ}\,\Omega} = 8\underline{/113.13^\circ}\,\text{A} \Leftarrow$$

With

$$V_\ell = 240\text{V}$$

$$I_\ell = \sqrt{3}I_p = 8\sqrt{3}\,\text{A}$$

and

$$\theta = 36.87^\circ$$

Thus

$$\begin{aligned}
P &= \sqrt{3}V_\ell I_\ell \cos\theta \\
&= \sqrt{3}(240\,\text{V})(8\sqrt{3}\,\text{A})\cos 36.87^\circ \\
&= (5760)(0.8000) = 4608\,\text{W} \Leftarrow
\end{aligned}$$

12.13: In Problem 12.12 (Fig 12.6), determine the phase currents and the power delivered to the balanced load if $Z = Z_1 = Z_2 = Z_3 = 50\underline{/16.26^\circ}\Omega$.

The line and phase voltages were developed in Problem 12.12. With

$$\mathbf{V}_{AB} = 240\underline{/30^\circ}\,\text{V} \Leftarrow$$
$$\mathbf{V}_{BC} = 240\underline{/-90^\circ}\,\text{V} \Leftarrow$$

and

$$\mathbf{V}_{CA} = 340\underline{/150^\circ}\,\text{V} \Leftarrow$$

the phase currents are

$$\mathbf{I}_{AB} = \frac{\mathbf{V}_{AB}}{Z} = \frac{240\underline{/30^\circ}\,\text{V}}{50\underline{/16.26^\circ}\,\Omega} = 4.80\underline{/13.74^\circ}\,\text{A} \Leftarrow$$

$$\mathbf{I}_{BC} = \frac{\mathbf{V}_{BC}}{Z} = \frac{240\underline{/-90°}\ \text{V}}{50\underline{/16.26°}\ \Omega} = 4.80\underline{/-106.26°}\ \text{A} \Leftarrow$$

and

$$\mathbf{I}_{CA} = \frac{\mathbf{V}_{CA}}{Z} = \frac{240\underline{/150°}\ \text{V}}{50\underline{/16.26°}\ \Omega} = 4.80\underline{/133.74°}\ \text{A} \Leftarrow$$

With

$$V_p = 240\text{V}$$

$$I_p = 4.80\,\text{A}$$

and

$$\theta = 16.26°$$

The power delivered to the load is

$$\begin{aligned}
P &= 3V_pI_p\cos\theta \\
&= 3(240\,\text{V})(4.80\,\text{A})\cos 16.26° \\
&= (3456)(0.9600) = 3317.8\,\text{W} \Leftarrow
\end{aligned}$$

12.14: In Fig 12.7, the rms amplitude of the line-to-line voltages in the three phase source is 240 V. The source is connected in the positive sequence and operates at 60 Hz. If the balanced load consists of $Z = Z_1 = Z_2 = Z_3 = 30\underline{/36.87°}\Omega$, determine the line voltages and currents and the power delivered to the load.

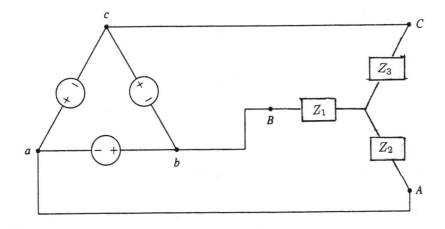

12.15: Rework Problem 12.14 (Fig 12.7) to determine the power delivered to the load with the balanced load consisting of $Z = Z_1 = Z_2 = Z_3 = 40\underline{/53.13°}\ \Omega$.

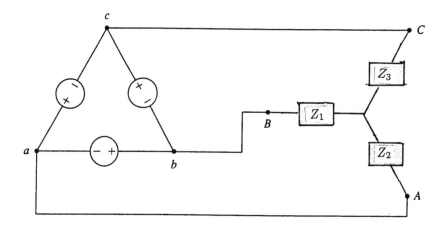

The line and phase voltages are identical to those in Problem 12.14. With $V_\ell = 240\,\text{V}$ the line currents are

$$I_p = \frac{240}{40} = 6\,\text{A}$$

With

$$\theta = 53.13°$$

$$
\begin{aligned}
P &= \sqrt{3}V_\ell I_\ell \cos\theta \\
  &= \sqrt{3}(240\,\text{V})(6\,\text{A})\cos 53.13° \\
  &= (2494.2)(0.6000) = 1496.5\,\text{W} \Leftarrow
\end{aligned}
$$

12.16: In Fig 12.8, the rms amplitude of the line-to-line voltages in the three phase source is 240 V. The source is connected in the positive sequence and operates at 60 Hz. If the balanced load consists of $Z = Z_1 = Z_2 = Z_3 = 30\underline{/36.87°}\ \Omega$, determine the phase voltages and currents and the power delivered to the load.

With $\mathbf{V}_{AB}$ chosen as the reference, the line voltages are

$$
\begin{aligned}
\mathbf{V}_{AB} = \mathbf{V}_{ab} &= 240\underline{/0°}\,\text{V} \Leftarrow \\
\mathbf{V}_{BC} = \mathbf{V}_{bc} &= 240\underline{/-120°}\,\text{V} \Leftarrow
\end{aligned}
$$

and

$$
\mathbf{V}_{CA} = \mathbf{V}_{ca} = 240\underline{/120°}\,\text{V} \Leftarrow
$$

and

$$
\mathbf{I}_{AB} = \frac{\mathbf{V}_{AB}}{Z} = \frac{240\underline{/0°}\,\text{V}}{30\underline{/36.87°}\,\Omega} = 8\underline{/-36.87°}\,\text{A} \Leftarrow
$$

$$
\mathbf{I}_{BC} = \frac{\mathbf{V}_{BN}}{Z} = \frac{240\underline{/-120°}\,\text{V}}{30\underline{/36.87°}\,\Omega} = 8\underline{/-156.87°}\,\text{A} \Leftarrow
$$

and

$$
\mathbf{I}_{CA} = \frac{\mathbf{V}_{CN}}{Z} = \frac{240\underline{/120°}\,\text{V}}{30\underline{/36.87°}\,\Omega} = 8\underline{/83.13°}\,\text{A} \Leftarrow
$$

With

$$
V_p = 240\,\text{V}
$$

$$
I_p = 8\,\text{A}
$$

and

$$
\theta = 36.87°
$$

Thus

$$
\begin{aligned}
P &= 3 V_p I_p \cos\theta \\
&= 3(240\,\text{V})(8\,\text{A})\cos 36.87° \\
&= (5760)(0.8000) = 4608\,\text{W} \Leftarrow
\end{aligned}
$$

12.17: Rework Problem 12.16 (Fig 12.8) to determine the power delivered to the load with the balanced load consisting of $Z = Z_1 = Z_2 = Z_3 = 50\underline{/16.26°}$.

The line voltages are identical to those in Problem 12.14. With $V_\ell = 240$ V and the phase currents are

$$I_p = \frac{V_\ell}{|Z|} = \frac{240\,\text{V}}{50\,\Omega} = 4.8\,\text{A}$$

With

$$\theta = 16.26°$$

$$
\begin{aligned}
P &= 3V_pI_p\cos\theta \\
&= 3(240\,\text{V})(4.80\,\text{A})\cos 16.26° \\
&= (3456)(0.9600) = 3317.8\,\text{W} \Leftarrow
\end{aligned}
$$

UNBALANCED SYSTEMS

12.18: In Fig 12.1, $\mathbf{V}_{AN} = \mathbf{V}_{NB} = 120\underline{/0°}$ V. If $Z_1 = 8 + j6\,\Omega$, $Z_2 = 8 - j6\,\Omega$ and $Z_3 = 4 - j20\,\Omega$, determine the total power delivered to the system of loads.

Here, with $I = V/|Z|$

$$I_1 = \frac{120\,\text{V}}{10\,\Omega} = 12\,\text{A}$$

$$I_2 = \frac{120\,\text{V}}{10\,\Omega} = 12\,\text{A}$$

and with

$$Z_3 = 4 - j20\,\Omega = 20.40\underline{/-78.69°}$$

then

$$I_3 = \frac{240\,\text{V}}{20.40\,\Omega} = 11.76\,\text{A}$$

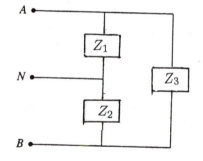

Then

$$
\begin{aligned}
P &= I_1^2 R_1 + I_2^2 R_2 + I_3^2 R_3 \\
&= (12\,\text{A})^2(8\,\Omega) + (12\,\text{A})^2(8\,\Omega) + (11.76\,\text{A})^2(4\,\Omega) \\
&= (144\,\text{A})^2(8\,\Omega) + (144\,\text{A})^2(8\,\Omega) + (138.3\,\text{A})^2(4\,\Omega) \\
&= 1152\,\text{W} + 1152\,\text{W} + 553.2\,\text{W} \\
&= 2857.2\,\text{W} \Leftarrow
\end{aligned}
$$

12.19: In Fig 12.1, $\mathbf{V}_{AN} = \mathbf{V}_{NB} = 120\underline{/0°}$ V. If $Z_1 = 20\underline{/30°}$, $Z_2 = 30\underline{/36.87°}$ and $Z_3 = 30\underline{/45°}\,\Omega$, determine the total power delivered to the system of loads.

Here, with $I = V/|Z|$

$$I_1 = \frac{120\,\text{V}}{20\,\Omega} = 6\,\text{A}$$

$$I_2 = \frac{120\,\text{V}}{30\,\Omega} = 4\,\text{A}$$

and

$$I_3 = \frac{240\,\text{V}}{30\,\Omega} = 4\,\text{A}$$

Then

$$
\begin{aligned}
P &= I_1^2 R_1 + I_2^2 R_2 + I_3^2 R_3 \\
&= (6\,\text{A})^2(17.32\,\Omega) + (4\,\text{A})^2(24\,\Omega) + (4\,\text{A})^2(15\sqrt{2}\,\Omega) \\
&= 623.5\,\text{W} + 384\,\text{W} + 339.4\,\text{W} \\
&= 1346.9\,\text{W} \Leftarrow
\end{aligned}
$$

12.20: In the system of Fig 12.9, the line voltages are 240 V and are in the positive sequence. Determine $\mathbf{I}_b$.

Here

$$
\begin{aligned}
\mathbf{V}_a &= 240\underline{/0°}\,\text{V.} = 240 + j0\,\text{V} \\
\mathbf{V}_b &= 240\underline{/-120°}\,\text{V} = -120 - j207.85\,\text{V}
\end{aligned}
$$

and

$$\mathbf{V}_c = 240\underline{/120°}\,\text{V} = -120 + j207.85\,\text{V}$$

Thus

$$\mathbf{I}_a = \frac{\mathbf{V}_a - \mathbf{V}_b}{30\underline{/0°}} = \frac{240 - (-120 - j207.85)}{30\underline{/0°}} = \frac{415.69\underline{/30°}}{30\underline{/0°}}$$

or

$$\mathbf{I}_a = 13.86\underline{/30°} = 12.00 + j6.93\,\text{A}$$

and

$$\mathbf{I}_c = \frac{\mathbf{V}_c - \mathbf{V}_b}{40\underline{/60°}} = \frac{(-120 + j207.85) - (-120 - j207.85)}{40\underline{/60°}} = \frac{415.70\underline{/90°}}{40\underline{/60°}}$$

or

$$\mathbf{I}_a = 10.39\underline{/30°} = 9.00 + j5.20° \text{ A}$$

By KCL

$$
\begin{aligned}
\mathbf{I}_b &= -(\mathbf{I}_a + \mathbf{I}_c) \\
&= -[(12.00 + j6.93) + (9.00 + j5.20)] \\
&= -21.00 - j12.13 \text{ A} = 24.25\underline{/-150°} \text{ A} \Leftarrow
\end{aligned}
$$

12.21: In Fig 12.10, determine the line currents. $\mathbf{I}_1, \mathbf{I}_2$ and $\mathbf{I}_3$ if the line-to-line voltages all have a magnitude of 300 V and are connected in the positive sequence.

Here

$$\mathbf{I}_1 = \frac{300\underline{/0°} \text{ V}}{20\underline{/36.87°} \text{ Ω}} = 15\underline{/-36.87°} \text{ A} = 12 - j9 \text{ A} \Leftarrow$$

$$\mathbf{I}_2 = \frac{300\underline{/-120°} \text{ V}}{30\underline{/53.13°} \text{ Ω}} = 10\underline{/-173.13°} \text{ A} = -9.93 - j1.20 \text{ A} \Leftarrow$$

and

$$\mathbf{I}_3 = \frac{300\underline{/120°} \text{ V}}{40\underline{/-45°} \text{ Ω}} = 7.5\underline{/165°} \text{ A} = -7.24 + j1.94 \text{ A} \Leftarrow$$

Then, by KCL

$$
\begin{aligned}
\mathbf{I}_n &= -(\mathbf{I}_1 + \mathbf{I}_2 + \mathbf{I}_3) \\
&= -[(12 - j9) + (-9.93 - j1.20) + (-7.24 + j1.94)] \\
&= 5.17 + j8.26 = 9.74\underline{/57.96°} \text{ A} \Leftarrow
\end{aligned}
$$

12.22: In Fig 12.3, determine the line currents. $\mathbf{I}_1, \mathbf{I}_2$ and $\mathbf{I}_3$ if the line-to-line voltages all have a magnitude of 300 V and are connected in the *negative sequence*.

Here

$$\mathbf{I}_1 = \frac{300\underline{/0°} \text{ V}}{20\underline{/36.87°} \text{ Ω}} = 15\underline{/-36.87°} = 12 - j9 \text{ A} \Leftarrow$$

$$\mathbf{I}_2 = \frac{300\underline{/120°} \text{ V}}{30\underline{/53.13°} \text{ Ω}} = 10\underline{/66.87°} = 3.93 + j9.20 \text{ A} \Leftarrow$$

and

$$\mathbf{I}_3 = \frac{300\underline{/-120°} \text{ V}}{40\underline{/-45°} \text{ Ω}} = 7.5\underline{/-75°} = 1.94 - j7.24 \text{ A} \Leftarrow$$

Then, by KCL

$$\begin{aligned} \mathbf{I}_n &= -(\mathbf{I}_1 + \mathbf{I}_2 + \mathbf{I}_3) \\ &= -[(12 - \jmath 9) + (3.93 + \jmath 9.20) + (1.94 - \jmath 7.94)] \\ &= -17.87 + \jmath 7.05 = 19.21\underline{/158.47^\circ}\,\text{A} \Leftarrow \end{aligned}$$

12.23: In Fig 12.11, the rms amplitude of the line-to-line voltages in the three phase source is 440 V. The source is connected in the positive sequence and operates at 60 Hz. If the unbalanced load consists of $Z_1 = 88\underline{/0^\circ}$, $Z_2 = 44\sqrt{2}\underline{/45^\circ}$ and $Z_3 = 44\underline{/36.87^\circ}$, determine the power delivered to the load.

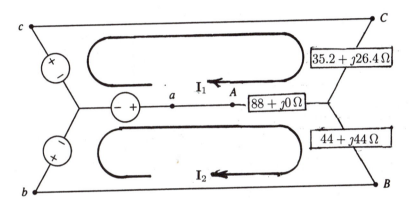

Set $\mathbf{V}_{an}$ as the reference and use mesh analysis. With

$$v_p = \frac{v_\ell}{\sqrt{3}} = 254\,\text{V}$$

the phase voltages are

$$\mathbf{V}_{an} = 254\underline{/0^\circ}\,\text{V} = 254 + \jmath 0\,\text{V}$$

$$\mathbf{V}_{bn} = 254\underline{/-120^\circ}\,\text{V} = -127 - \jmath 220\,\text{V}$$

$$\mathbf{V}_{cn} = 254\underline{/120^\circ}\,\text{V} = -127 + \jmath 220\,\text{V}$$

Then

$$\begin{aligned} \mathbf{V}_{ab} &= \mathbf{V}_{an} - \mathbf{V}_{bn} \\ &= (254 + \jmath 0) - (-127 - \jmath 220) \\ &= 381 + \jmath 220\,\text{V} = 440\underline{/30^\circ}\,\text{V} \end{aligned}$$

and

$$\begin{aligned} \mathbf{V}_{ca} &= \mathbf{V}_{cn} - \mathbf{V}_{an} \\ &= (-127 + j220) - (254 - j0) \\ &= -381 + j220 \text{ V} = 440\underline{/150°} \text{ V} \end{aligned}$$

The two clockwise mesh currents in the network diagram may be written using two applications of KVL. With $Z_1 = 88 + j0\,\Omega$, $Z_2 = 44 + j44\,\Omega$ and $Z_3 = 35.2 + j26.4\,\Omega$

$$\begin{aligned} (123.2 + j26.4)\mathbf{I}_1 \quad - (88 + j0)\mathbf{I}_2 &= 440\underline{/150°} \text{ V} \\ -(88 + j0)\mathbf{I}_1 \quad + (132 + j44)\mathbf{I}_1 &= 440\underline{/30°} \text{ V} \end{aligned}$$

These may be written in matrix form

$$\begin{bmatrix} (123.2 + j26.4) & -(88 + j0) \\ -(88 + j0) & (132 + j44) \end{bmatrix} \begin{bmatrix} \mathbf{I}_1 \\ \mathbf{I}_2 \end{bmatrix} = \begin{bmatrix} -381 + j220 \text{ V} \\ 381 + j220 \text{ V} \end{bmatrix}$$

and a matrix inversion can be employed to find the current vector

$$\begin{bmatrix} \mathbf{I}_1 \\ \mathbf{I}_2 \end{bmatrix} = \begin{bmatrix} (123.2 + j26.4) & -(88 + j0) \\ -(88 + j0) & (132 + j44) \end{bmatrix}^{-1} \begin{bmatrix} -381 + j220 \text{ V} \\ 381 + j220 \text{ V} \end{bmatrix}$$

Here,

$$\begin{bmatrix} (123.1 + j26.4) & (-88 + j0) \\ (-88 + j0) & (132 + j44) \end{bmatrix}^{-1} = \begin{bmatrix} (0.0102 - j0.0064) & (0.0049 - j0.0059) \\ (0.0049 - j0.0059) & (0.0086 - j0.0068) \end{bmatrix}$$

so that

$$\begin{bmatrix} \mathbf{I}_1 \\ \mathbf{I}_2 \end{bmatrix} = \begin{bmatrix} 0.653 + j3.509 \text{ A} \\ 4.192 + j2.609 \text{ A} \end{bmatrix} = \begin{bmatrix} 3.569\underline{/79.46°} \text{ A} \\ 4.938\underline{/31.90°} \text{ A} \end{bmatrix}$$

Then, with

$$\mathbf{I}_a = \mathbf{I}_1 - \mathbf{I}_2 = (0.653 + j3.509) - (4.192 + j2.609) = -3.539 + j0.900 = 3.651\underline{/165.72°} \text{ A}$$

$$\begin{aligned} P &= 35.2(I_1)^2 + 44(I_2)^2 + 88(I_a)^2 \\ &= (35.2\,\Omega)(3.569 \text{ A})^2 + (44\,\Omega)(4.938 \text{ A})^2 + (88\,\Omega)(3.651 \text{ A})^2 \\ &= 2694.3 \text{ W} \Leftarrow \end{aligned}$$

12.24: Determine the overall power factor for the system treated in Problem 12.23.

Reference to Problem 12.23 shows that

$$\begin{aligned}
\mathbf{I}_1 &= 3.569\underline{/79.46°}\text{ A} \\
\mathbf{I}_2 &= 4.938\underline{/31.90°}\text{ A} \\
\mathbf{I}_a &= 3.651\underline{/165.72°}\text{ A}
\end{aligned}$$

The power was obtained as $P = 2245.9\,\text{W}$ and the reactive power will be

$$\begin{aligned}
Q &= 26.4(I_1)^2 + 44(I_2)^2 + 0(I_a)^2 \\
&= (26.4\,\Omega)(3.569\,\text{A})^2 + (44\,\Omega)(4.938\,\text{A})^2 \\
&= 1409.2\,\text{VAR} \Leftarrow
\end{aligned}$$

Thus

$$\mathbf{S} = P + jQ = 2694.3 + j1409.2 = 3040.55\underline{/27.61°}\text{ VA}$$

The power factor angle is

$$\theta = 27.61°$$

and the power factor is

$$PF = \cos\theta = 0.8861 \Leftarrow$$

12.25: In Fig 12.12, the rms amplitude of the line-to-line voltages in the three phase source is 300 V. The source is connected in the positive sequence and operates at 60 Hz. If the unbalanced load consists of $Z_1 = 60\underline{/0°}$, $Z_2 = 75\sqrt{2}\underline{/-45°}$ and $Z_3 = 50\underline{/36.87°}$, determine the power delivered to the load.

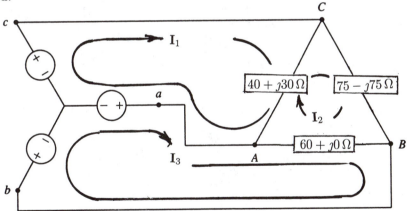

Set $\mathbf{V}_{an}$ as the reference and use mesh analysis. With

$$v_p = \frac{v_\ell}{\sqrt{3}} = \frac{300\,\text{V}}{\sqrt{3}} = 173.2\,\text{V}$$

then

$$\mathbf{V}_{ab} = 300\underline{/30°}$$

and

$$\mathbf{V}_{ca} = 300\underline{/150°}$$

The three clockwise mesh currents are shown in the system diagram and the three mesh equations may be written using three applications of KVL. With $Z_1 = 60+j0\,\Omega$, $Z_2 = 75-j75\,\Omega$ and $Z_3 = 40 + j30\,\Omega$

$$
\begin{array}{rll}
(40 + j30)\mathbf{I}_1 \quad - (40 + j30)\mathbf{I}_2 & = & -259.8 + j150\,\text{V} \\
- (40 + j30)\mathbf{I}_1 \quad + (175 - j45)\mathbf{I}_2 \quad - (60 + j0)\mathbf{I}_3 & = & 0 \\
- (60 + j0)\mathbf{I}_2 \quad + (60 + j0)\mathbf{I}_3 & = & 259.8 + j150\,\text{V}
\end{array}
$$

These may be written in matrix form

$$
\begin{bmatrix}
(40 + j30) & -(40 + j30) & 0 \\
-(40 + j30) & +(175 - j45) & -(60 + j0) \\
0 & -(60 + j0) & +(60 + j0)
\end{bmatrix}
\begin{bmatrix}
\mathbf{I}_1 \\ \mathbf{I}_2 \\ \mathbf{I}_3
\end{bmatrix}
=
\begin{bmatrix}
-259.8 + j150\,\text{V} \\
0\,\text{V} \\
259.8 + j150\,\text{V}
\end{bmatrix}
$$

and a matrix inversion can be employed to find the current vector

$$
\begin{bmatrix}
\mathbf{I}_1 \\ \mathbf{I}_2 \\ \mathbf{I}_3
\end{bmatrix}
=
\begin{bmatrix}
(40 + j30) & -(40 + j30) & 0 \\
-(40 + j30) & +(175 - j45) & -(60 + j0) \\
0 & -(60 + j0) & +(60 + j0)
\end{bmatrix}^{-1}
\begin{bmatrix}
-259.8 + j150\,\text{V} \\
0\,\text{V} \\
259.8 + j150\,\text{V}
\end{bmatrix}
$$

With

$$
\begin{bmatrix}
(40 + j30) & -(40 + j30) & 0 \\
-(40 + j30) & +(175 - j45) & -(60 + j0) \\
0 & -(60 + j0) & +(60 + j0)
\end{bmatrix}^{-1}
$$

$$
=
\begin{bmatrix}
(0.0227 - j0.0053) & (0.0067 + j0.0067) & (0.0067 + j0.0067) \\
(0.0067 + j0.0067) & (0.0067 + j0.0067) & (0.0067 + j0.0067) \\
(0.0067 + j0.0067) & (0.0067 + j0.0067) & (0.0233 + j0.0067)
\end{bmatrix}
$$

the current vector is

$$
\begin{bmatrix}
\mathbf{I}_1 \\ \mathbf{I}_2 \\ \mathbf{I}_3
\end{bmatrix}
=
\begin{bmatrix}
-4.36 + j7.52\,\text{A} \\
-2.00 + j2.00\,\text{A} \\
2.33 + j4.50\,\text{A}
\end{bmatrix}
=
\begin{bmatrix}
8.693\,\underline{/120.10°}\,\text{A} \\
2\sqrt{2}\,\underline{/-45°}\,\text{A} \\
5.067\,\underline{/62.63°}\,\text{A}
\end{bmatrix}
$$

Then, with

$$\begin{aligned}
\mathbf{I}_a &= \mathbf{I}_1 - \mathbf{I}_2 \\
&= (-4.36 + j7.52) - (-2.00 + j2.00) \\
&= -2.36 + j5.52 = 6.00\underline{/113.12°}\,\text{A}
\end{aligned}$$

and

$$\begin{aligned}
\mathbf{I}_b &= \mathbf{I}_3 - \mathbf{I}_2 \\
&= (2.33 + j4.50) - (-2.00 + j2.00) \\
&= 4.33 + j2.50 = 5.00\underline{/30°}\,\text{A}
\end{aligned}$$

Then

$$\begin{aligned}
P &= 40(I_a)^2 + 60(I_b)^2 + 75(I_2)^2 \\
&= (40\,\Omega)(6.00\,\text{A})^2 + (60\,\Omega)(5.00\,\text{A})^2 + (75\,\Omega)(2\sqrt{2}\,\text{A})^2 \\
&= 3540\,\text{W} \Leftarrow
\end{aligned}$$

12.26: Determine the overall power factor for the system treated in Problem 12.25.

Reference to Problem 12.23 shows that

$$\begin{aligned}
I_a &= 6.00\,\text{A} \\
I_2 &= 2\sqrt{2}\,\text{A} \\
I_b &= 5.00\,\text{A}
\end{aligned}$$

The power was obtained as $P = 3540\,\text{W}$ and the reactive power will be

$$\begin{aligned}
Q &= 30(I_a)^2 - 75(I_2)^2 + 0(I_b)^2 \\
&= (30\,\Omega)(6.00\,\text{A})^2 - (75\,\Omega)(2\sqrt{2}\,\text{A})^2 \\
&= 480\,\text{VAR}
\end{aligned}$$

Thus

$$\mathbf{S} = P + jQ = 3540 + j480 = 3572.4\underline{/7.72°}\,\text{VA}$$

The power factor angle is

$$\theta = 7.72°$$

and the power factor is

$$PF = \cos\theta = 0.9909 \Leftarrow$$

12.27: Figure 12.13 shows a wye-connected generator delivering power to a wye-connected load in a four wire system. The rms amplitude of the line-to-line voltages in the three phase source is 240 V. The source is connected in the positive sequence and operates at 60 Hz. If the unbalanced load consists of three impedances, $Z_1 = 48\underline{/0°}\ \Omega$, $Z_2 = 24\sqrt{2}\underline{/-135°}\ \Omega$ and $Z_3 = 60\underline{/36.87°}\ \Omega$, determine the power delivered to the load.

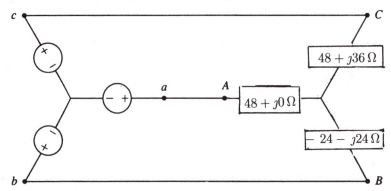

Notice the short circuit between n and N which means that

$$v_n = V_N$$

Moreover

$$V_p = \frac{V_\ell}{\sqrt{3}} = \frac{240\ \text{V}}{\sqrt{3}} = 138.6\ \text{V}$$

Then

$$\mathbf{I}_{aA} = \frac{\mathbf{V}_{an}}{48 + \jmath0} = \frac{138.6\underline{/0°}}{48} = 2.887 + \jmath0 = 2.887\underline{/0°}\ \text{A}$$

$$\mathbf{I}_{bB} = \frac{\mathbf{V}_{bB}}{-24 - \jmath24} = \frac{138.6\underline{/-120°}}{24\sqrt{2}\underline{/-135°}} = 3.943 + \jmath1.057 = 4.082\underline{/15°}\ \text{A}$$

$$\mathbf{I}_{cC} = \frac{\mathbf{V}_{cn}}{48 + \jmath36} = \frac{138.6\underline{/120°}}{60\underline{/36.87°}} = 0.276 + \jmath2.293 = 2.309\underline{/83.13°}\ \text{A}$$

Then

$$
\begin{aligned}
P &= 48(I_{aA})^2 + 24(I_{bB})^2 + 48(I_{cC})^2 \\
&= (48\,\Omega)(2.887\,\text{A})^2 + (24\,\Omega)(4.082\,\text{A})^2 + (48\,\Omega)(2.309\,\text{A})^2 \\
&= 1055.9\,\text{W} \Leftarrow
\end{aligned}
$$